A NOTE FROM THE AUTHORS

Congratulations on your decision to take the AP Biology exam! Whether or not you're completing a yearlong AP Biology course, this book can help you prepare for the exam. In it you'll find information about the exam as well as Kaplan's test-taking strategies, a targeted review that highlights important concepts on the exam, and practice tests. Take the diagnostic test to see which subjects you should review most, and use the two full-length exams to get comfortable with the testing experience. The review chapters include summaries of AP Biology investigations so that even if you haven't completed them in class, you won't be surprised on Test Day. Don't miss the strategies for answering the free-response questions: You'll learn how to cover the key points AP graders will want to see.

To help you manage your content review, we've designed an intuitive rating system. The star system takes numerous factors into account, most notably how closely the content aligns with knowledge required for the AP exam and the level of detail with which the content is covered. Topics that are not explicitly tested on the exam will have a lower rating, as will sections of the book that provide additional supplementary details on a required topic.

Take a few minutes to flip through the book to see a few instances of the rating system in action. By getting familiar with the star ratings, you will also get a better idea of the test maker's expectations.

★★★★ A section with a four-star rating means you should know essentially all of the terms and facts in that section. Given the College Board's emphasis on "doing science," all of the investigations also have four-star ratings.

★★★ A three-star section indicates that the majority of the content is essential knowledge, but there are some extraneous details included primarily to enhance your comprehension. For example, the section on mitosis is given a three-star rating, because, while the concept is important, some of the details (including the names of the phases) are not required knowledge for the exam.

★★ A two-star section contains content and examples designed to enrich your understanding of a required concept. For instance, the section on the respiratory system has a two-star rating because this section is primarily an example used to support the required concepts that "organisms have various mechanisms of eliminating waste" and "organisms exhibit specialization of organs." Read these sections with a focus on the "big picture," rather than the specifics.

★ One-star sections include facts that are relatively low-yield on the AP exam. Don't waste precious time memorizing these sections. However, reading to understand even these less-important sections can help round out your science expertise, especially for free-response questions.

By studying college-level biology in high school, you've placed yourself a step ahead of other students. You've developed your critical-thinking and time-management skills, as well as your understanding of the practice of biological research. This book will help you show off what you've learned!

Best of luck,

Allison Ann Wilkes

Linda Brooke Stabler

Mark Metz

RELATED TITLES

AP® BIOLOGY

2016

Allison Ann Wilkes

Linda Brooke Stabler

Mark Metz

KAPLAN

PUBLISHING

New York

This publication is designed to provide accurate and authoritative information in regard to the subject matter covered. It is sold with the understanding that the publisher is not engaged in rendering legal, accounting, or other professional service. If legal advice or other expert assistance is required, the services of a competent professional should be sought.

© 2015 by Kaplan, Inc.

Published by Kaplan Publishing, a division of Kaplan, Inc.
750 Third Avenue
New York, NY 10017

Printed in the United States of America

10 9 8 7 6 5 4 3 2 1

ISBN-13: 978-1-62523-146-8

Kaplan Publishing books are available at special quantity discounts to use for sales promotions, employee premiums, or educational purposes. For more information or to order books, please call the Simon & Schuster special sales department at 866-506-1949.

TABLE OF CONTENTS

PART ONE: THE BASICS

PART TWO: DIAGNOSTIC TEST

PART THREE: PRACTICE TESTS

ABOUT THE AUTHORS

Allison Ann Wilkes holds a MS in biology from the University of West Florida and has taught for Kaplan since 2005.

Linda Brooke Stabler holds a PhD in biology from Arizona State University and teaches biology at Redlands Community College in Oklahoma.

Mark Metz works for the U.S. Department of Agriculture and holds a PhD in environmental science from the University of Illinois. He taught biology at the university level for four years.

KAPLAN PANEL OF AP EXPERTS

Congratulations—you have chosen Kaplan to help you get a top score on your AP exam.

Kaplan understands your goals, and what you're up against—achieving college credit and conquering a tough test—while participating in everything else that high school has to offer.

You expect realistic practice, authoritative advice, and accurate, up-to-the-minute information on the test and that's exactly what you'll find in this book, as well as every other in the AP series. To help you (and us!) reach these goals, we have sought out leaders in the AP community. Allow us to introduce our experts:

AP BIOLOGY EXPERTS

Franklin Bell has taught AP Biology since 1994, most recently at Saint Mary's Hall in San Antonio, Texas, where he is the science department cochair. He has been an AP Biology reader for six years and a table leader for two years. In addition, he has been an AP Biology consultant at College Board workshops and has received the Southwestern Region's College Board Special Recognition Award.

Larry Calabrese has taught biology for 37 years and AP Biology for 20 years at Palos Verdes Peninsula High School in California. He also teaches anatomy and physiology at Los Angeles Harbor College in Wilmington, California. He has been an AP Biology reader since 1986 and a table leader since 1993. For the past 10 years, he has taught College Board workshops.

Cheryl G. Callahan has been teaching biology for over 20 years, first at the college level and now at the high school level at Savannah Country Day School in Savannah, Georgia. She has been an AP Biology reader and table leader since 1993. She also moderates the AP Biology Electronic Discussion Group.

Ruth Moss has taught biology at Grimsley High School in Greensboro, North Carolina, since 1985. She has taught AP Biology for 20 years and has been an AP reader for 11 years.

Part One

THE BASICS

CHAPTER 1: INSIDE THE AP BIOLOGY EXAM

IF YOU ONLY LEARN FIVE THINGS IN THIS CHAPTER . . .

1. The AP Biology exam is three hours long and includes a 90-minute multiple-choice and grid-in section (69 questions), a 10-minute reading period, and an 80-minute free-response section (8 questions).

2. The AP Biology exam is scored on a scale of 1–5, with 5 being the highest possible score.

3. Use Kaplan's AP Biology guide for a better understanding of what will be covered on the test. In addition, the Educational Testing Service (ETS) releases a list of topics that will be covered on the exam each year, along with detailed information about how each topic will be tested.

4. Remember that it's better to understand concepts rather than trying to memorize facts. Having a thorough understanding of the concepts covered in your AP Biology course will help you with the test as a whole.

5. Rereading your textbook and class notes isn't always enough. Use this guide to assess your test strengths and weaknesses and to practice for the exam. The more you prep, the more confident you will feel on Test Day.

INTRODUCTION

There's a good way and a bad way to skip the Introduction to Biology class in college. Many students take the bad way, which consists of going to sleep ridiculously late every night with the Xbox controller still wedged in their sweaty hands, setting the alarm for 1:30 pm, then waking up and asking a roommate, "What did I miss?" This is not exactly the sort of behavior that will land you on the dean's list.

Then there's the good way: Skip the whole Introduction to Biology experience entirely—hundreds of students crammed into an auditorium, the tiny dot that is the professor just visible down in front of an ocean of seats—by getting a good score on the Advanced Placement (AP) Biology exam.

YOUR AP SCORE

Depending on the college, a score of 4 or 5 on the AP Biology exam will allow you to leap over the freshman intro course and jump right into more advanced classes.

These advanced classes are usually smaller in size, better focused, more intellectually stimulating, and simply put, just more interesting than a basic course. If you are just concerned about fulfilling your science requirement so that you can get on with your study of pre-Columbian art or Elizabethan music or some such nonbiological area, the AP exam can help you there, too. Ace the AP Biology exam and, depending on the requirements of the college you choose, you may never have to take a science class again.

TEST PREP ≠ STUDYING

If you're holding this book, chances are you are already gearing up for the AP Biology exam and probably nearing completion of the AP Biology course. Your teacher has spent the year cramming your head full of the biology know-how you will need to have at your disposal. There is more to the AP Biology exam than biology know-how, however. You have to be able to work around the challenges and pitfalls of the test—and there are many—if you want your score to reflect your abilities. You see, studying biology and preparing for the AP Biology exam are not the same thing. Rereading your textbook is helpful, but it's not enough.

That's where this book comes in. We'll show you how to marshal your knowledge of biology and put it to brilliant use on Test Day. We'll explain the ins and outs of the test structure and question format so you won't experience any nasty surprises. We'll even give you answering strategies designed specifically for the AP Biology exam.

THE POWER OF PREPARING

Preparing effectively for the AP Biology exam means doing some extra work. You need to review your text *and* master the material in this book. Is the extra push worth it? If you have any doubts, keep in mind that you can always sleep until 1:30 pm with the Xbox controller in your hand on the weekend.

OVERVIEW OF THE TEST STRUCTURE

Advanced Placement exams have been around for half a century. While the format and content have changed over the years, the basic goal of the AP program remains the same: to give high school students a chance to earn college credit or advanced placement. To do this, a student needs to do two things:

- Find a college that accepts AP scores
- Score well enough on the exam

The first part is easy, because most colleges accept AP scores in some form or another. The second part requires a little more effort. If you have worked diligently all year in your course work, you've laid the groundwork. The next step is familiarizing yourself with the test.

WHAT'S ON THE TEST

Two main goals of the College Board are (1) to help students develop a conceptual framework for modern biology, and (2) to help students gain an appreciation of science as a process. To this end, the AP Biology course is designed to expose the student to four main ideas.

Big Idea 1: The process of evolution drives the diversity and unity of life.

Big Idea 2: Biological systems utilize free energy and molecular building blocks to grow, reproduce, and maintain dynamic homeostasis.

Big Idea 3: Living systems store, retrieve, transmit, and respond to information essential to life processes.

Big Idea 4: Biological systems interact, and these systems and their interactions possess complex properties.

These four big ideas are referred to as evolution, cellular processes, genetics and information transfer, and interactions, respectively. The four big ideas encompass the core principles, theories, and scientific processes that guide the study of life. Each of these big ideas is broken down into enduring understandings and learning objectives that will help you to organize your knowledge.

This approach to scientific discovery is about thinking, not just memorization. It's about learning concepts and how they relate, not just facts. Because of this, **the College Board is increasing the emphasis on themes and concepts and placing less weight on specific facts in both the AP Biology course and exam**. The chapters in the review section of this book are intended to take advantage of this design by focusing on concepts and synthesizing information from different concepts to better understand, and learn, the AP Biology course and exam content.

Now that you know what's on the test, let's talk about the test itself. The AP Biology exam consists of two sections, or, more precisely, two sections and one intermission. Section I has a Part A and a Part B. In Part A, there are typically 63 multiple-choice questions with four answer choices each. (However, depending on the form of the exam, the exact number of questions may vary.)

In Part B, there are six grid-in items that integrate scientific thinking and mathematical skills. For each grid-in item, you will need to calculate the correct answer and then enter it into a grid section on the answer sheet. You will have 90 minutes to complete all 69 questions. This section is worth 50 percent of your total score.

After this section is completed, there will be a 10-minute break. During the break you will not be able to consult teachers, other students, or textbooks. Furthermore, you may not access any electronic or communication device, which means cell phones, computers and calculators are off limits.

After the break, there's a 10-minute, recommended "reading period." This doesn't mean you get to pull your favorite novel out of your backpack and finish that chapter you started earlier. Instead, you're given 10 minutes to pore over Section II of the exam, which consists of two long free-response questions and six short free-response questions that are worth 50 percent of your total score. You then have 80 minutes to answer all of these questions. The term "long free-response" means roughly the same thing as "large, multistep, and involved." Although the two long free-response questions are worth a significant amount each and are often broken into multiple parts, they usually don't cover an obscure topic. Instead, they take a fairly basic biology concept and ask you several questions about it. Sometimes diagrams are required, or experiments must be set up properly. It's a lot of biology work, but it is fundamental biology work. The short free-response items are typically illustrative examples or concepts that you are expected to explain or analyze, providing appropriate scientific evidence and reasoning. A typical response for the short free-response items will be a few sentences to a paragraph in length. You have approximately 20 minutes for each of the long free-response questions and six minutes for each of the short-response questions.

HOW THE EXAM IS SCORED

AP scores are based on the number of questions answered correctly. **No points are deducted for wrong answers**. No points are awarded for unanswered questions. Therefore, you should **answer every question, even if you have to guess.**

When the 180 minutes of testing are up, your exam is sent away for grading. The multiple-choice part is handled by a machine, while qualified graders—a group that includes biology teachers and professors, both current and former—grade your responses to Section II. After an interminable wait, your composite score will arrive by mail. (For information on rush score reports and other grading options, visit collegeboard.com or ask your AP Coordinator.) Your results will be placed into one of the following categories, reported on a five-point scale:

5 = Extremely well qualified

4 = Well qualified

3 = Qualified

2 = Possibly qualified

1 = No recommendation

Some colleges will give you credit for a score of 3 or higher, but it's much safer to get a 4 or a 5. If you have an idea of where you will be applying to college, check out the schools' websites or call the admissions offices to find out their particular rules regarding AP scores.

REGISTRATION AND FEES

You can register for the exam by contacting your guidance counselor or AP Coordinator. If your school doesn't administer the exam, contact the Advanced Placement Program for a list of schools in your area that do. At the time of this book's publication, the fee for each AP exam is $91 ($121 at schools outside of the United States). For students with acute financial need, the College Board offers a $29 fee reduction. In addition, most states offer exam subsidies to cover all or part of the remaining cost for eligible students. To learn about other sources of financial aid, contact your AP Coordinator.

For more information on all things AP, contact the Advanced Placement Program:

Phone: (888) 225-5427 or (212) 632-1780

Email: apstudents@info.collegeboard.org

Website: https://apstudent.collegeboard.org/home

ADDITIONAL RESOURCES

Books

Biology (Tenth Edition)

Neil A. Campbell, Jane B. Reece, Lisa A. Urry, Steven A. Wasserman, Peter V. Minorsky, and Robert B. Jackson, 2013

Benjamin Cummings, San Francisco, CA

ISBN: 978-0321775658

Introduction to Organic and Biochemistry (Eighth Edition)

Frederick A. Bettelheim, William H. Brown, and Jerry March, 2013

Brooks/Cole, Pacific Grove, CA

ISBN: 1-133-10976-4

Biology (Fourth Edition)

Karen Arms and Pamela S. Camp, 1995

Saunders College Publishing, Philadelphia, PA

ISBN: 0-0301-5434-0

Plant Physiology (Fourth Edition)

Frank Salisbury and Cleon Ross, 1991

Wadsworth Publishing, Inc., Belmont, CA

The Cartoon Guide to Genetics (Revised Edition)

Larry Gonick and Mark Wheelis, 1991

HarperResource (HarperCollins), New York, NY

ISBN: 0-0627-3099-1

Insect Molecular Genetics: An Introduction to Principles and Applications (Third Edition)

Marjorie A. Hoy, 2013

Academic Press/Elsevier, San Diego, CA

ISBN: 0-1241-5874-9

Evolutionary Biology (Third Edition)

Douglas J. Futuyma, 1997

Sinauer Associates, Inc., Sunderland, MA

ISBN: 0-8789-3189-9

Human Physiology (Twelfth Edition)

Stuart Ira Fox, 2010

McGraw-Hill Science/Engineering/Math, New York, NY

ISBN: 0-0773-5006-5

Silent Spring (Anniversary Edition)

Rachel Carson, 2002 (1962)

Mariner Books (Houghton-Mifflin), New York, NY

ISBN: 0-6182-4906-0

WEBSITES

The pH Scale, Dr. Paul Decelles, staff.jccc.net/pdecell/chemistry/phscale.html

Enzymes, Mark Rothery's Biology Website, mrothery.co.uk/enzymes/enzymenotes.htm

The Cell Cycle, James A. Sullivan, cellsalive.com cellsalive.com/cell_cycle.htm

Mitosis, James A. Sullivan, cellsalive.com cellsalive.com/mitosis.htm

Cellular Respiration, Regina Bailey, Biology on About.com, biology.about.com/library/weekly/aa090601a.htm

Photosynthesis—Autotrophic Metabolism, Dr. Charles Mallery, University of Miami fig.cox.miami.edu/~cmallery/150/phts/phts.htm

Photosynthesis, Western Kentucky University, Biology 120 course tutorial, bioweb.wku.edu/courses/Biol120/Web/Photosynth1x.asp

Plant Physiology Online, Sinauer Associates, Inc., Sunderland, MA, 5e.plantphys.net

Mendelian Genetics, The Biology Project, The University of Arizona, biology.arizona.edu/mendelian_genetics/mendelian_genetics.html

A Monk's Flourishing Garden: The Basics of Molecular Biology Explained, BioTeach, scq.ubc, ca/a-monks-flourishing-garden-the-basics-of-molecular-biology-explained/

Gel Electrophoresis Simulator Physlets, webpages by John Cowan, revised and modified by Wolfgang Christian, webphysics.davidson.edu/applets/biogel/biogel.html

Origins of Life Seminar Series, New York Center for Astrobiology, origins.rpi.edu/seminars/

The Origin of Species, by Charles Darwin, Online Literature Library, literature.org/authors/darwin-charles/the-origin-of-species

Understanding Evolution, The University of California Museum of Paleontology, Berkeley, evolution.berkeley.edu

Tree of Life Web Project, David R. Maddison and Katja S. Schulz (eds.), 2007, tolweb.org

The Biosphere: Life on Earth, The University of California Museum of Paleontology, Berkeley, ucmp.berkeley.edu/alllife/threedomains.html

Welcome to the Dynamics of Development! Dr. Jeff Hardin, University of Wisconsin, Department of Zoology, worms.zoology.wisc.edu/dd2/

Human Population Growth, Kimball's Biology Pages, John W. Kimball, users.rcn.com/jkimball.ma.ultranet/BiologyPages/P/Populations2.html

Introduction to Biomes, Susan L. Woodward, Department of Geospatial Science, Radford University, radford.edu/swoodwar/CLASSES/GEOG235/biomes/intro.html

CHAPTER 2: STRATEGIES FOR SUCCESS: IT'S NOT ALWAYS HOW MUCH YOU KNOW

IF YOU ONLY LEARN FIVE THINGS IN THIS CHAPTER . . .

1. Remember the keys to Kaplan's general test-taking strategies:
   - Pacing
   - Process of Elimination
   - Knowing When to Guess
   - Recognizing Patterns and Trends
   - Taking the Right Approach

2. Know how to manage your stress. You can beat anxiety the same way you can beat the AP Biology exam—by knowing what to expect beforehand and developing strategies to deal with it.

3. Take Kaplan's diagnostic test to learn your test-taking strengths and weaknesses. Knowing these will allow you to focus on your problem areas as you prepare for Test Day.

4. Get organized! Make a study schedule between now and Test Day, and give yourself plenty of time to prepare. Waiting until the last minute to cram an entire semester's worth of information is not only unwise but also exhausting.

5. Make a Test Day game plan. Have everything you need to bring to the exam ready the night before. On the morning of the test, make it a priority to eat a good breakfast. Avoid an overindulgence of caffeine. Read something to warm up your brain. And finally, get to the test site early.

INTRODUCTION

Even nonbiologists know that the world constantly changes. Fifteen years ago there weren't many cell phones. There weren't many standardized tests, either. Nowadays, you can't go a semester of school without taking some letter-jumble exam like the PSAT, SAT, ACT, BLAM, ZORK, or FWOOSH (some of those tests are fake, some aren't). And right after leaving school, many students call their friends on their cell phones to talk about the test they all just took.

Rampant cell phone usage is a problem for another book; this one will concentrate on standardized testing. Because everyone reading this has taken a standardized test of one kind or another, you are all probably familiar with some of the general strategies that help students increase their scores on a standardized exam. Let's review some of these.

GENERAL TEST-TAKING STRATEGIES

1. **Pacing.** Because many tests are timed, proper pacing allows someone to attempt every question in the time allotted. Poor pacing causes students to spend too much time on some questions to the point where they run out of time before completing all the questions.

2. **Process of Elimination.** On every multiple-choice test you ever take, the answer is given to you. If you can eliminate answer choices you know are incorrect and only one choice remains, then that must be the correct answer.

3. **Knowing When to Guess.** The AP Biology exam does not deduct points for wrong answers, while questions left unanswered receive zero points. That means you should always guess on a question you can't answer any other way.

4. **Recognizing Patterns and Trends.** The AP Biology exam doesn't change greatly from year to year. Sure, each question won't be the same, and different topics will be covered from one administration to the next, but there will also be a lot of overlap from one year to the next. Because of this, certain patterns can be uncovered regarding any standardized test. Learning about these trends and patterns can help students taking the test for the first time.

5. **Taking the Right Approach.** Having the right mindset plays a large part in how well people do on a test. Those students who are nervous about the exam and hesitant to make guesses often fare much worse than students with an aggressive, confident attitude.

These points are all valid for every standardized test, but they are quite broad in scope. The rest of this chapter will discuss how these general ideas can be modified to apply specifically to the AP Biology exam. These test-specific strategies and the factual information covered in your course as well as in this book's review section are the one-two punch that will help you succeed on the exam.

HOW TO APPROACH THE MULTIPLE-CHOICE QUESTIONS

The multiple-choice questions are numbered, but that does not mean you must answer the questions in the given order. In fact, it's highly unlikely that the questions will be presented to you in a confidence-inspiring, point-building, time-saving order. There is good news, though: You are free to navigate the section in a manner that highlights your strengths and downplays your weaknesses. All of the multiple-choice questions carry the same weight so you don't get extra credit for correctly answering a hard question. In the end, colleges will never know which questions you answered correctly. All they will know is that you were smart enough to spend your time where it was more likely to turn into points.

Of the 63 multiple-choice questions, there are two distinct question types.

1. **The Stand-Alones.** These questions typically make up a little over half of the AP Biology exam. Each Stand-Alone question covers a specific topic, and then the next Stand-Alone hits a different topic. The question stem may be as short as a single sentence, but it is not uncommon to see multiple paragraphs for a single question! In addition to words, many question stems will be accompanied by an equation, table, graph or figure. The next question is a typical Stand-Alone.

    22. AB + energy → A + B

        Which of the following best characterizes the reaction represented above?

        (A) Anabolism
        (B) Endergonic reaction
        (C) Exergonic reaction
        (D) Hydrolysis

You get some information to start with, and then you're expected to answer the question. The number of the question, 22, makes no difference because there's no order of difficulty on the AP Biology exam. Tough questions are scattered between easy and medium questions.

2. **Data Questions.** Just as the name suggests, a group of two to five questions is preceded by data in one form or another. The data might be a simple sentence or two, but usually it is something more complex, such as:

    • A description of an experiment (50–200 words), often with an accompanying illustration

    • A graph or series of graphs

    • A large table

    • A diagram

The next question is a sample Data question.

Questions 61–63 refer to the following graphs. These graphs present the frequency of size classes for tail length of *Felis domesticus* as measured at four different sites around the world.

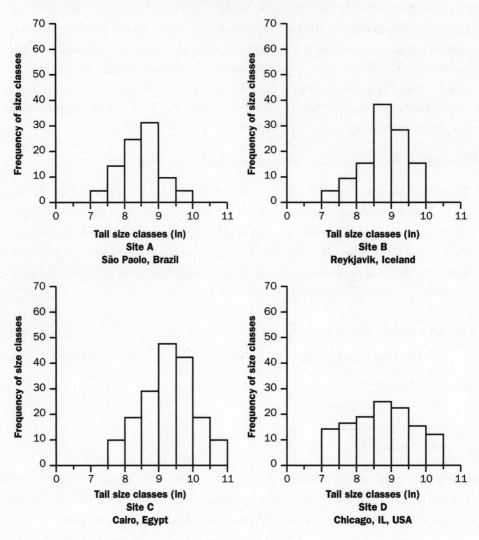

61. According to the data, at Site C, the most common size class of tails among *Felis domesticus* is

(A)   7–8 inches.

(B)   8–8.5 inches.

(C)   9–9.5 inches.

(D)   9.5–10 inches.

Some Stand-Alone questions will require you to analyze data; the difference between Stand-Alone and Data questions is the number of questions associated with the data. In a Stand-Alone question, you will analyze the data then answer a single question. On Data questions you'll get multiple questions referencing the same data.

Those are the two types of questions you'll see in Part A. Combine this knowledge with the fact that you must answer 63 multiple-choice *and* 6 grid-in questions in 90 minutes, and you'll see why it is imperative to take control of the test. Section management may not come to mind when you think about Biology, but it will be tested on the AP Biology exam!

KEY POINT

You don't need to get every multiple-choice question right on the AP Biology exam. To get a 4 or 5, you need to get a large portion, but not all, of the questions right. If you don't have enough time to get to every question, make sure that the questions you skip are the longest, most involved ones. That's a great use of your limited resource: time.

There's more to it than just tackling questions in the right order, however. The more you know about the question types, the better equipped you will be to handle them.

TAKE AN EDUCATED GUESS

Many times you can eliminate at least one answer choice from a problem. It may seem insignificant, but it gets you closer to the correct answer and it can significantly increase your chances of guessing correctly. You won't get every guess right, but over the course of the test, this form of educated guessing will improve your score.

THE STAND-ALONES

It's easier to talk about what isn't in the Stand-Alone questions than what is there.

- There's no order of difficulty; that is, questions don't start out easy and gradually become tougher.
- There are no two questions connected to each other in any way.
- There's no pattern as to what biology concepts appear when.

The Stand-Alones look like a bunch of disconnected biology questions, one following another, and that's just what they are. A genetics question may follow a taxonomy question, which may follow a question about the Krebs cycle.

There's no overall pattern, so don't bother looking for one. Nevertheless, just because the section is randomly ordered doesn't mean you have to approach it on the same random terms. Instead, draw two lists right now using the topics covered in the table of contents. Label one list "Concepts I Enjoy and Know" and label the other list "Concepts that Are Not My Strong Points."

When you get ready to tackle the Stand-Alones, keep these two lists in mind. On your first pass through the section, answer all the questions that deal with concepts you like and know a lot about. Quickly glance over any tables, figures or images and look for terminology in the question stem to clue you into the concept being tested. It should not take very long for you to figure out whether or not you have the factual chops needed to answer it. If you do, answer that problem and move on. If the question is on a subject that's not one of your strong points, skip it and come back later.

The overarching goal is to answer correctly the greatest possible number of questions in the time available. To do this, focus on your strengths during the first pass through the section. Some questions might be very difficult, even in a subject you're familiar with. Take a minute or so on a tough question, and if you can't come up with an answer, make a mark by the question number in your test booklet and move on. The first pass is about picking up easy points.

Once you've swept through and snagged all the easy questions, take a second pass and try the tougher ones. These tougher questions might cover subjects you're not strong in, or they might just be very difficult questions on subjects you are familiar with. Odds are high that you won't know the answer to some of these questions, but don't leave them blank. You should always take a stab at eliminating some answer choices, and then make an educated guess.

To select the correct answer on the AP Biology exam you will need to know the relevant science, but even if certain facts elude you, you can still increase your odds of choosing correctly by keeping the following two key ideas in mind.

COMPREHENSIVE, NOT SNEAKY

Some tests are sneakier than others. They have convoluted writing, questions designed to trip you up mentally, and a host of other little tricks. Students taking a sneaky test often have the proper facts, but get the question wrong because of a trap in the question itself.

DON'T FORGET!

The AP Biology test is NOT a sneaky test. The test works hard to be as comprehensive as it can be so that students who only know one or two biology topics will soon find themselves struggling.

Understanding these facts about how the test is designed can help you answer questions on it. The AP Biology exam is comprehensive, not sneaky. You've probably taken an AP Biology course, so trust your instincts when guessing. If you think you know the right answer, chances are you dimly remember the topic being discussed in your AP course. The test is about science, not traps, so trusting your instincts will help more often than not.

You don't have much time to ponder every tough question, so trusting your instincts can help keep you from getting bogged down and wasting time on a problem. You might not get every educated guess correct, but again, the point isn't about getting a perfect score. It's about getting a good score, and surviving hard questions by going with your gut feelings is a good way to achieve this.

On other problems, though, you might have no inkling of what the correct answer should be. In that case, turn to the second key idea.

THINK "GOOD SCIENCE"!

The AP Biology test rewards good biologists. The test wants to foster future biologists by covering fundamental topics and sound laboratory procedure. What the test doesn't want is bad science. It doesn't want answers that are factually incorrect, too extreme to be true, or irrelevant to the topic at hand.

Yet these "bad science" answers invariably appear, because it's a multiple-choice test and you must have three incorrect answer choices around the one right answer. So, if you don't know how to answer a question, look at the answer choices and think "good science." This may lead you to find some poor answer choices that can be eliminated.

44. $AB + energy \rightarrow A + B$

    Which of the following best characterizes the reaction represented above?

    (A) Anabolism
    (B) Endergonic reaction
    (C) Exergonic reaction
    (D) Hydrolysis

Here's our trusty Stand-Alone question from before. Even if you don't know what the problem is asking, look at choice (D), *hydrolysis*. The prefix *hydro-* means "water." You don't even have to be a biologist to know that. Even if you don't know exactly what hydrolysis is, you should know that it has something to do with water.

Nothing in the question stem has anything to do with water, does it? The AP Biology exam is comprehensive, not sneaky, so how could choice (D) be the correct answer? It just can't. That makes basic good science sense. Thinking about good science, you can cross out (D). You have a one-in-three shot, so take a guess.

You would be surprised how many times the correct answer on a multiple-choice question is a simple, blandly worded fact like, "Cells come in a variety of sizes and shapes." No breaking news there, but it is good science: a carefully worded statement that is factually accurate.

THE POWER OF GOOD SCIENCE

Thinking about good science in terms of the AP Biology exam can help you in two ways:

1. It helps you cross out extreme answer choices or choices that are untrue or out of place.

2. It can occasionally point you toward the correct answer, because the correct answer will be a factual piece of information sensibly worded.

Neither the "good science" nor the "comprehensive, not sneaky" strategy is 100-percent effective every time, but they do help more often than not. On a tough Stand-Alone question, these techniques can make the difference between an unanswered question and a good guess.

DATA QUESTIONS

Data questions require a slightly greater initial time investment, but don't let that intimidate you! Once the data is understood you may find that you can answer the questions rather quickly. Because most of the new information is in the shared introduction, you'll probably notice that the question stems are actually a little shorter than those of the average Stand-Alone. As you navigate through the multiple-choice questions, treat the Data questions in much the same way you would the Stand-Alone questions, with awareness of your strengths and weaknesses and your overall goal of getting more questions correct. If you see a Punnett square and you love heredity, then dive right in. However, if the topic is one that is more likely to induce anxiety than correct answers, then skip it and return after your first pass through the section.

The key to getting through the Data Questions section of the exam is to be able to analyze quickly and draw a conclusion from data presented.

KNOW YOUR EXPERIMENTS

At least one—and most likely several—of the Data Questions will deal with experiments. Make sure you understand all the basic points of an experiment: testing a hypothesis, setting up an experiment properly to isolate a particular variable, and so on so that you will be able to breeze through this section when you come to it.

QUESTIONS WITH GRAPHS

Most graph questions usually require a bit of biology knowledge to determine what the right answer is, but some graph questions only test whether or not you can read a graph properly. If you can make sense of the vertical and horizontal axes, then you can determine what the correct answer is. Granted, very, very few graph questions are this easy, but even so, it's nice to have a slam-dunk question or two. Therefore, if you see a graph, look at the problem and see if you can answer the question just by knowing how to read a graph.

That's all that can be said for the multiple-choice section of the AP Biology exam. Be sure to practice these strategies on the practice test in this book so that you'll actually use them on the real test. Once you implement these techniques, your mindset, approach, and score should benefit.

HOW TO APPROACH THE GRID-IN ITEMS

Grid-in items appear in Part B of Section I after the multiple-choice questions. Six questions are presented that require you to apply your scientific thinking and mathematical skills to calculate a response. Then you "grid in" your responses in the grid provided on your answer sheet, which will resemble the one shown here.

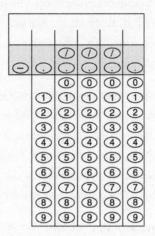

The next question is a sample grid-in item.

>   **Part B Directions:** Part B consists of questions requiring numeric answers. Calculate the correct answer for each question.
>
>   In a certain species of fruit fly, the allele for red eyes R is dominant to the white allele r.
>
>   A scientist performed a cross between a red-eyed fruit fly and a white-eyed fly. When crossed, 62 offspring result. Of these 62 offspring, 43 have red eyes and 19 have white eyes.
>
>   Calculate the chi-squared value for the null hypothesis that the red-eyed parent was heterozygous for the eye-color gene. Give your answer to the nearest tenth.

Just like Data Questions, Grid-In items require you to analyze the presented information carefully. You must identify the given information, determine the correct formula for solving the problem, make your calculations, and then correctly grid in the answer. Obviously, there is more room for error on the Grid-In questions and this should be factored in to your decision to "answer now" or "save for later." Keep in mind, all things being equal, if you need to guess, you'll have better odds guessing on a multiple-choice question, where the answer is actually given to you. As a case in point, would you rather pick the answer to the chi-squared question above from four answer choices, or grid in your own answer?

To help with the more quantitative problems on the AP Biology exam, you will be allowed to use basic four-function calculators (with square root). In addition, because these questions focus on applying math and not just recalling information, a formula list is provided for you. A sample formula list is provided in Appendix A of this book.

Of course, the multiple-choice questions only account for 50 percent of your total score. To get the other 50 percent, you have to tackle the free-response questions. Before jumping to the free-response question strategies, take a few minutes to read about the 10-minute reading period.

HOW TO APPROACH THE 10-MINUTE READING PERIOD

There is a 10-minute reading period sandwiched between the multiple-choice section and the free-response (essay) portion of the AP Biology exam. Notice that this is a "reading" period, not a "nap" period. Ten minutes isn't much time, but it does give you an opportunity to read the essay questions. You can write notes in the question booklet, but these will not be seen by graders.

This gives you just over one minute per question to plan your response. Take at least 30 seconds to read and reread each question. Make sure you understand what is being asked. Next, you should jot down any thoughts you have about the answer on a piece of scratch paper or in the test booklet. Write down key words you want to mention. Undoubtedly, many of you have brainstormed ideas when writing an essay for your English class. This is exactly what you want to do here as well: Brainstorm ideas about the best way to answer each free-response question.

With the remaining time, make a quick outline of how you would answer each question. You don't need to write complete sentences; just jot down notes that you can understand. If drawings or diagrams are requested, make a brief, crude version of what they will look like.

The following is an example of a free-response question and some notes that could be taken. The actual answer should not be in this outline form, but in a more coherent essay, with more detailed descriptions of each concept where appropriate.

Transcription is the process of generating RNA from a DNA blueprint. It occurs in both prokaryotes and eukaryotes, but the mechanism and results are different. **Describe** the differences between prokaryotic and eukaryotic transcription for the following cases.

(A) Where does transcription occur, and how does this affect translation?

(B) What is the structure of a messenger RNA?

(C) What enzymes and other molecules and sequences are necessary for transcription?

Things to Cover:

Prok — cytoplasm, translation → Christmas-tree structure

Euk — nucleus (rRNA in nucleolus), no translation until exits

Prok — many genes on one mRNA (operon/polycistronic), no introns, cap, or tail

Euk — single gene, introns, cap, poly A tail

Prok — same RNA pol for all RNA (sigma factor), –10 and –35 region of promoter, maybe rho for termination

Euk — RNA pol I, II, and III (rRNA, mRNA, tRNA), transcription factors, enhancers, TATA box or something like it, spliceosome

By the time you've written that, your one-to-two minutes on that question should be up. Move on to the next question. When the time comes to start writing your answers, you'll have a good set of notes on which to base your answer to this question.

HOW TO APPROACH THE FREE-RESPONSE QUESTIONS

For Section II, you'll have 80 minutes (after the reading period) to answer eight questions. That's an average of 10 minutes per question, which gives you an idea of how much work each question may take. You will likely spend more time on each of the two long free-response questions than on each of the six short-response questions. Take the time to make your answers as precise and detailed as possible while managing the allotted time.

IMPORTANT DISTINCTIONS

Each free-response question will, of course, be about a distinct topic. However, this is not the only way in which these questions differ from one another. Each question will also need a certain kind of answer, depending on the type of question it is. Part of answering each question correctly is understanding what general type of answer is required. There are five important signal words that indicate the rough shape of the answer you should provide:

- Describe
- Discuss
- Explain
- Compare
- Contrast

Each of these words indicates that a specific sort of response is required; none of them mean the same thing. Questions that ask you to *describe*, *discuss*, or *explain* are testing your comprehension of a topic. A description is a detailed verbal picture of something; a description question is generally asking for "just the facts." This is not the place for opinions or speculation. Instead, you want to create a precise picture of something's features and qualities. A description question might, for example, ask you to describe the results you would expect from an experiment. A good answer here will provide a rich, detailed account of the results you anticipate.

A question that asks you to discuss a topic is asking you for something broader than a mere description. A discussion is more like a conversation about ideas, and—depending on the topic—this may be an appropriate place to talk about tension between competing theories and views. For example, a discussion question might ask you to discuss which of several theories offers the best explanation for a set of results. A good answer here would go into detail about why one theory does a better job of explaining the results, and it would talk about why the other theories cannot cope with the results as thoroughly.

A question that asks you to explain something is asking you to take something complicated or unclear and present it in simpler terms. For example, an explanation question might ask you to explain why an experiment is likely to produce a certain set of results, or how one might measure a certain sort of experimental result. A simple description of an experimental setup would not be an adequate answer to the latter question. Instead, you would need to describe that setup *and* talk about why it would be an effective method of measuring the result.

COMPARE VS. CONTRAST QUESTIONS

Questions that ask you to *compare* or *contrast* are asking you to analyze a topic in relation to something else. A question about comparison needs an answer that is focused on similarities between the two things. A question that focuses on contrast needs an answer emphasizing differences and distinctions.

THREE POINTS TO REMEMBER ABOUT THE FREE-RESPONSE QUESTIONS

1. **Most Questions Are Stuffed with Smaller Questions**. You usually won't get one broad question like, "Are penguins really happy?" Instead, you'll get an initial setup followed by questions labeled (a), (b), (c), and so on. Expect to spend a paragraph writing about each lettered question.

2. **Writing Smart Things Earns You Points**. For each subquestion on a free-response question, points are given for saying the right thing. The more points you score, the better off you are on that question. Going into the details about how points are scored would make your head spin, but in general, the AP Biology people have a rubric, which acts as a blueprint for what a good answer should look like. Every subsection of a question has

two to five key ideas attached to it. If you write about one of those ideas, you earn yourself a point. There's a limit to how many points you can earn on a single subquestion, and there are other strange regulations, but it boils down to this: Writing smart things about each question will earn you points toward that question.

So don't be terse or in a hurry. You have about 10 minutes to answer each free-response question. Use the time to be as precise as you can be for each subquestion. Part of being precise is presenting your answer in complete sentences. Do not simply make lists or outlines. Sometimes doing well on one subquestion will earn you enough points to cover up for another subquestion you're not as strong on. When all the points are tallied for that free-response question, you come out strong on total points, even though you didn't ace every single subquestion.

3. **Mimic the Data Questions.** Data often describe an experiment and provide a graph or table to present the information in visual form. On at least one free-response question, you will be asked about an experiment in some form or another. To score points on this question, you must describe the experiment well and perhaps present the information in visual form.

So, look over the sample Data Questions you see in this book and on the actual test, because you can use knowledge of this format when tackling the free-response questions. In a way, this is just another aspect of the good science idea. The AP Biology test wants to show you what good science looks like on the Data Questions. You can then use that information when crafting your free-response answers.

Beyond these points, there's a bit of a risk in the free-response section because there are only eight questions. If you get a question on a subject you're weak in, things might look grim. Still, take heart. Quite often, you'll earn some points on every question because there will be some subquestions or segments that you are familiar with.

Remember, the goal is not perfection. If you can ace four of the questions and slug your way to partial credit on the other four, you will put yourself in a position to get a good score on the entire test. That's the Big Picture, so don't lose sight of it just because you don't know the answer to one subquestion.

FREE-RESPONSE POINTS

Don't forget—you *only* receive points for relevant correct information; you receive *no points* for incorrect information or for restating the question, which also eats up valuable time!

MAXIMIZE YOUR FRQ SCORE

1. Only answer the number of subsections the long free-response questions call for. For example, if the question has four sections (a, b, c, and d) and says to choose three parts, then choose *only* three parts.

2. There are almost always easy points that you can earn. State the obvious and provide a brief but accurate explanation for it.

3. In many instances, you can earn points by defining relevant terms. (Example: Writing *osmosis* would not get you a point, but mentioning "movement of water down a gradient across a semipermeable membrane" would likely get the point).

4. While grammar and spelling are not assessed on the free-response portion, correct spellings of words and legible sentences will increase your chances of earning points.

5. You do not have to answer free-response questions in the order in which they appear on the exam. It's a good strategy to answer the questions you are most comfortable with first, and then answer the more difficult ones.

6. The length of your response does not determine your score—a one-page written response containing accurate, succinct, yet detailed information can score the maximum amount of points, while other essays spanning three to four pages of vague, inaccurate materials may not earn any.

7. Be careful that you do not over-explain a concept. Where the initial explanation gets you points, contradictions cause points to be taken away.

8. Keep personal opinions out of free responses. Base your response on factual researched knowledge.

9. Relax and do your best. You know more than you think!

Be sure to use all the strategies discussed in this chapter when taking the practice exams. Trying out the strategies there will get you comfortable with them, and you should be able to put them to good use on the real exam.

STRESS MANAGEMENT

You can beat anxiety the same way you can beat the AP Biology exam—by knowing what to expect beforehand and developing strategies to deal with it.

SOURCES OF STRESS

In the space provided, write down your sources of test-related stress. The idea is to pin down any sources of anxiety so you can deal with them one by one. Here are some common examples to get you started:

- I always freeze up on tests.

- I'm nervous about the molecular biology (or the evolutionary biology, ecology, etc.).

- I need a good/great score to get into my first-choice college.

- My older brother/sister/best friend/girlfriend/boyfriend did really well. I must match that score or do better.

- My parents, who are paying for school, will be quite disappointed if I don't do well.

- I'm afraid of losing my focus and concentration.

- I'm afraid I'm not spending enough time preparing.

- I study like crazy but nothing seems to stick in my mind.

- I always run out of time and get panicky.

- The simple act of thinking, for me, is like wading through refrigerated honey.

Now list your own.

MY SOURCES OF STRESS

_____

_____

_____

_____

_____

_____

Read through the list. Cross out things or add things. Now rewrite the list in order of most disturbing to least disturbing.

MY SOURCES OF STRESS, IN ORDER

_____

_____

_____

_____

_____

_____

Chances are, the top of the list is a fairly accurate description of exactly how you react to test anxiety, both physically and mentally. The later items usually describe your fears (disappointing mom and dad, looking bad, etc.). Taking care of the major items from the top of the list should go a long way toward relieving overall test anxiety. That's what we'll do next.

STRENGTHS AND WEAKNESSES

Take 60 seconds to list the areas of biology that you are good at. They can be general (evolution) or specific (protein synthesis). Put down as many as you can think of, and if possible, time yourself. Write for the entire time; don't stop writing until you've reached the one-minute stopping point. Go.

STRONG TEST SUBJECTS

_____

_____

_____

_____

_____

Now, take one minute to list areas of the test you're not so good at, just plain bad at, have failed at, or keep failing at. Again, keep it to one minute, and continue writing until you reach the cutoff. Go.

WEAK TEST SUBJECTS

_____

_____

_____

_____

_____

Taking stock of your assets and liabilities lets you know the areas you don't have to worry about and the ones that will demand extra attention and effort. It helps a lot to find out where you need to spend extra effort. We mostly fear what we don't know and are nervous to face. You can't help feeling more confident when you know you're actively strengthening your chances of earning a higher overall score.

Now, go back to the "good" list, and expand on it for two minutes. Take the general items on that first list and make them more specific; take the specific items and expand them into more general conclusions. Naturally, if anything new comes to mind, jot it down. Focus all of your attention and effort on your strengths. Don't underestimate yourself or your abilities. Give yourself full credit. At the same time, don't list strengths you don't really have; you'll only be fooling yourself.

Expanding from general to specific might go as follows. If you listed ecology as a broad topic you feel strong in, you would then narrow your focus to include areas of this subject about which you are particularly knowledgeable. Your areas of strength might include population analysis, energy flow in communities, and so on. Whatever you know well goes on your "good" list. OK. Check your starting time. Go.

STRONG TEST SUBJECTS: AN EXPANDED LIST

_____

_____

_____

_____

_____

After you've stopped, check your time. Did you find yourself going beyond the two minutes allotted? Did you write down more things than you thought you knew? Is it possible you know more than you've given yourself credit for? Could that mean you've found a number of areas in which you feel strong?

CONGRATULATIONS!

You just took an active step toward helping yourself. Enjoy your increased feelings of confidence, and use them when you take the AP Biology exam.

STRESS-MANAGEMENT ACTIVITY

Sit in a comfortable chair in a quiet setting. Close your eyes and breathe in a deep, satisfying breath of air. Then exhale the air completely. Imagine you're blowing out a candle with your last little puff of air. Do this two or three more times, filling your lungs to their maximum and emptying them totally. Let your body sink deeper into the chair as you become even more comfortable.

1. With your eyes shut, you can notice something very interesting. You're no longer dealing with the worrisome stuff going on in the world outside of you. Now you can concentrate on what happens inside you. The more you recognize your own physical reactions to stress and anxiety, the more you can do about them.

2. Visualize a relaxing situation. Make it as detailed as possible and notice as much as you can.

3. Stay focused on the images as you sink deeper into your chair. Breathe easily and naturally. You might have the sensations of any stress or tension draining from your muscles and flowing downward, out of your feet and away from you.

4. Take a moment to check how you're feeling. Notice how comfortable you've become. Imagine how much easier it would be if you could take the test feeling this relaxed and in this state of ease.

5. Close your eyes and start remembering a real-life situation in which you did well on a test. Make the memory as detailed as possible. Think about the sights, the sounds, the smells, even the tastes associated with this remembered experience. Now start thinking about the AP Biology exam. Just imagine taking the upcoming test with the same feelings of confidence and relaxed control.

This exercise is a great way to bring the test down to earth. You should practice this exercise often, especially when you feel burned out on test preparation. The more you practice it, the more effective the exercise will be for you.

EXERCISE

Whether it's jogging, walking, biking, mild aerobics, push-ups, or a pickup basketball game, physical exercise is a very effective way to stimulate both your mind and body and to improve your ability to think and concentrate. Lots of students get out of the habit of regular exercise when they're prepping for the exam. Also, sedentary people get less oxygen to the blood, and hence to the brain, than active people. You can watch TV fine with a little less oxygen; you just can't think as well.

Any big test is a bit like a race. Finishing the race strong is just as important as being quick early on. If you can't sustain your energy level in the last sections of the exam, you could blow it. Along with a good diet and adequate sleep, exercise is an important part of keeping yourself in fighting shape and thinking clearly for the long haul.

There's another thing that happens when students don't make exercise an integral part of their test preparation. Like any organism in nature, you operate best if all your "energy systems" are in balance. Studying uses a lot of energy, but it's all mental. When you take a study break, do something active. Take a 5-to-10-minute exercise break for every 50 or 60 minutes that you study. The physical exertion helps keep your mind and body in sync. This way, when you finish studying for the night and go to bed, you won't lie there unable to sleep because your head is wasted while your body wants to run a marathon.

IMPORTANT

One warning about exercise: It's not a good idea to exercise vigorously right before you go to bed. This could easily cause sleep-onset problems. For the same reason, it's also not a good idea to study right up to bedtime. Make time for a "buffer period" before you go to bed. Take 30 to 60 minutes to take a long, hot shower, to meditate, or to watch a TV show.

STAY DRUG FREE

Using drugs to prepare for or take a big test is not a good idea. Don't take uppers to stay alert. Amphetamines make it hard to retain information. Mild stimulants, such as coffee, cola, or over-the-counter caffeine pills can help you study longer because they keep you awake, but they can also lead to agitation, restlessness, and insomnia. Some people can drink a pot of coffee and sleep like a baby. Others have one cup and start to vibrate. It all depends on your tolerance for caffeine. Remember, a little anxiety is a good thing. The adrenaline that gets pumped into your bloodstream helps you stay alert and think more clearly.

You can also rely on your brain's own endorphins. Endorphins have no side effects and they're free. It just takes some exercise to release them. Running, bicycling, swimming, aerobics, and power walking all cause endorphins to occupy the happy spots in your brain's neural synapses. In addition, exercise develops your mental stamina and increases the oxygen transfer to your brain.

STRESS-BUSTING TIP

To reduce stress, you should eat fruits and vegetables; low-fat protein such as fish, skinless poultry, beans and legumes; or whole grains such as brown rice, whole wheat bread, and pastas. Don't eat sweet, high-fat snacks. Simple carbohydrates like sugar make stress worse, and fatty foods lower your immunity. Don't eat salty foods either. They can deplete potassium, which you need for nerve functions. You can go back to your Combos-and-Dew diet after the AP Biology exam.

ISOMETRICS

Here's another natural route to relaxation and invigoration. You can do it whenever you get stressed out, including during the test. Close your eyes. Start with your eyes and—without holding your breath—gradually tighten every muscle in your body (but not to the point of pain) in the following sequence:

- Close your eyes tightly.

- Squeeze your nose and mouth together so that your whole face is scrunched up. (If it makes you self-conscious to do this in the test room, skip the face-scrunching part.)

- Pull your chin into your chest, and pull your shoulders together.

- Tighten your arms to your body, then clench your fists.

- Pull in your stomach.

- Squeeze your thighs and buttocks together, and tighten your calves.

- Stretch your feet, then curl your toes (watch out for cramping in this part).

At this point, every muscle should be tightened. Now, relax your body, one part at a time, in reverse order, starting with your toes. Let the tension drop out of each muscle. The entire process might take five minutes from start to finish (maybe a couple of minutes during the test). This clenching and unclenching exercise will feel silly at first, especially the buttocks part, but if you get good at it, you will feel very relaxed.

COUNTDOWN TO THE TEST

STUDY SCHEDULE

The schedule presented here is the ideal. Compress the schedule to fit your needs. Do keep in mind, though, that research in cognitive psychology has shown that the best way to acquire a great deal of information about a topic is to prepare over a long period of time. Because you may have several months to prepare for this exam, it makes sense for you to use that time to your advantage. This book, along with your text, should be invaluable in helping you prepare for this test.

If you have two semesters to prepare, use the following schedule:

September:

Take the diagnostic test in this book and isolate areas in which you need help. The diagnostic will serve to familiarize you with the type of material you will be asked about on the AP Biology exam.

Begin reading your biology textbook along with the class outline.

October–February:

Continue reading this book and use the star system to help guide you to the most salient information for the exam.

March and April:

Take the two practice tests and get an idea of your score. Also, identify the areas in which you need to brush up. Then go back and review those topics in both this book and your biology textbook.

May:

Do a final review and take the exam.

If you have only one semester to prepare, you'll need a more compact schedule:

January:

Take the diagnostic test in this book.

February–April:

Begin reading this book and identify areas of strengths and weaknesses.

Late April:

Take the two practice tests and use your performance results to guide you in your preparation.

May:

Do a final review and take the exam.

PACE YOURSELF!

Pace yourself throughout the school year as you prep for the exam. Try to allow at least 30 minutes nightly to review. Use your textbook, class handouts, and Internet resources for content mastery.

THREE DAYS BEFORE THE TEST

It's almost over. Eat a PowerBar, drink some soda—do whatever it takes to keep going. Here are Kaplan's strategies for the three days leading up to the test.

Take a full-length practice test under timed conditions. Use the techniques and strategies you've learned in this book. Approach the test strategically, actively, and confidently.

WARNING

DO NOT take a full-length practice test if you have fewer than 48 hours left before the test. Doing so will probably exhaust you and hurt your score on the actual test.

TWO DAYS BEFORE THE TEST

Go over the results of your practice test. Don't worry too much about your score or about whether you got a specific question right or wrong. The practice test doesn't count, but examine your performance on specific questions with an eye to how you might get through each one faster and better on the test to come.

THE NIGHT BEFORE THE TEST

DO NOT STUDY. Get together an "AP Biology Exam Kit" containing the following items:

PROBLEM SHOOTING

As you study, highlight challenging terms and keep a list of their meanings so they won't trip you up on Test Day.

- A watch (as long as it doesn't beep)
- A few No. 2 pencils (pencils with slightly dull points fill the ovals better), mechanical pencils are NOT permitted
- A pen with black or dark blue ink (for the free-response questions)
- Erasers
- Photo ID card
- Your admission ticket from ETS

Know exactly where you're going, exactly how you're getting there, and exactly how long it takes to get there. It's probably a good idea to visit your test center sometime before the day of the test so that you know what to expect—what the rooms are like, how the desks are set up, and so on.

Relax the night before the test. Do the relaxation and visualization techniques. Read a good book, take a long, hot shower, watch something you'll enjoy. Get a good night's sleep. Go to bed early and leave yourself extra time in the morning.

THE MORNING OF THE TEST

First, wake up. After that:

- Eat breakfast. Make it something substantial, but not anything too heavy or greasy.

- Don't drink a lot of coffee if you're not used to it. Bathroom breaks cut into your time, and too much caffeine is a bad idea.

- Dress in layers so that you can adjust to the temperature of the test room.

- Read something. Warm up your brain with a newspaper or a magazine. You shouldn't let the exam be the first thing you read that day.

- Be sure to get there early. Allow yourself extra time for traffic, mass transit delays, and/or detours.

STAY ORGANIZED!

Staying organized is a big part of AP success. Keep your class notes and handouts together, and keep an easy-to-follow schedule—so you know just where you are in the program and what material you still need to cover.

DURING THE TEST: PANIC PLAN

Don't be shaken. If you find your confidence slipping, remind yourself how well you've prepared. You know the structure of the test; you know the instructions; you've had practice with—and have learned strategies for—every question type.

If something goes really wrong, don't panic. If you accidentally misgrid your answer page or put the answers in the wrong section, raise your hand and tell the proctor. He or she might be able to arrange for you to regrid your test after it's over, when it won't cost you any time.

AFTER THE TEST

You might walk out of the AP Biology exam thinking that you blew it. This is a normal reaction. Lots of people—even the highest scorers—feel that way. You tend to remember the questions that stumped you, not the ones that you knew. We're positive that you will have performed well and scored your best on the exam because you followed the Kaplan strategies outlined in this section. Be confident in your preparation, and celebrate the fact that the AP Biology exam is soon to be a distant memory.

Now, continue your exam prep by taking the diagnostic test that follows this chapter. This short test will give you an idea of the format of the actual exam, and it will demonstrate the scope of topics covered. After the diagnostic test, you'll find answers with detailed explanations. Be sure to read these explanations carefully, even if you got the question right, as you can pick up bits of knowledge from them. Use your score to learn which topics you need to review more carefully. Of course, all the strategies in the world can't save you if you don't know anything about biology. The chapters following the diagnostic test will help you review the primary concepts and facts that you can expect to encounter on the AP Biology exam.

Part Two

DIAGNOSTIC TEST

AP BIOLOGY DIAGNOSTIC TEST

Enough said about strategies; on to the review portion of our program. The following chapters contain a wealth of information and review questions about all the main topics covered on the exam. Ideally, you will have the time to go through every chapter and try every review question while working at a steady pace. You'll finish with enough time to take a practice test each week, and then follow that up by taking the real AP Biology exam.

This is the ideal scenario, but one thing often prevents students from following it. That one thing is the real world. The fact is, many students have schedules that are already chock-full of activities. Finding large chunks of time to devote to studying for a test—one that isn't even part of your regular schoolwork—isn't just difficult, it's next to impossible.

If this is the case with you, do not feel bad. Most students feel this way. To help you decide where to start studying, take the following diagnostic test. The questions in this diagnostic test are designed to cover most of the topics you will encounter on the AP Biology exam. After you take it, you can use the results to give yourself a broad idea of what subjects you are strong in and what topics you need to review more. You can use this information to tailor your approach to the following review chapters. Hopefully, you'll still have time to read all the chapters, but if pressed, you can start with the chapters and subjects you know you need to work on.

Time yourself, and take the entire test without interruption—you can always call your friend back after you finish. Also, no TV! You won't get to watch your favorite show while taking the real AP Biology exam, so you may as well get used to it now.

DON'T CRAM!

Mastering the enormous amount of material in AP Biology will be much easier if you learn it in small chunks rather than cramming the night before the exam!

DON'T FORGET!

Be sure to read the answer explanations for all questions, even those you answered correctly. Even if you got the question right, reading another person's answer can give you insights that will prove helpful on the real exam.

Good luck on the diagnostic test!

HOW TO TAKE PRACTICE TESTS

The next section of this book consists of a diagnostic practice test. Taking a practice AP exam gives you an idea of what it's like to answer these test questions for a longer period of time, one that approximates the real test. You'll find out which areas you're strong in and where additional review may be required. Any mistakes you make now are ones you won't make on the actual exam, as long as you take the time to learn where you went wrong.

For the most accurate results you should approximate real test conditions as closely as possible. Before taking a practice test, find a quiet place where you can work uninterrupted for three hours. Time yourself according to the time limit at the beginning of each section. The full-length diagnostic test includes 63 multiple-choice questions, six grid-in items, and two long and six short free-response questions. You are allotted 90 minutes for the multiple-choice questions and grid-in items, a ten-minute break, and 90 minutes to answer the free-response questions, which begins with a recommended 10-minute reading period. Use the reading period to plan your answers for the free-response questions, although you may begin writing your responses before the 10 minutes are up.

As you take the practice tests, remember to pace yourself. Train yourself to be aware of the time you are spending on each question. Try to be aware of the general types of questions you encounter, and alert to certain strategies or approaches that help you to handle the various question types more effectively.

After taking a practice exam, be sure to read the detailed answer explanations that follow. These will help you identify areas that could use additional review. Even when you've answered a question correctly, you can learn additional information by looking at the answer explanation.

Finally, it's important to approach the test with the right attitude. You're going to get a great score because you've reviewed the material and learned the strategies in this book.

Good luck!

HOW TO COMPUTE YOUR SCORE
SCORING THE MULTIPLE-CHOICE QUESTIONS AND GRID-IN ITEMS

To compute your score on this portion of the diagnostic test, calculate the number of questions you got right, then divide by 69 to get the percentage score for the multiple-choice portion of that test.

SCORING THE FREE-RESPONSE QUESTIONS

Reviewers will have key bits of information that they are looking for in response to each free-response question. Each piece of information that they are able to check off in your essay is a point toward a better score.

To figure out your approximate score for the free-response questions, look at the key points found in the sample response for each question. For each key point you included, add a point. Tally the number of points you earned for each question. Add the points earned for each question to get your total number of points. Divide your sum by the total number of points available for all of the free-response questions. This will give you a percentage score for the free-response portion of that test.

CALCULATING YOUR COMPOSITE SCORE

Your score on the AP Biology exam is a combination of your scores on the multiple-choice portion of the exam and the free-response section. The free-response section is worth 50 percent of the exam score, and the multiple-choice is worth 50 percent.

To determine your score, multiply the percentage score for the free-response section by 0.5, and for the multiple-choice section, multiply by 0.5. Add these numbers together and multiply by 100 to get your final score. Remember, however, that much of this depends on how well all of those taking the AP test do. If you do better than average, your score would be higher. The numbers here are just approximations.

The approximate score range is as follows:

5 = 65 –100 (extremely well qualified), **4** = 55–64 (well qualified), **3** = 45–54 (qualified),
2 = 40–44 (possibly qualified), **1** = 0–39 (no recommendation)

If your score falls between 55 and 100, you're doing great, keep up the good work! If your score is lower than 54, there's still hope—keep studying and you will be able to obtain a much better score on the exam before you know it.

Diagnostic Test Answer Grid

1. Ⓐ Ⓑ Ⓒ Ⓓ
2. Ⓐ Ⓑ Ⓒ Ⓓ
3. Ⓐ Ⓑ Ⓒ Ⓓ
4. Ⓐ Ⓑ Ⓒ Ⓓ
5. Ⓐ Ⓑ Ⓒ Ⓓ
6. Ⓐ Ⓑ Ⓒ Ⓓ
7. Ⓐ Ⓑ Ⓒ Ⓓ
8. Ⓐ Ⓑ Ⓒ Ⓓ
9. Ⓐ Ⓑ Ⓒ Ⓓ
10. Ⓐ Ⓑ Ⓒ Ⓓ
11. Ⓐ Ⓑ Ⓒ Ⓓ
12. Ⓐ Ⓑ Ⓒ Ⓓ
13. Ⓐ Ⓑ Ⓒ Ⓓ
14. Ⓐ Ⓑ Ⓒ Ⓓ
15. Ⓐ Ⓑ Ⓒ Ⓓ
16. Ⓐ Ⓑ Ⓒ Ⓓ
17. Ⓐ Ⓑ Ⓒ Ⓓ
18. Ⓐ Ⓑ Ⓒ Ⓓ
19. Ⓐ Ⓑ Ⓒ Ⓓ
20. Ⓐ Ⓑ Ⓒ Ⓓ
21. Ⓐ Ⓑ Ⓒ Ⓓ

22. Ⓐ Ⓑ Ⓒ Ⓓ
23. Ⓐ Ⓑ Ⓒ Ⓓ
24. Ⓐ Ⓑ Ⓒ Ⓓ
25. Ⓐ Ⓑ Ⓒ Ⓓ
26. Ⓐ Ⓑ Ⓒ Ⓓ
27. Ⓐ Ⓑ Ⓒ Ⓓ
28. Ⓐ Ⓑ Ⓒ Ⓓ
29. Ⓐ Ⓑ Ⓒ Ⓓ
30. Ⓐ Ⓑ Ⓒ Ⓓ
31. Ⓐ Ⓑ Ⓒ Ⓓ
32. Ⓐ Ⓑ Ⓒ Ⓓ
33. Ⓐ Ⓑ Ⓒ Ⓓ
34. Ⓐ Ⓑ Ⓒ Ⓓ
35. Ⓐ Ⓑ Ⓒ Ⓓ
36. Ⓐ Ⓑ Ⓒ Ⓓ
37. Ⓐ Ⓑ Ⓒ Ⓓ
38. Ⓐ Ⓑ Ⓒ Ⓓ
39. Ⓐ Ⓑ Ⓒ Ⓓ
40. Ⓐ Ⓑ Ⓒ Ⓓ
41. Ⓐ Ⓑ Ⓒ Ⓓ
42. Ⓐ Ⓑ Ⓒ Ⓓ

43. Ⓐ Ⓑ Ⓒ Ⓓ
44. Ⓐ Ⓑ Ⓒ Ⓓ
45. Ⓐ Ⓑ Ⓒ Ⓓ
46. Ⓐ Ⓑ Ⓒ Ⓓ
47. Ⓐ Ⓑ Ⓒ Ⓓ
48. Ⓐ Ⓑ Ⓒ Ⓓ
49. Ⓐ Ⓑ Ⓒ Ⓓ
50. Ⓐ Ⓑ Ⓒ Ⓓ
51. Ⓐ Ⓑ Ⓒ Ⓓ
52. Ⓐ Ⓑ Ⓒ Ⓓ
53. Ⓐ Ⓑ Ⓒ Ⓓ
54. Ⓐ Ⓑ Ⓒ Ⓓ
55. Ⓐ Ⓑ Ⓒ Ⓓ
56. Ⓐ Ⓑ Ⓒ Ⓓ
57. Ⓐ Ⓑ Ⓒ Ⓓ
58. Ⓐ Ⓑ Ⓒ Ⓓ
59. Ⓐ Ⓑ Ⓒ Ⓓ
60. Ⓐ Ⓑ Ⓒ Ⓓ
61. Ⓐ Ⓑ Ⓒ Ⓓ
62. Ⓐ Ⓑ Ⓒ Ⓓ
63. Ⓐ Ⓑ Ⓒ Ⓓ

64.

65.

66.

67.

68.

69.

Section I: Multiple-Choice Questions and Grid-In Items

Time: 90 Minutes
63 Multiple-Choice Questions and 6 Grid-In Items

1. Amylase is an enzyme important in the digestion of starches. What kind of organic macromolecule is amylase?

   (A) A protein
   (B) A carbohydrate
   (C) A lipid
   (D) A nucleic acid

2. Which two subcellular organelles contain unique DNA similar to that of bacteria and are thought to have evolved from prokaryotic symbionts of the first eukaryotic cells?

   (A) The nucleus and endoplasmic reticulum
   (B) The chloroplast and mitochondrion
   (C) The nucleolus and mitochondrion
   (D) The chloroplast and ribosomes

3. Which of the following structures is present in all cells?

   (A) Nucleus
   (B) Cell wall
   (C) Plasma membrane
   (D) Mitochondria

Questions 4 and 5 refer to the following information.

$$6CO_2 + 6H_2O + energy \rightarrow C_6H_{12}O_6 + O_2$$
$$C_6H_{12}O_6 + 6O_2 \rightarrow 6CO_2 + 6H_2O + ATP$$

4. During aerobic cellular respiration, the C–H bond in glucose is broken and the electrons are ultimately transferred to oxygen. Which of the following best explains why the concentration of electron poor hydrogens (H^+) does not drastically change as a result?

   (A) Protons are used as building blocks for macromolecules.
   (B) Protons associate with ATP, which normally carries a negative charge.
   (C) Protons are transported out of the cell, where they are removed via diffusion.
   (D) Protons combine with oxygen anions to form water.

5. Which molecule is reduced to form glucose?

   (A) CO_2
   (B) ATP
   (C) O_2
   (D) H_2O

6. Production of purple kernels in *Zea maize* (corn) is dominant over yellow kernels, and smooth kernels are dominant over wrinkled. The two traits are passed on independently of one another. Two heterozygous corn plants with purple, smooth kernels are crossed. Out of 160 of their offspring plants, how many would you expect to have yellow and smooth kernels?

   (A) 120
   (B) 90
   (C) 60
   (D) 30

GO ON TO THE NEXT PAGE

7. Sex-linked recessive disorders are usually carried on the X chromosome and most often affect males. This is because

(A) mothers always pass the disorders on to their sons

(B) it takes only one copy of the gene to affect males

(C) it takes only one copy of the gene to affect females

(D) it takes two copies of the gene to affect males

Questions 8 and 9 refer to the following information.

Normal DNA strand:

5′ TAC ACA GAA GGA GAG GGA ACA ATT 3′
Methionine Cystine Leucine Proline Leucine Proline Cystine Stop

Mutated DNA strand:
5′ TAC GAC AGA AGG AGA GGG AAC AAT 3′
Methionine Leucine Serine Serine Serine Proline Leucine Phenylalanine

8. What type of mutation is shown?

(A) Substitution

(B) Deletion

(C) Insertion

(D) Translocation

9. What effect does the mutation have on the codon sequence?

(A) Transposition

(B) Trinucleotide repeat

(C) Inversion

(D) Frame shift

10. While on the Galápagos Islands, Charles Darwin noted that several distinct kinds of finches had beak characteristics well suited to the various kinds of foods that they ate. This is an example of

(A) analogous structures.

(B) Hardy-Weinberg equilibrium.

(C) genetic drift.

(D) adaptive radiation.

11. Animals in the phylum *Echinodermata*, such as sea stars and sand dollars, are thought to be more closely related to the phylum *Chordata* (which includes humans and other vertebrates) than to other animal phyla. Which of the following observations provides the best justification for this conclusion?

(A) *Echinodermata* and *Chordata* are the most abundant animal phyla.

(B) During development, the anus forms prior to the mouth in both phyla.

(C) All species in both phyla have a common ancestor.

(D) Neither phyla includes obligate anaerobes.

12. *Naeglaria fowleri* causes a fatal form of meningitis. The infectious form of this organism inhabits fresh water in warm climates, often in the sediment of lakes. It can infect humans when they swim in infested lakes, allowing entry through the nose. *N. fowleri* has a true membrane-bound nucleus and cellular organelles. It is a unicellular, heterotrophic organism that lacks a cell wall and moves via pseudopodia. What type of organism is it?

(A) Bacteria

(B) Virus

(C) Protozoan

(D) Fungus

GO ON TO THE NEXT PAGE ⟩

13. Which organelle is principally involved in protein degradation and cellular digestion?

(A) Mitochondrion

(B) Lysosome

(C) Smooth endoplasmic reticulum

(D) Rough endoplasmic reticulum

14. What is the probability that a mother who is a carrier for cystic fibrosis, an autosomal recessive disorder, will have an affected child if the father is genotypically normal?

(A) 0%

(B) 25%

(C) 50%

(D) 100%

15. The sodium potassium pump is an ATPase that pumps 3 Na^+ out of the cell and 2 K^+ into the cell for each ATP hydrolyzed. Cells can use the pump to help maintain cell volume. Which of the following would most likely happen to the rate of ATP consumption immediately after a cell is moved to a hypotonic environment?

(A) It would remain the same.

(B) It would decrease.

(C) It would increase.

(D) It would increase, then decrease.

Questions 16–19 refer to the following images.

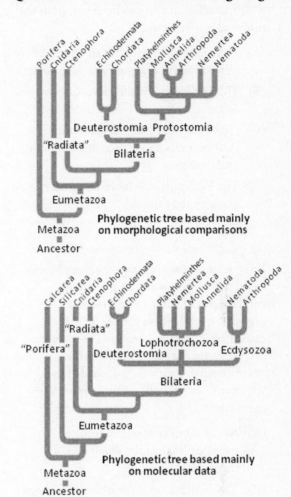

Phylogenetic tree based mainly on morphological comparisons

Phylogenetic tree based mainly on molecular data

16. The main difference between the two phylogenetic trees is best explained by which of the following?

(A) Molecular differences among *Porifera* undetected by visual comparison

(B) Subdivision of *Protostomia* into two sister taxa

(C) The relative timeline of *Deuterostomia* divergence

(D) The convergence of the entire animal kingdom on a common ancestor

GO ON TO THE NEXT PAGE

17. The molecular-based tree does NOT support which conclusion that can be drawn from the morphology-based tree?

(A) *Annelida* and *Arthropoda* are more closely related than *Arthropoda* and *Nematoda*.

(B) The *Porifera* are divided into *Calcarea* and *Silicarea*.

(C) *Nemertea* and *Nematoda* have more common DNA sequences than *Mollusca* and *Nemertea*.

(D) The *Radiata* have a more distant common ancestor than the *Bilateria*.

18. If the scientists who created the morphology-based tree relied mainly on phenotypic comparisons of adult and developing organisms, while those who created the molecular-based tree compared homologous *hedgehog* genes (a gene important for development), then which of the following statements would both scientists most likely agree with?

(A) As the number of shared features increases, so does the likelihood of independent evolution.

(B) The fossil record is the ultimate authority in phylogenetics.

(C) The molecular-based tree is the result of modern technology and is unlikely to be altered.

(D) Neither phylogenetic tree completely and accurately describes actual evolutionary history.

19. The phylum *Cnidaria* consists of over 10,000 aquatic species, with cnidocytes (specialized cells used mainly for capturing prey) being the distinguishing characteristic. Which of the following is most likely to be classified as a Cnidarian?

(A) An organism lacking digestive, circulatory and nervous systems that feeds by drawing in water through pores

(B) An organism with clearly defined sides (i.e., top/bottom; left/right)

(C) A planktonic organism capable of responding to the environment equally from all directions

(D) An organism that actively stalks its prey with coordinated complex movements

20. Which of the following is the correct sequence of the passage through which air travels during inhalation?

(A) Trachea—pharynx—bronchi—alveoli

(B) Pharynx—trachea—bronchioles—bronchi—alveoli

(C) Pharynx—larynx—trachea—bronchi—alveoli

(D) Larynx—trachea—bronchi—alveoli

21. When organisms that are homozygous for a particular beneficial allele gradually replace those that possess more harmful alleles within a population, this is known as

(A) disruptive selection.

(B) Hardy-Weinberg equilibrium.

(C) stabilizing selection.

(D) directional selection.

GO ON TO THE NEXT PAGE

Questions 22 and 23 refer to the following information.

Consider the following blood group data taken from a population in Hardy-Weinberg equilibrium with respect to the alleles responsible for different blood factors. All individuals in the population possess two different blood factors, each coded for by a dominant allele and a recessive allele. For the first blood factor, allele *R* is dominant to allele *r*, so both *RR* and *Rr* individuals test as blood type R, while *rr* individuals test as blood type r. For the second blood factor, allele *F* is dominant to allele *f*, so both *FF* and *Ff* individuals test as blood type F, while ff individuals test as blood type f. The frequencies observed for blood type in the population are as follows:

Type	Frequency
RF	0.60
Rf	0.15
rF	0.24
rf	0.01

22. The frequency of the *r* allele in this population is _____, while the frequency of the *F* allele in the population is _____.

    (A) 0.60, 0.50
    (B) 0.25, 0.25
    (C) 0.50, 0.60
    (D) 0.25, 0.84

23. Given the information in the previous question, all of the following statements concerning this population are true EXCEPT

    (A) there is no mutation between the blood factor alleles.
    (B) there is significant migration between this population and others.
    (C) the population is large in size.
    (D) there is no positive selection for the *R* allele.

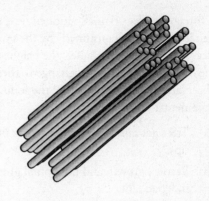

24. The structure pictured above shows the arrangement of microtubules that can be found within

    (A) cilia.
    (B) flagella.
    (C) basal bodies.
    (D) sarcomeres.

25. The only type of RNA that does not leave the nucleus once it is assembled is

    (A) hnRNA.
    (B) mRNA.
    (C) rRNA.
    (D) tRNA.

26. The membrane protein that actually makes ATP is

    (A) the electron transport chain.
    (B) ATP synthase.
    (C) cytochrome c oxidase.
    (D) hexokinase.

GO ON TO THE NEXT PAGE

27. A foreign object originally located next to a tree is eventually "consumed" by the growing tree. As the tree grows taller, the height of the object in the tree is unchanged. These observations support which of the following statements?

(A) Trees get taller by growing at the branch tips.

(B) Vertical growth and horizontal growth are independent.

(C) Once a certain maximum is attained, vertical height is relatively constant.

(D) Height is increased via cell proliferation in the root system.

28. It is hypothesized that negative pressure in the phloem and cohesion among water molecules are responsible for the bulk movement of water against gravity in vascular plants. Which of the following fluid-filled cylinders best illustrates cohesion?

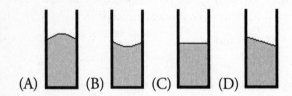

29. A certain plant is grown in darkness and observed to produce tall stems with non-expanded leaves. After being transported into daylight, the same plant develops broad, green leaves, short sturdy stems and long roots. Which of of following provides the best explanation for these observations?

(A) There is minimal energy expenditure until sunlight is detected.

(B) Sunlight stimulates the elongation of stems and proliferation of leaves.

(C) The plant is exhibiting the triple response to mechanical stress.

(D) The plant normally sprouts underground.

30. Nerve impulses move along the axons of neurons via action potentials caused by

(A) the movement of sodium across the cell membrane.

(B) an electron transport chain.

(C) the release of calcium into the sarcomeres.

(D) an ATP proton pump.

31. Which of the following is NOT part of the digestive tract?

(A) Esophagus

(B) Spleen

(C) Stomach

(D) Duodenum

Questions 32–34 refer to the following information.

A scientist studying the ecology of cities found that in developed landscapes, plant roots were not colonized by mycorrhizal fungi to the same degree that they were in a nearby nature preserve. In addition, she found that rates of photosynthesis and root respiration were much higher in plants in the preserve than for plants in city landscapes. She conducted a controlled greenhouse experiment to see what effects mycorrhizal colonization had on plant photosynthesis and respiration. Her experimental design involved growing 10 plants in soil rich in mycorrhizal fungal elements and 10 in the same soil that had been sterilized to remove the fungi. She made periodic measurements of plant photosynthesis and root respiration and calculated the mean rates for each experimental treatment. Her results are shown here.

GO ON TO THE NEXT PAGE

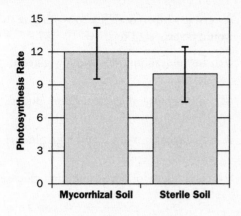

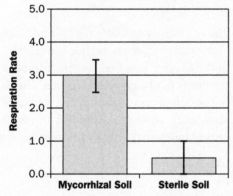

34. Assuming that further experimentation showed conclusively that plants in cities had reduced rates of photosynthesis and respiration due to lack of colonization by mycorrhizae, what important biogeochemical cycle of an ecosystem would be most affected?

(A) The nitrogen cycle

(B) The water cycle

(C) The hydrological cycle

(D) The carbon cycle

35. A patient's parents both have a disease that is caused by a sex-linked dominant allele. This disease was passed down to the patient's mother from the patient's grandfather. The chance that the patient will have this disease is

(A) 25%.

(B) 50%.

(C) 75%.

(D) 100%.

32. What conclusion can be drawn from the data?

(A) The presence of mycorrhizae increased photosynthesis and respiration rates significantly.

(B) The presence of mycorrhizae increased photosynthesis but not respiration.

(C) The presence of mycorrhizae increased respiration but not photosynthesis.

(D) The presence of mycorrhizae had no effect on photosynthesis or respiration.

33. Why might mycorrhizae influence photosynthesis and/or respiration?

(A) Mycorrhizae are important plant pathogens.

(B) Mycorrhizae are important plant parasites.

(C) Mycorrhizae are important plant predators.

(D) Mycorrhizae are important plant symbionts.

36. The terminal electron acceptor in the electron transport chain of the mitochondrion is

(A) NADH.

(B) O_2.

(C) H_2O.

(D) NAD^+.

37. The composition of chyme that empties from the stomach into the small intestine, relative to pure water, is normally

(A) hypotonic and acidic.

(B) hypertonic and basic.

(C) hypertonic and acidic.

(D) isotonic and acidic.

GO ON TO THE NEXT PAGE

38. What organism exists as a protein capsule surrounding nucleic acids, and is, perhaps, the simplest and oldest form of life on Earth?

(A) Virus

(B) Bacterium

(C) Protist

(D) Fungal spore

39. Plasmids (circular genetic elements present in bacteria)

(A) can act like transposons.

(B) are found in all eukaryotic cells.

(C) replicate only when other plasmids replicate.

(D) never move from cell to cell.

Questions 40–42 refer to the following image.

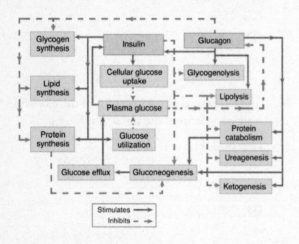

40. Based on the data in the figure above, the most reasonable conclusion is that

(A) protein is the body's preferred energy source.

(B) insulin and glucagon act antagonistically.

(C) insulin and glucagon are produced by beta and alpha cells, respectively.

(D) brain cells are able to uptake glucose without insulin.

41. A person with an inability to synthesize insulin would be expected to show

(A) low glucagon levels and low glucose levels.

(B) high glucagon levels and low glucose levels.

(C) low glucagon levels and high glucose levels.

(D) high glucagon levels and high glucose levels.

42. The respiratory quotient (RQ) is calculated as the ratio of carbon dioxide produced to the oxygen consumed for the complete combustion of a given fuel source. The RQ for carbohydrates is around 1.0, while the respiratory quotient for lipids is around 0.7. In resting individuals, the RQ is most likely to be

(A) 1.2.

(B) 1.0.

(C) 0.8.

(D) 0.6.

43. Which of the following is an essential building block of phospholipids?

(A) Glycerol

(B) Nucleic acids

(C) Amino acids

(D) Cholesterol

44. All of the following can affect enzyme function EXCEPT

(A) pH.

(B) temperature.

(C) specificity.

(D) transition state stabilization.

GO ON TO THE NEXT PAGE

45. In some organisms, features that have no function become vestigial and are ultimately lost. In many cave-dwelling animals, organs such as the eyes have been lost while other sense organs have increased in size. Which of the following hypotheses to explain the loss of nonfunctioning organs would NOT be considered correct?

    (A) Mutations causing the reduction in size of nonfunctional traits become fixed by genetic drift.

    (B) Natural selection against organs that are not used exists because the organs interfere with other, more important functions.

    (C) The development of the organ requires energy expenditures that are better spent on building other tissues or maintaining other traits.

    (D) All organs are maintained or eliminated as a result of how much they are used.

46. Rotifers are tiny, pseudocoelomate animals that have complete digestive systems and other specialized organ systems. Which of the following is also true of rotifers?

    (A) They possess both a mouth and an anus.

    (B) Their internal body cavity is entirely enclosed by mesoderm.

    (C) Their dorsal nerve cord runs superior to their notochord.

    (D) All of the above

47. Which of the following is NOT a characteristic of viruses in general?

    (A) Contain DNA and/or RNA

    (B) Do not require a host

    (C) Affect plants and animals

    (D) Reproduce using a lytic cycle

48. The DNA of a mouse is analyzed to reveal the presence of alleles resulting in brown fur color. The mouse is heterozygous, *Bb*, yet is completely colorless in phenotype. This can be explained as an occurrence of

    (A) epigenesis.

    (B) epistasis.

    (C) incomplete dominance.

    (D) pleiotropy.

49. Tetrodotoxin, an extremely potent poison produced by the puffer fish, binds tightly to voltage-gated sodium channels and blocks the flow of sodium ions but does not affect either potassium or chloride ion channels. Tetrodotoxin directly blocks which phase of the action potential?

    (A) Depolarization

    (B) Repolarization

    (C) Hyperpolarization

    (D) Neurotransmitter release

50. The punctuated equilibrium hypothesis claims that

    (A) cataclysmic events (e.g., asteroid strikes) have shaped the history of life on Earth.

    (B) most speciation occurs sympatrically.

    (C) species go through long periods of time during which they do not change markedly in genotype or phenotype.

    (D) new species arise through mutations that have large effects on phenotype.

51. In a food chain that consists of grass → grasshoppers → spiders → mice → snakes → hawks, the organism(s) that possess the most biomass within the community is (are) the

    (A) grass.

    (B) grasshoppers.

    (C) mice.

    (D) snakes.

GO ON TO THE NEXT PAGE

52. The earliest forms of life were most likely

    (A) unicellular autotrophs.

    (B) multicellular autotrophs.

    (C) unicellular heterotrophs.

    (D) multicellular heterotrophs.

53. Many neighboring animal cells have connections between them that serve as direct passageways between their cytoplasms, allowing the movement of ions and small molecules back and forth. These connections also couple electrical responses in one cell with electrical responses of adjacent cells. These cell-to-cell linkages are known as

    (A) plasmodesmata.

    (B) gap junctions.

    (C) hemidesmosomes.

    (D) lamella.

54. Skin-associated lymphatic tissue (SALT), which is located just underneath the epidermal layer of the skin, is associated with which of the following organ systems?

    (A) Digestive

    (B) Endocrine

    (C) Immune

    (D) Secretory

55. Reactivity of a skeletal muscle cell to the neurotransmitter acetylcholine may be increased by

    (A) increasing the concentration of acetylcholinesterase at the neuromuscular junction.

    (B) increasing the amount of myosin-II present within the skeletal muscle fibers.

    (C) downregulation of acetylcholine receptors on the surface of the T-tubule system.

    (D) increasing the number of calcium ions stored within the sarcoplasmic reticulum.

56. Nitrogen fixation can be carried out by free-living bacteria or by bacterial symbionts living in the roots of plants. The bacteria *E. coli* can use nitrate as an electron acceptor in the electron transport chain that is part of its cell membrane. The nitrate, therefore, serves the same purpose as which molecule in eukaryotes?

    (A) Oxygen

    (B) NADH

    (C) Carbon dioxide

    (D) Hydrogen

57. Which of the following INCORRECTLY pairs a metabolic process with its site of occurrence?

    (A) Glycolysis—cytosol

    (B) Citric acid cycle—mitochondrial membrane

    (C) Electron transport chain—mitochondrial membrane

    (D) ATP phosphorylation—mitochondria

58. The primary function of fermentation is to

    (A) generate ATP for the cell.

    (B) synthesize glucose.

    (C) regenerate NAD^+.

    (D) synthesize ethanol or lactic acid.

GO ON TO THE NEXT PAGE

59. The Cdk inhibitor *p16* binds to Cdk4/cyclin D complexes, which are normally responsible for allowing cells to pass through the restriction point from G_1 into S phase. Underexpression of *p16* protein could lead to

(A) uncontrolled cell division.

(B) cessation of mitosis.

(C) increased inhibition of Cdk4/cyclin D complexes.

(D) overexpression of *p53* protein.

60. In the diagram pictured above, the letter *X* represents

(A) Glucose.

(B) $NADP^+$.

(C) ATP.

(D) ADP.

61. Two plants, X and Y, are grown as potential food crops. Plant X is able to maintain a high rate of photosynthesis as the oxygen level in the air around it increases from a low of 10 percent to a high of 50 percent. Yet plant Y's rate of photosynthesis drops drastically under these circumstances. The best conclusion to draw from this data is that

(A) plant X is a CAM plant.

(B) plant Y is performing only the Calvin cycle in higher oxygen partial pressures.

(C) plant X is a CAM plant and plant Y is a C4 plant.

(D) Plant X must be a C4 plant.

62. The highest primary productivity in kilojoules per year would be found in which of the following ecosystems?

(A) Desert

(B) Tundra

(C) Taiga

(D) Tropical rain forest

63. Two plants growing together in the same pot are separated and planted in different pots. One plant dies while the other grows much higher. The plants most likely had what kind of relationship?

(A) Parasitic

(B) Commensalism

(C) Mutualistic

(D) Mycorrhizal

GO ON TO THE NEXT PAGE

GRID-IN ITEMS

The following questions require numeric answers. Calculate the correct answer for the question and then enter the answers on the provided grids. You may use a simple four-function calculator and the included formula sheet in Appendix A.

64. You complete an experiment in which a piece of potato is placed in an open container of 0.32 M sucrose solution. After a few minutes, you measure the mass of the potato again and determine that no change in mass has occurred.

    Calculate the water potential of the solutes within the piece of potato.

65. A cell has 48 chromosomes after DNA replication but before undergoing meiosis. During meiosis I, the homologs for a particular chromosome do not segregate and all the genetic material for that chromosome stays inside one of the two daughter cells. After undergoing meiosis II, what is the minimum possible number of chromosomes that will be present in one of the cells?

66. In a given population, 7,192 people have attached earlobes and 4,841 people have unattached earlobes. The allele for unattached earlobes is recessive. Assuming that the population is in Hardy-Weinberg equilibrium, how many of the individuals with attached earlobes are heterozygous?

67. A swab of bacteria is taken and placed in a petri dish containing the growth medium agar. Twenty bacteria were transferred into the petri dish by the swab. If the bacteria have a growth rate of 0.5, how many bacteria will be present in the petri dish after 4 steps of reproduction? Use 2.72 for e and round your answer to the nearest whole number.

68. In a rare breed of African cichlids, yellow fins (Y) are dominant to orange fins (y). Yellow and orange cichlids were bred in captivity to sell into the aquarium trade. The breeders counted 426 yellow babies and 328 orange babies. Calculate the chi-square value for the null hypothesis that the yellow parent was heterozygous for yellow fins. Round your answer to the nearest tenth.

69. A cross between a wild-type (Ww) and a white-eyed (ww) fruit fly will yield what percentage of white-eyed offspring? Give your answer as a decimal.

TEN-MINUTE READING PERIOD

Take the next 10 minutes to glance over the four questions that comprise Section II of this test. You can take notes in the margins, and begin writing your responses, but it is advisable that you use this time to develop your responses.

When the 10 minute period is over, you will have 80 minutes to complete the eight free-response questions.

Section II: Free-Response Questions

Time: 90 Minutes

2 Long Questions

6 Short Questions

Part A Directions: Answer each part of the following free-response questions with complete sentences. Answers should be in essay form. Make sure to provide a detailed response to each part of each question. Diagrams and other figures may be included to demonstrate knowledge of a topic, but they should not be the only answer you provide.

1. Some scientists believe that for ecosystems to maintain a balanced or steady state, biodiversity must be conserved. Using both classical Darwinian thought and the modern synthesis of evolutionary theory, answer the following questions as they relate to biodiversity.

   (A) **Compare** current biodiversity with the biodiversity that existed during the late Triassic and early Jurassic periods.

   (B) Insects are the most diverse group of organisms on Earth. Give some possible **explanations** for why this diversity exists.

   (C) **Discuss** an example in which human interaction significantly affected the selective forces acting on an organism in a natural system as well as the results of that interaction.

2. Proteins are complex molecules important to the function of every living organism, and their chemical properties make them susceptible to conditions in the environment.

   (A) **Discuss** the chemical composition of proteins and the chemical reactions involved in protein synthesis.

   (B) **Discuss** protein structure and the structure and function of proteins in relation to environmental factors.

   (C) **Discuss** the role of proteins in the intracellular and extracellular exchange of information.

GO ON TO THE NEXT PAGE

Part B Directions: Answer each part of the following free-response questions with complete sentences. Answers should be in essay form. Make sure to provide a detailed response to each part of each question. Diagrams and other figures may be included to demonstrate knowledge of a topic, but they should not be the only answer you provide.

3.   If a plant cell with a low water potential is placed into an open water bath with a higher water potential, what will happen to the cell? If solute is added to the open water bath, what will happen to the cell?

4.   A scientist conducts an experiment with penicillin-sensitive bacteria in which he adds a plasmid containing a gene that confers penicillin resistance. Following a protocol that elicits normal growth and uptake of the plasmid DNA, the scientist then adds bacteria to four new plates, as shown here.

	Glucose Medium With No Antibiotic	Glucose Medium Penicillin Added
Bacterial Strain Without Plasmid	#1	#2
Bacterial Strain With Plasmid Added	#3	#4

What type of protocol did the scientist likely use to encourage the uptake of the plasmid DNA? Describe the growth patterns scientists should expect to see on the plates.

5.   Refer to the pedigree diagrammed here.

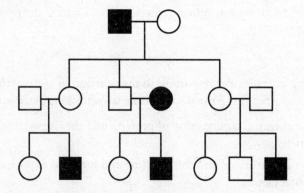

What inheritance pattern can best explain the allele(s) that are represented in the pedigree shown? Based on your decision about inheritance pattern, what can you conclude about the grandmother at the top of the tree for this trait?

GO ON TO THE NEXT PAGE

6. The graph shown here depicts the population growth curves for three different populations of an organism.

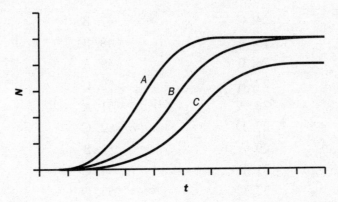

Compare and contrast the growth rates and carrying capacities of the three populations.

7. DNA makes up the genetic code for all living organisms. Although it is omnipresent in all living organisms, DNA cannot transmit this code on its own. A series of processes and functions must occur within the cell to carry out the genetic instructions encoded in DNA. Describe the process of transcription in protein synthesis. Include in your answer: mRNA, DNA, complementary base pairs, terminator, promoter, nucleus, introns, exons, and cytoplasm.

8. Geneticists can determine which genes will be expressed in offspring by tracking inheritance patterns and using Punnett squares. Explain the role of alleles in determining the genotype and phenotype of offspring.

DIAGNOSTIC TEST: ANSWER KEY

1. A	24. C	47. B
2. B	25. A	48. B
3. C	26. B	49. A
4. D	27. A	50. C
5. A	28. A	51. A
6. D	29. D	52. C
7. B	30. A	53. B
8. C	31. B	54. C
9. D	32. C	55. D
10. D	33. D	56. A
11. B	34. D	57. B
12. C	35. C	58. C
13. B	36. B	59. A
14. A	37. C	60. C
15. C	38. A	61. D
16. B	39. A	62. D
17. A	40. B	63. A
18. D	41. C	64. −7.776
19. C	42. C	65. 23
20. C	43. A	66. 0.47
21. D	44. D	67. 148
22. C	45. D	68. 12.7
23. B	46. A	69. 0.5

ANSWERS AND EXPLANATIONS

SECTION I

1. A

Amylase is a typical enzyme and like the vast majority of enzymes it is a protein.

2. B

Both chloroplasts and mitochondria have unique circular strands of DNA similar to that of bacteria, which are not found in the nucleus of the cell. It is thought that the earliest eukaryotic cells similar to unicellular protists formed symbioses with these bacteria. Endosymbiotic theory holds that energy-producing bacteria came to reside symbiotically within eukaryotic cells and ultimately formed the mitochondria, while photosynthetic bacteria inhabited the cells that later evolved into algae and plants by forming chloroplasts.

3. C

All cells are bound by a plasma membrane. Prokaryotic cells lack a true nucleus and organelles, ruling out (A) and (D). Animal cells and many protist cells lack a cell wall, ruling out (B).

4. D

While neither electrons nor protons are *directly* transferred from glucose to oxygen, both are indirectly combined with oxygen to form water. The second equation shows the "big picture," wherein glucose loses hydrogens and oxygen gains them (and electrons) to form water.

5. A

The first equation shown is photosynthesis, the process by which plants harvest the sun's energy to fix carbon dioxide and produce glucose. An important step in the light-independent reactions of photosynthesis, the Calvin cycle, involves reduction of fixed carbon dioxide to the three-carbon precursors of glucose. Water is oxidized during the light-dependent reactions of photosynthesis.

6. D

In a dihybrid cross such as the one described in the question, the expected phenotypic outcome is: 9/16 individuals will show both dominant traits, 3/16 will show one dominant and one recessive trait, 3/16 will show the *other* dominant and the *other* recessive trait, and 1/16 will show both recessive traits. The question asks about individuals showing one dominant and one recessive trait, so out of 160 offspring, 30 would be expected to fit the bill.

7. B

Sex-linked recessive disorders that are carried on the X chromosome are not always passed to offspring, because the mother has two copies of the X chromosome, ruling out (A). Because males have one X and one Y chromosome, they can only have one copy of an X-linked gene, ruling out (D). For females to be affected, two copies of the gene must be present; because they are recessive disorders, one copy in a female would not be expressed, ruling out (C). In males, because there is only one X chromosome, one copy of the recessive gene is expressed because there is no dominant copy to mask its expression.

8. C

The mutated gene sequence has the single base guanine (G) inserted in the fourth position in the sequence; all of the previous and subsequent bases are the same as the normal strand. A substitution, (A), involves replacement of one base by another, and deletion, (B), involves loss of a single base. Translocation, (D), involves movement of a segment of DNA.

9. D

Both insertions and deletions cause frame shifts that can affect all of the codons that occur after the error, as in the example shown. Both are point mutations involving single nucleotides. Transposition, (A), involves movement of a gene position on a chromosome. A trinucleotide repeat, (B), involves sequences of three nucleotides repeated in a series on the same chromosome a number of times. Inversion, (C), occurs when a DNA sequence is flipped, such that ATT would become TTA.

10. D

The beaks of Darwin's finches represent homologous structures, not analogous ones, ruling out (A). Choices (B) and (C) describe how allele frequencies contribute to microevolution. Hardy-Weinberg equilibrium describes populations with stable allele frequencies. Genetic drift is a phenomenon associated with statistically unexpected changes in allele frequency, especially in small populations. Darwin's finches are examples of adaptive radiation, an evolutionary process that results from adaptation to specific ecological niches.

11. B

Of the nine major animal phyla, only *Echinodermata* and *Chordata* have deuterostome embryonic development; the anus forms prior to the mouth in the gut cavity. (A) is not true, even so, a relative abundance of species would not support a close relationship. (C) and (D) are characteristics common among all *Animalia* and would not explain why *Echinodermata* are thought to be more closely related to *Chordata* than to other animals.

12. C

The key to this question is in the last sentence before the question. Any organism that is a unicellular heterotrophic eukaryote without a cell wall must be a protist. Bacteria, (A), and viruses, (B), lack a nucleus and cellular organelles, and fungi, (D), have cell walls containing chitin.

13. B

The lysosome is responsible for cellular digestion and protein degradation. None of the other organelles listed have functions of digestion and degradation.

14. A

If the mother is a carrier for an autosomal recessive disorder, then her genotype would be represented with one dominant allele and one recessive allele, *Aa*. The father is genotypically normal (*AA*), so 50% of the children will be carriers *Aa* and 50% will be homozygous dominant, *AA*. The likelihood of having an affected child is therefore, 0%.

15. C

A cell placed in a hypotonic environment will swell, unless action is taken. The sodium potassium pump can be used to maintain cell volume. During each cycle of the pump, the *net* effect is the removal of one solute particle, so activating the pump (and consuming ATP) would reduce the solute gradient between the cell and the environment and mitigate the swelling. (D) is wrong because the question specifically asks about an *immediate* effect.

16. B

While there are some subtle differences between the two phylogenetic trees, the *main* difference is in the relationship among the bilaterians. The morphology-based tree divides the bilaterians into two clades: deuterostomes and protostomes. In contrast, the tree based on molecular data assigns two sister taxa to the protostomes. (A) and (C) do represent differences, but they are fairly small in comparison. Both trees show a convergence on a single ancestor, ruling out (D).

17. A

This one requires close examination of the two trees. In the morphology-based tree, *Annelida* and *Arthropoda* have a more recent common ancestor than *Arthropoda* and *Nematoda*. However, in the molecular data-based tree, *Arthropoda* and *Nematoda* have a more recent ancestor than *Annelida* and *Arthropoda*, therefore (A) is correct. (B) is wrong because it does not represent a conclusion that can be drawn from the morphology-based tree. (C) can be eliminated because the morphology-based tree does not use DNA sequencing. (D) is wrong because both trees show a more distant common ancestor for the *Radiata* than the *Bilateria*.

18. D

Phylogenetics, by its very nature—looking back into prehistory—is an imperfect science, one that will never be completely accurate. The best answer is therefore, (D). (A) is unlikely to be a true statement and less likely to be supported by both scientists. The fossil record may be "set in stone," but it is far from the ultimate authority—the most robust phylogenetic hypotheses are those evidenced by numerous lines of molecular and morphological incidence as well as by fossil evidence, so rule out (B). (C) is unlikely to be supported by either scientist.

19. C

The question stem mentions *Cnidaria* and the figure implies that these organisms exhibit radial symmetry (they are "Radiata" as opposed to "Bilateria"). The correct answer will describe radial symmetry. (C) is the best answer.

20. C

This is a basic question about the flow of air during inhalation. Air enters via the pharynx and travels through the larynx into the trachea. From there it is divided into the bronchi, where it then travels to the alveoli.

21. D

Movement in one phenotypic direction over time is defined as directional selection, where one extreme phenotype is favored over all other possible phenotypes given a particular environment. Disruptive selection, (A), weeds out intermediate phenotypes, favoring extremes; (C), stabilizing selection, weeds out extreme phenotypes allowing the majority of members within the population to converge on the intermediate physical feature (i.e., very light and very dark fur color may be less advantageous than a grayish, intermediate color).

22. C

This is a fairly difficult question with some tricky answer choices. You might have been tempted to choose (D) if you simply added the frequency of individuals expressing blood type r (0.24 + 0.01) and did the same for blood type F (0.24 + 0.60). This type of addition works for the recessive r allele, but it does not work for the F allele. The frequency of the r type, q^2, is ($rF + rf$), or 0.24 + 0.01 = 0.25; thus, the frequency of the r allele, q, is 0.5. However, individuals with blood type F can be either homozygous or heterozygous; therefore, the frequency of RF individuals in the table includes both FF individuals as well as Ff ones. Therefore, you cannot simply add 0.60, the RF frequency, to the rF frequency to calculate the frequency of the F allele. You must first find the frequency of the recessive f allele and then subtract that from 1.00 to find the frequency of the F allele. The frequency of the f trait is 0.15 + 0.01 = 0.16, so the frequency of the f allele is 0.4. The allele frequency of F is $1.00 - \text{freq}(f) = 0.60$. Therefore, (C) is correct.

23. B

Hardy-Weinberg equilibrium requires no mutation, large population size, no natural selection, and random (panmictic) mating—but no migration from other populations with different allele frequencies. Thus, (B) is the correct answer.

24. C

The structure pictured, showing nine microtubule triplets, can be found in basal bodies, which form the base of eukaryotic cilia and flagella. Within the cilia and flagella are nine microtubule doublets surrounding a central pair of microtubules. Sarcomeres, the functional units of muscle tissue, are composed of many parallel filaments (thick and thin) that slide past each other.

25. A

The only type of RNA that does not leave the nucleus is the primary transcript of DNA, which is called heterogeneous nuclear RNA, or hnRNA. This type of RNA is formed and processed wholly within the nucleus; its noncoding regions are excised and the resultant mRNA strand exits the nucleus through a nuclear pore.

26. B

ATP synthase is the protein that actually makes ATP from ADP and P.

27. A

The tree grows taller, but the height of the foreign object remains constant. This implies that the tree does not grow taller by adding building blocks at the base, but rather by extending the ends. This matches (A). Choice (B) is not a valid conclusion based on

the stated observations, because both types of growth are observed. Choice (C) is not supported by the observations, because the tree continues to grow taller. If the tree grew via the mechanism described in (D), then the foreign object would be expected to rise over time.

28. A

The best example of cohesion would illustrate the preference of the fluid molecules for other fluid molecules—meaning the correct answer will show the fluid sticking together instead of sticking to the container (which would exhibit *adhesion*). The correct answer is, therefore, (A).

29. D

The plant in question is grown in darkness, certain traits are observed, then the same plant is transported into the light and the previously observed traits are "reversed." Without light, the plant produces tall stems and non-expanded leaves—as if it were simply trying to reach something. When presented with light, the same plant develops broad, green leaves and short sturdy stems and long roots (i.e., it begins to look like a "normal plant"). The best explanation is that the plant is exhibiting normal behavior, which is explained if the plant normally sprouts underground. (A) can be eliminated because the plant is observed to expend energy (tall stems and leaves) in darkness. (B) is wrong because the sunlight actually thickens the stems—the elongation is more evident in the darkness. (C) may be tempting, but cannot be concluded because there is no evidence of mechanical stress.

30. A

The transmission of nerve impulses occurs when gated sodium channels are stimulated to open, depolarizing the membrane of the axon and setting off a chain reaction of depolarization/repolarization events. Electron transport chains, (B), are associated with ATP production via the action of proton pumps, (D), in respiration and photosynthesis. Calcium functions in sarcomeres, (C), to allow muscle contraction.

31.   B

The spleen is an organ associated with the lymphatic and immune systems. All of the other structures listed are part of the digestive tract, a continuous tube that extends from the mouth to the anus.

32.   C

Although both figures show that rates of photosynthesis and respiration tended to be higher when mycorrhizae were present, the error bars indicate that there really was no difference in terms of photosynthesis, because the bars overlap the mean values for the two treatments. This rules out (A) and (B). The figures do show a difference between the two treatments in terms of respiration, ruling out (D).

33.   D

Most land plants form mutualistic symbioses with mycorrhizae; it is thought that the two groups evolved together. Both organisms typically benefit from the association.

34.   D

Both photosynthesis and respiration are important components of the carbon cycle. The nitrogen cycle is predominantly moderated by bacteria. The water cycle and hydrological cycle are basically the same thing and would not necessarily be influenced by change in photosynthesis and respiration.

35.   C

Call the disease allele X^A. The corresponding normal allele is X^a. If the patient's mother inherited the disease from her father (the patient's grandfather), her genotype is $X^A X^a$. Because the patient's father is also affected, his genotype is $X^A Y$. The cross is, therefore, $X^A X^a \times X^A Y$. Using a Punnett square, the offspring can have the following genotypes:

	X^A	X^a
X^A	$X^A X^A$	$X^A X^a$
Y	$X^A Y$	$X^a Y$

The patient's chances of having the disease are 75 percent, because from the Punnett square, 75 percent of the offspring will inherit the dominant X^A allele.

36.   B

Oxygen is the terminal electron acceptor in the electron transport chain of the mitochondrion; thus, choice (B) is the correct answer.

37.   C

As chyme enters the small intestine from the stomach, it contains many dissolved ions and molecules, and is therefore hypertonic, not hypotonic or isotonic to pure water; thus, (A) and (D) are incorrect. The acidity of the stomach leaves the chyme acidic; thus, (C), hypertonic and acidic, is the correct answer.

38.   A

Because viral and bacterial research has played such a major part in our understanding of molecular genetics, the College Board has determined that these concepts should be part of the AP Biology curriculum. You have already learned about bacteria and the relative sizes of organisms as seen under a microscope. The smallest of the organisms studied are viruses. Smaller doesn't always indicate simpler or older, but in this case it does. Viruses are very simple structures, and scientists still argue as to whether they are alive or not. Viral infections of bacteria and the actions of retroviral RNA have been crucial to developing new techniques for studying the DNA molecule and have led to advances in genomic research.

39. A

Plasmids are able to reintegrate themselves back into the main bacterial chromosome, much as transposable elements in eukaryotic genomes can move from chromosome to chromosome using enzymes and insertion sequences. Plasmids are generally not found in eukaryotic cells, ruling out (B). Plasmids can replicate themselves independently of the main bacterial chromosome, and they can easily be transferred from cell to cell, so (C) and (D) are also incorrect.

40. B

The most reasonable conclusion that can be drawn from the figure, is that glucagon and insulin act antagonistically. Most processes in the figure that are stimulated by insulin are inhibited by glucagon (and vice versa). (A) is untrue and not a reasonable conclusion that may be drawn from the figure. (C) and (D) are both true statements, but neither can be concluded from the data in the figure.

41. C

Insulin lowers plasma glucose levels. Plasma glucose inhibits glucagon. An inability to synthesize insulin would lead to increased plasma glucose levels and decreased glucagon levels.

42. C

In resting individuals both carbohydrates and fat are utilized for energy, so the RQ would be expected to be somewhere between the RQ for carbohydrates and the RQ for lipids.

43. A

The building blocks of a phospholipid include: glycerol, fatty acids, a phosphate group and a small, variable molecule linked to the phosphate group. Therefore, (A) is the best answer.

44. D

pH, temperature, and enzyme specificity can all have an effect on the function of an enzyme. Transition state stabilization, while important to the concept of enzymes, is not a major factor affecting enzyme function.

45. D

Only (D) is a Lamarckian explanation of why a vestigial organ would disappear. All other explanations are Darwinian in nature and can be considered valid reasons that a nonfunctional organ may be reduced in size to the point of disappearing.

46. A

Being invertebrates and not chordates, rotifers do not have a notochord or dorsal nerve cord. In addition, being pseudocoelomate means that their internal body cavity is only partially covered with mesoderm. The question stem states that rotifers have a complete digestive system, so that is equivalent to their having a mouth and an anus. While the specifics of a given phylum are beyond the scope of the test, be sure to know the basic vocabulary of animal phyla for the AP Biology exam.

47. B

Although viruses have a lot of unknowns and diversity, it is believed that most viruses need some form of host to complete their full life cycle.

48. B

Epistasis occurs when one gene alters the expression of another gene that is independently inherited. In this instance, the brown fur color allele is most likely affected by the presence of recessive alleles on a different chromosome that code for the expression of "no color." Don't confuse epistasis with epigenesis, the phenomenon whereby genes are expressed differently depending upon which parent they are inherited from.

49. A

The toxin blocks sodium channels; therefore, depolarization, or the initial rise in positive charge within the neuron, is blocked. The opening of sodium channels is what causes a rush of positive sodium ions into the nerve cell when a stimulus arrives, and it is this initial rush of sodium that causes the action potential to start. The phase after depolarization, when potassium channels open to restore the resting potential of −70 mV, is called repolarization.

50. C

Punctuated equilibrium postulates that most evolutionary change happens quickly in small isolated populations that break off from larger groups, so this eliminates (B). Between bursts of change, phenotypes in a population hover around some mean value, not changing very much. Even though change is rapid in punctuated equilibrium, it relies on variation in the population, not on the sudden production of novel features. In the fossil record, two different phenotypes would mark one of

these punctuation events and would, therefore, not be arbitrary.

51. A

Organisms at the top of the food chain generally have the least biomass, while organisms at the bottom have the greatest biomass. In this food chain, grass is at the bottom and has the greatest biomass.

52. C

According to the heterotroph hypothesis, the earliest forms of life were probably unicellular organisms that used organic molecules as their sources of food.

53. B

Gap junctions are connections between cells that enable ions and other small material to move between adjacent cells. They are essential for rapid electrical conduction across large tissues or organs, such as the heart. Plasmodesmata, (A), occur only in plants, and lamella, (D), are inner sections of plant cell wall. Try to eliminate answer choices quickly if they are not relevant to the information presented in the question stem—here, the stem mentions "animal cells."

54. C

The word *lymphatic* should tip you off here that SALT is associated with the immune system. Although the lymphatic system is generally part of the circulatory system because it removes excess fluids from the tissues and returns them to the general circulation, it is a repository for immune system cells. B- and T-lymphocytes fill the lymphatic system, particularly the lymph nodes. In SALT, immune system cells known as Langerhans cells are phagocytic guardians that engulf and destroy pathogens that enter through wounds.

55. D

The neurotransmitter acetylcholine (ACh) causes the rapid release of Ca^{2+} ions from the sarcoplasmic reticulum in muscle fibers. This calcium allows myosin and actin filaments to slide past each other and causes the muscle to contract. Of the answer choices given, only (D) would allow for a greater response to the neurotransmitter. Increasing acetylcholinesterase or downregulating ACh receptors would facilitate faster breakdown of ACh and cause less muscle stimulation, so (A) and (C) are incorrect; increasing the amount of myosin would not help muscle contraction unless entire sarcomere units were also produced in greater quantities, (B).

56. A

Neither NADH nor NAD^+ serves as electron acceptors within the electron transport chain (ETC) of eukaryotes. However, oxygen does. Recall that oxygen is the terminal electron acceptor in the ETC of most organisms, allowing water to be formed from hydrogen and electrons coming off the ETC. Without the presence of oxygen, ATP production would rapidly stop.

57. B

The citric acid cycle, otherwise known as the Krebs cycle, takes place within the matrix (or inner fluid) of the mitochondria. All other answer choices are correctly placed within their proper areas of occurrence.

58. C

The key function of fermentation is to regenerate NAD^+. The reason this is so essential is that as glucose gets broken down within the cytoplasm of a cell, electrons fly out of the broken bonds. These electrons are normally caught by NAD+ and ferried into the mitochondria and ETC. Yet, fermentation occurs in the absence of the mitochondria and ETC. Thus, to keep a ready supply of NAD^+ around to catch electrons coming off the broken glucose molecules, the NADH must be reoxidized into NAD^+ by the process of fermentation. This allows the breakdown of glucose to continue and the subsequent production of small amounts of ATP via glycolysis.

59. A

Removal of cyclin D inhibition would allow cells to progress unhindered from G_1 into S phase. This could cause, in the absence of other controls, a more rapid onset of mitosis and uncontrolled cell division.

60. C

This is a diagram of a mitochondrion. It is here that the reactions of cellular respiration take place. Pyruvate, produced from the breakdown of glucose in glycolysis, enters the mitochondrial matrix along with oxygen and NADH, the Krebs cycle takes place (as well as electron transport and oxidative phosphorylation), and ATP is produced. In fact, the letter X represents ATP, whose production is the most essential aspect of the entire cellular respiration process.

61. D

Plant X must be a C4 plant. The enzyme used by most plants (C3 plants) to capture carbon dioxide for the Krebs cycle is called rubisco. Yet, rubisco tends to accept oxygen rather than carbon dioxide when oxygen concentrations are high. This process, known as photorespiration, consumes oxygen and generates no ATP. However, C4 plants, known for living in hot and dry climates, use the enzyme PEP carboxylase, which has a much higher affinity for carbon dioxide than rubisco. This allows these plants to continue accepting carbon dioxide, and thus photosynthesizing, long after oxygen concentrations rise around them.

62. D

The highest primary productivity, or production of food from light, would occur in the tropical rainforest because of its year-round growing season, ample light, and huge plant biomass. None of the other biomes listed could achieve the same productivity due to shorter growing seasons or lower plant biomass.

63. A

If one plant dies while the other is strengthened because of this separation, it is likely that one plant was preying on the other as a parasite. Only after removal does the host species grow taller, but the parasite will die off because of lack of nutrition. It is not clear that the separation would affect plants in a commensalistic relationship, and those in a competitive relationship would both be strengthened by the separation.

GRID-IN ITEM ANSWER EXPLANATIONS

64. Water potential is equal to the sum of the solute potential and pressure potential. To solve this problem, first you must calculate the solute potential.

$$\psi_s = -iCRT$$
$$\psi_s = (-1)(0.32 \text{ mol/L})(0.0831 \text{ Lbar/mol K})(300 \text{ K})$$
$$\psi_s = -7.776 \text{ bars}$$

The pressure potential of the solution is 0 bars because the container is open to the atmosphere. Therefore,

$$\psi = \psi_s + \psi_p$$
$$\psi = 0 \text{ bars} + -7.776 \text{ bars}$$
$$\psi = -7.776 \text{ bars}$$

The correct answer is –7.776 bars.

65. The cell starts out with 48 chromosomes before meiosis. During meiosis I, if everything goes properly, both daughter cells should have 24 chromosomes. During meiosis II, the daughter cells should split, producing two daughter cells with 24 chromosomes. If a chromosome does not segregate properly from its homolog, then one cell will have 25 chromosomes and one will have 23 chromosomes. Then, during meiosis II, the identical sister chromotids will separate from each other, so the cell with 25 chromosomes will produce two cells with 25 chromosomes. Additionally, the cell with 23 chromosomes will produce two cells with 23 chromosomes. Therefore, the minimum possible number of chromosomes in these cells is 23.

66. The frequencies of alleles in a Hardy-Weinberg population can be determined using the following equations:

$$p^2 + 2pq + q^2 = 1$$
$$p + q = 1$$

You can use A and a to represent the alleles for attached and unattached earlobes. The frequency of the aa phenotype is $4{,}841/(4{,}841 + 7{,}192) = 0.402$. Therefore, $q^2 = 0.402$ and $q = 0.63$ because $p + q = 1$, $p = 0.37$.

The frequency of the Aa genotype is equal to $2pq$, or $2(0.63)(0.37) = 0.47$.

67. Bacteria in the petri dish will grow at a steady exponential rate, provided there is enough food for them to consume. Because the rate of population growth is the difference of reproduction and death rates, the rate will continue to grow as long as the bacteria are able to continue to rapidly reproduce.

Using the equation $N_t = N_0\, e^{rt}$ where $N_0 = 20$ (the number of bacteria at the beginning), $r = 0.5$ (the rate of population growth), and $t = 4$ (steps of reproduction), the number of bacteria in the petri dish after 4 steps of reproduction will be:

$$N_t = (20)(2.72)^{(0.5)(4)} = (20)(2.72)^{(2)} = (20)(2.72)(2.72) \approx 148 \text{ bacteria.}$$

68. A cross between a heterozygous yellow fish (Yy) and an orange fish (yy) would yield an offspring ratio of 1:1 between yellow- and orange-finned fish. Of the 754 fish, it would be expected that 377 would be yellow and 377 would be orange. The chi-squared values are calculated as follows:

phenotype	observed	expected	obs-exp	$(\text{obs-exp})^2$	$(\text{obs-exp})^2/\text{exp}$
yellow	426	377	49	2401	6.368
orange	328	377	−49	2401	6.368
					$X^2 = 12.736$

Rounded to the nearest tenth, the answer is 12.7.

69. A cross between a wild-type (Ww) and a white-eyed (ww) fruit fly will yield 50 percent Ww and 50 percent ww. Therefore, the answer is 0.5.

SECTION II

PART A: ANSWER EXPLANATION #1

Remember, your free-response answers must be written in complete sentences and formulated as a cohesive essay. You are allowed to use diagrams and figures, but they must be explained. Feel free to compose a quick outline, but keep in mind that you only have 80 minutes to complete two long questions (20 minutes each) and six short questions (6 minutes each). Readers are looking more for key bits of information in complete sentences than they are looking at the overall composition of the paper. While composition and coherence are important in the answer, the content of the answer is what is most important to the reader.

This question, rather than asking you to recall particular facts or even to apply the facts you know about evolution, asks you to apply an appreciation and synthesis of the principles of evolution. There are additional biological concepts from other course topics that you will have to recall to answer parts of the question as well.

The first part of the question asks you to compare the conditions that exist today to those of an easily recognizable geological time span, the age of the dinosaur. To answer the question, you need to recognize the period, the dominant organisms that supposedly lived during that time, and the environmental conditions of the period. You then need to make some comparisons between the organisms and environmental conditions that existed during the late Triassic and early Jurassic period, and the organisms and environmental conditions that exist today. There is no right or wrong answer—just be explicit without being wordy.

Part (B) requires a knowledge of the patterns of biodiversity that exist today. The obvious answer is that either arthropods or insects are the most diverse group of organisms. Either answer should be acceptable as long as you support it. You then need to make some statements that address the high amount of adaptive radiation that took place in the case of arthropods/insects.

The last part allows you to apply a particular piece of information that you provide. Providing your own example makes it easier for you to answer the question but more difficult for the grader to give you points. The key to answering this part of the question is to be very explicit and pay close attention to the question you are answering with your explanation. You will not want to "wax poetic" or be political about an environmental disaster caused by humans for part (C). Give your example and explain the cause-and-effect relationship within the context of evolutionary biology.

Here is a possible response:

(A)　The transition between the Triassic and Jurassic periods marks the time in Earth's history when life on Earth became dominated by dinosaurs. The major continental plates were still massed into the supercontinent Pangaea and were just beginning to separate. The major limitation to the distribution of organisms was environmental conditions that existed at the time, as most of the major landmasses were contiguous and only local geographic isolations caused speciation. The majority of the terrestrial organisms that existed on Earth were centered around a warm, dry climate with a centralized region near the equator. There was no glaciation or polar ice caps present. The marine systems were warm and full of both simple and complex organisms.

The limited variation in climate during the late Triassic and early Jurassic periods made a great impact on diversity. There were fewer niches available for an organism to take advantage of, so there were fewer organisms. Flowering plants, mammals, and birds were just starting to evolve, and the land was dominated by reptiles. The Earth today is much more diverse in terms of climate and species. After the Jurassic period, the major tectonic plates separated, isolating several large lines of organismal groups. Modern flowering plants, insects, mammals, and birds all have major adaptive radiations that are much more recent than the late Jurassic period. Geographic isolation and the availability of a large number of new ecological niches allowed the diversification of organisms that could take advantage of these niches.

(B) The most diverse group of organisms by far is the Class Insecta. Their diversity is explained by a combination of certain morphological advantages and ecological associations. First of all, insects evolved wings that allowed them to escape predation and take advantage of several new ecological niches. Insects are closely associated with flowering plants and their fast radiation is closely linked to the radiation of the angiosperms as they colonized a changing land climate. Insects also have short generations, allowing them to produce many offspring in short periods of time. These factors combined in traits that conferred selective advantages and provided insects with higher fitness. Also, the higher rate of reproduction and gamete production allowed more chances for viable mutations to occur. Since mutations are the ultimate source of genetic variation, the greater the rate of maintaining a mutation, the higher the rate of accruing variation.

(C) Both biotic and abiotic (nonliving) factors in the local environment act as selective agents on the fitness of each species of organism. When these selective factors are stable, population numbers are stable. When human intervention impacts the selective forces affecting an organism, it changes the pressure for a population to remain constant. A good example of this is the effect of global warming on the seasonal activity of migrating birds. The northern temperate regions are becoming warmer sooner in the spring, but the migratory birds are arriving at the same time. The insect communities the birds use to feed their nestlings are emerging earlier in the year, so the food resource is becoming limited for the birds that had previously adapted to the later arrival time. Some birds have a genetic predisposition to arrive a little sooner, so they will have a selective advantage over the later-arriving birds. This should cause directional selection away from the birds that arrive at the previously adapted time toward birds that arrive earlier. After several generations, the bird populations with earlier arrival times should be the dominant phenotype. This likely will not cause speciation, but will probably change the behavioral phenotype of some bird species. The bird species will either have to adapt to the new ecological niche or suffer extinction because of extremely low fitness.

PART A: ANSWER EXPLANATION #2

Read and reread the question over again until you understand it. Parts (A) and (B) are about chemical and physical properties of proteins. Part (C) requires synthesis of information and a working knowledge of other themes and hierarchical levels of biology.

To answer the first two parts of the question, you need to organize information starting with tiny structures and ending with broader themes. Be sure to touch on all of the important points for each level of organization. Each level of organization has different chemical and physical properties that are affected by environmental conditions. These properties need to be discussed for part (B). Part (C) should include examples of information exchange and a general explanation of how it is accomplished.

Here is a (good) potential answer to the free-response question:

(A)   A chemical compound composed of a series of amino acids joined by peptide bonds is called a polypeptide or, when the polypeptide is long, a protein. Amino acids are composed of a backbone of a centralized α-carbon atom bonded to a carboxyl group and an amine group. The fourth bond on the α-carbon is to a side chain that gives each amino acid its unique identity and chemical and physical properties. For example, glycine has the simple side chain of only an H atom; alanine has a side chain of a methyl group (CH_3). The charges on the amine and carboxyl groups depend on the pH of the surrounding environment. Some side chains contain S atoms, which allow disulfide bonds between some amino acids along the length of the protein.

(B)   Protein structure is uniquely coded by regions of DNA through an mRNA intermediary in which triplet nucleic acids code for specific amino acids. Amino acids are added sequentially at the end of the polypeptide chain through a peptide bond between a carboxyl group on one amino acid and an amine group from the next amino acid. The addition of amino acids is mediated by a tRNA and is performed by a ribosome. The peptide bond is formed through dehydration synthesis, generating one molecule of H_2O.

The conformation of a protein is dictated by its primary structure, the sequence of amino acids that make the polypeptide. Bond angles and hydrogen bonding between atoms on amino acids contribute to the secondary structure that is the twisting and turning that occurs around the amino acid backbone. A tertiary structure develops from interactions between different side chains on the amino acids (e.g., disulfide bonds). A quaternary structure is also possible between different or similar polypeptides, which form one functional macromolecule. Protein conformation is affected by environmental factors such as pH and temperature. The pH of the environment affects the electrical charges on the amine, carboxyl, and side chain groups of amino acids, which in turn affect hydrogen and tertiary bonding.

Temperature can also cause a protein to unravel as free energy surpasses bond energy.

(C)   The unique coding of the primary structure of proteins causes a great deal of specificity of attachment sites in the tertiary and quaternary structures. These highly specific proteins can exist on the external membrane of cells, allowing cellular recognition by the particular proteins that occur there. Transmembrane proteins can exchange information from the intracellular matrix to the extracellular matrix and vice versa via temporary changes in the protein conformation due to the presence or absence of a messenger at a receptor site (e.g., acetyl CoA-mediated messenger system). Proteins in the cellular matrix can relay the same information from the cell nucleus to the cell membrane through specificity at a receptor site and changing bond affinities with other proteins.

PART B

3. If the plant cell has a lower water potential than its surrounding environment, that means the cell is hypertonic to its environment. When a cell is hypertonic, it means that water will move into the cell by osmosis. Water follows its concentration gradient from a place of high concentration inside the cell to a place of low concentration outside the cell. The cell will expand with water until the concentrations of solute equalize. The cell wall provides a barrier to too much water moving into the cell.

    Adding solute to the water bath would increase the solute potential and therefore decrease the water potential. Since the pressure potential is 0 because it is an open water bath, I would expect water to move back out of the cell by osmosis. This will depend on the exact concentrations in the cell and the surrounding environments. Water will move by osmosis down its concentration gradient until it reaches a balance.

4. Here is a possible response:

    There are two ways scientists promote transformation in bacteria: increasing temperature and adding chemicals such as $CaCl_2$. The scientist could have used either treatment to promote the uptake of the plasmid DNA.

    There will be no growth on the plate that has penicillin and bacteria that have not acquired resistance from the plasmid because the penicillin will kill the bacteria. This means that there will be no growth on plate 2. There will likely be bacterial growth on plate 4 because some of the bacteria will be transformed with the plasmid and will acquire penicillin resistance, but there should be considerably more growth on the plates where penicillin resistance is not necessary. There should be the most growth on plates 1 and 3.

5. You need to be familiar with pedigrees. Pedigrees are family trees with all the males indicated by squares and the females by circles. The shapes that are filled in express the trait in question. If you are given a pedigree, look first at the expression of the trait in the males, especially if sex-linkage is an option among the possible answers. The typical inheritance pattern in sex-linkage is for a trait to be more common in males and to skip generations. Both of these occur in this pedigree. The grandfather at the top of the tree expresses the trait. If the gene in question were sex-linked, the father would pass on his X chromosome to all of his daughters. If the trait were dominant, all of his daughters would express the trait. The daughters do not express the trait, so the answer is sex-linked recessive. (This pedigree could also exist for an autosomal recessive allele. Both answers are acceptable.)

    If the pedigree is showing the inheritance of the sex-linked recessive trait, the grandmother *cannot* be homozygous recessive. This is about all you can conclude from the available information, though. The grandmother could be either a heterozygous carrier or homozygous dominant and still produce the pedigree.

6. All of the populations start with the same number of individuals at the far left of the curves. Logistic growth is typified by an ultimate plateau that is equal to the carrying capacity: The higher the plateau, the higher the carrying capacity. Carrying capacity is the maximum number of individuals an environment will support. Based on the plateaus of the three curves, populations A and B have the same carrying capacity, which is higher than it is for population C.

Growth rate affects the steepness of the exponential portion of the curve, the incline before the plateau. The quicker the population gets to its carrying capacity, the higher the growth rate. Population A has the highest growth rate, followed by B, then C.

7. Since DNA is found in the nucleus of a cell and protein synthesis takes place in the ctyoplasm, base-pair coding information in the DNA has to be moved to the cytoplasm before protein synthesis can commence. The DNA is "transcribed" by mRNA, which mirrors the original strand of DNA with complementary base pairs, to be translated later on into proteins in the cytoplasm. In mRNA, uracil replaces thymine as a complementary base pair for adenine.

Transcription of the DNA starts at a region called the "promoter" and ends at a region called the "terminator." All base pairs within this region will be coded for with complementary base pairs in the mRNA. For an unknown reason, areas of the original strand of DNA, called introns, are not coded for in protein synthesis. Since these areas are not necessary for protein synthesis, they are cut out before the mRNA enters the cytoplasm. The areas that do code for protein synthesis, called exons, are transcribed to the mRNA to be moved to the cytoplasm.

8. Alleles come in pairs and compose the genetic makeup of a gene. One allele comes from the mother in a sexually reproducing species, and one allele comes from the father. The combination of alleles codes for the phenotype, or gene expression, when placed in the genotype (gene location) for a particular trait. A sample answer could be as follows:

Alleles play an important role in the expression of genes in offspring. Since two alleles make up a gene, the combination of two alleles leads to the expression or repression of genetic outcomes. Since the phenotype is the physical expression of a gene, the phenotype is what you will see when two alleles are combined to form a gene. If, for example, a codes for pink eyes and A codes for black eyes, with a as the recessive allele and A as the dominant allele, an allelic combination of Aa would produce a black-eyed offspring. In terms of genotype, the alleles are found on a specific gene and chromosomal location that codes for the expression of a trait.

Normally, lethal traits or traits that lead to disease are recessive, which means that the alleles that code for a phenotypic trait will only be expressed if both alleles are recessive. For example, Tay-Sachs is a recessive autosomal disease. There are exceptions, however—some alleles that code for disease are dominant, as in Huntington's disease.

Some genes have two different alleles, as in sickle cell anemia. An individual will only express symptoms of sickle cell anemia if he or she is homozygous for the alleles AA. An individual with the alleles AS is resistant to malaria, which is beneficial in certain areas of the world. An individual with an SS gene is neither resistant to malaria, nor will he or she express symptoms of sickle cell anemia. Since alleles produce the phenotype for a gene, some phenotypic outcomes are beneficial, as in resistance to a disease; some alleles have no genetic repercussions (e.g., eye color); and some allelic combinations are detrimental, even lethal, to an organism.

DIAGNOSTIC TEST CORRELATION CHART

Use the following table to determine which AP Biology topics you need to review most. After scoring your test, check to find out the areas of study covered by the questions you answered incorrectly. **Please note, some of the questions are listed under multiple topics because they can be related to multiple areas of study.**

Area of Study	Question Number
Evolution	10, 38, 45, 50, 52
Diversity of organisms	11, 12, 16, 17, 18, 19, 46, 56
Biological molecules	1, 42, 43
Cell	2, 3, 13, 24, 39
Membrane traffic	15, 53
Photosynthesis	5, 57, 61
Cellular respiration	4, 36, 42, 56, 57, 58, 60
Thermodynamics and homeostasis	26
Immune response	54
Molecular genetics, gene regulation, mutation	6, 7, 8, 9, 21, 25, 48
Cell cycle	24, 59
Chromosomal basis of inheritance	25
Mendelian genetics	7, 14, 21, 22, 23, 35
Viruses	38, 47
Cell communication	53
Hormones and the endocrine system	40, 41, 54
Neurons and the nervous system	30, 49, 55
Enzymes	1, 44
Plant vascular and leaf systems	27, 28, 29, 61
Respiration and the circulatory system	20, 44
Digestion and excretion	31, 37
Musculoskeletal system	43, 55
Ecology	32, 33, 34, 46, 51, 62, 63

CHAPTER 3: EVOLUTION

BIG IDEA 1: The process of evolution drives the diversity and unity of life.

IF YOU ONLY LEARN SIX THINGS IN THIS CHAPTER . . .

1. It is thought that chemical components of life on Earth originated through radiation and storms. These compounds became increasingly complex, forming protobionts and, ultimately, living organisms.

2. According to Darwin, species evolve via natural selection, where animals with certain traits are more likely to survive and reproduce, passing on those traits. Selection can be stabilizing (median is encouraged), directional (the norm shifts toward an extreme), or disruptive (extremes are favored over the norm).

3. Evidence for evolution comes from comparative anatomy (homologous and analogous structures), biogeography, embryology, the fossil record, biological classification, and molecular biology (relatives share DNA).

4. Biological species concept: A species is a reproductively isolated population able to interbreed and produce fertile offspring.

5. In allopatric speciation, geographically separated populations develop into different species. Sympatric speciation occurs when populations in the same environment adapt to fill different niches. Parapatric speciation occurs with limited interbreeding between two groups.

6. Hardy-Weinberg equilibrium states that genetic distribution remains constant in large, isolated, randomly mating populations with no mutation and no natural selection. These conditions rarely (if ever) occur together.

A LITTLE CHANGE CAN ADD UP: ORIGINS OF LIFE AND EVOLUTION

There is no biological concept more controversial and more misunderstood than biological evolution. The concept can be simply defined with the phrase "descent with modification." This means that over time, populations of organisms exhibit changes in characteristics that are passed on through inheritable (i.e., genetic) means. The important distinction between this simple definition and what is commonly accepted as biological evolution is mechanism. Mechanisms will be covered on the exam. You should also clarify the semantic distinction between evolution and biological evolution. Personalities evolve. Societies evolve. But only populations of organisms undergo biological evolution by means of a modification of inheritable characteristics. In this chapter, the term *evolution* will always refer to biological evolution.

EARLY ORIGINS OF LIFE ★★★★

Scientists estimate that the Earth is about 4.5 billion years old and that in the beginning, it was a very inhospitable place. When Earth first came into existence, there was very little or no atmospheric oxygen, and the surface of the Earth was bombarded by intense ultraviolet radiation. Around 39 billion years ago, there were heavy rains and violent storms, which led to the production of basic inorganic chemical building blocks from the soil and the energy needed to drive reactions for producing simple organic molecules. The prevailing theory of the origin of life is that these organic molecules became more and more complex until amino acids and nucleic acids were formed. Once strings of nucleic acids formed, they could self-replicate within the organic "soup." These self-replicating structures organized into **protobionts**, which were droplets of segregated chemicals. Chemicals continued to organize, until the first identifiable organism came into being. The RNA World Hypothesis theorizes that these early protobionts were ribonucleic acids (RNA) that eventually evolved into the DNA, RNA, and protein-based life that exists today. The origins of life have been theorized based on biochemistry, some actual laboratory experiments, and models of early Earth environments. At this time, it cannot be substantiated that this theory of the origin of life is true. Fossil evidence of early life is the only way the origins of life can be directly substantiated. The first fossil records of prokaryotes exist from geological deposits thought to be 3.5 billion years old. The oldest eukaryotic cells are from deposits that are about 2 billion years old. The oldest multicellular fossils are from deposits that are 1.25 billion years old.

Molecular and genetic evidence demonstrates that all life on Earth shares a common ancestor. For example, all eukaryotic cells share common traits such as the presence of a cytoskeleton, a nucleus, membrane-bound organelles, linear chromosomes, and endomembrane systems, elements that have likely been conserved from the earliest life-forms.

EARLY EVOLUTIONARY IDEAS AND DARWINISM ★★★★

Although some early scientists believed that species are immutable and do not evolve or change, others believed that evolution occurs. Theories of evolution developed long before the time of

Charles Robert Darwin (1809–1882). Prior to Darwin, even the scientists who accepted the idea that species change over time did not have a solid idea for a mechanism that could explain the changes they observed. The most well-known hypothesis prior to Darwin was that of **Jean-Baptiste de Lamarck** (1744–1829), who suggested that organisms pass on acquired traits in an attempt to reach a more perfect form. For example, by Lamarck's logic, if a mother works out at the gym and becomes strong and healthy, she will pass her acquired health to her children and so on. We know today that Lamarck's ideas are incorrect, as parents only pass on genetically controlled traits and not those they have acquired from the environment they live in.

Darwin presented his postulates for his **theory of evolution** by **natural selection** in his work *The Origin of Species* (1859). There have been slight modifications to the theory based on more up-to-date knowledge about genetics and molecular biology, but for the most part, Darwin's theories are accepted today. Virtually all scientists consider evolution by means of natural selection to be established fact.

DARWIN AND NATURAL SELECTION

Darwin had several original postulates for his theory of evolution by natural selection:

1. Individuals vary in their characteristics within a population. This means that all giraffes have long necks, but their necks aren't all the exact same length.

2. The variations observed in populations are inherited. When a big dog has puppies, they tend to be big, and a little dog's puppies tend to be little.

3. A considerable number of individuals in a population seem to die as they compete for limited resources in the environment. This is where the term "survival of the fittest" emerged; it simply means that some individuals have characteristics that make them more likely to survive. These characteristics could include being bigger or stronger, but they also could include being smaller and smarter. It simply means that the characteristic(s) suit(s) the environment better.

4. Individuals who have more resources because of their particular characteristics tend to produce more offspring that survive. For example, if a bird with a long beak can get more food from holes in trees because of its long beak, it will be more likely to survive and provide more food for its offspring. If beak length is a result of genetics, that bird's offspring are more likely to have long beaks, and the following generations of offspring are more likely to have long beaks, until every bird in the population has a long beak.

The selection for more "adaptive" traits tends to narrow a population of individuals down to those who are best suited for a particular environment. If changes occur in the environment, selection favors individuals best suited to the new environment. The theory of natural selection seemed to explain the differences Darwin observed in species he studied and helps to explain the biodiversity in organisms today.

EVIDENCE USED TO SUPPORT EVOLUTION ★★★★

THE FOSSIL RECORD

The geological layers in the Earth's crust stack on top of each other, with the oldest layers deeper and the youngest layers closer to the surface. Older geological layers hold more "primitive" fossils. Following is a table displaying different major biological events that have occurred through geologic time, the period in which these events occurred, and approximately how many millions of years ago they happened. You are not required to know the dates of these events, but be prepared to analyze given data and develop scientific inquiries to investigate certain claims.

THE FOSSIL RECORD OVER GEOLOGIC TIME

Period	Millions of Years Ago	Events(s)
Precambrian	> 3,500	First prokaryotes
Precambrian	> 1,000	Earliest eukaryotes
Cambrian	540–490	Origin of all extant and some extinct animal phyla, including chordates
Ordovician	489–446	Continued evolution of ocean life
Silurian	445–415	First terrestrial organisms
Devonian	415–360	Diversification of bony fishes; first insects; first seed plants
Carboniferous	360–300	First gymnosperms
Permian	300–250	Diversification of reptiles
Triassic	250–200	First mammals and dinosaurs
Jurassic	200–145	Diversification of dinosaurs; first birds
Cretaceous	145–65	Origin and diversification of angio-sperms; extinction of dinosaurs at end of period
Paleogene and Neogene*	65–1.8	Diversification of all major living groups of birds and mammals, including hominids
Quaternary	1.8–Present	Extinction of large land mammals; rise of humans

*The Paleogene and Neogene were traditionally grouped into a time period called the Tertiary, but this name no longer has an official place in the Geologic Time Scale.

BIOGEOGRAPHY

Organisms are more like other organisms in their geographic vicinity. Organisms in adjacent dissimilar environments are more similar than organisms in similar environments on opposite sides of the Earth. This suggests that organisms in adjacent dissimilar environments are descended from recent common ancestors and have not magically come about in the environment in which they currently live.

COMPARATIVE ANATOMY

Organisms have very different structures that are composed of the same basic components. For example, the human arm has the same bones as the wing of a bat. These structures are called homologous structures because they are considered to have arisen from a common ancestor. Analogous structures are structures that may perform a similar function, but have not arisen from the same ancestral condition.

EMBRYOLOGY OR ONTOGENY

Organisms that share a more recent common ancestor have similar modes of development. A classic example is that all vertebrate embryos have a stage of development in which they possess gills, whether they are aquatic or terrestrial. The presence of these more "primitive" characters in the embryos of "advanced" organisms suggests that these organisms share genetically controlled developmental physiologies that have been passed on from their common ancestors. The process through which an organism develops from an embryo to an adult is called ontogeny.

TAXONOMY

Organisms are classified into smaller and smaller subgroups based on similar and dissimilar characters. This hierarchy is an implicit illustration of the tree of life, leading to common ancestry by linkage to a superceding group. For example, a plant in the family Euphorbiaceae is more closely related to other plants in Euphorbiaceae than it is to plants in the family Cactaceae. The housefly, *Musca domestica*, is more closely related to other flies in the genus *Musca* than it is to flies in the genus *Stomoxys*.

Phylogenetic trees and cladograms are illustrations that can be used to represent the relationships between similar and dissimilar organisms. They are constructed using morphological, molecular, or DNA evidence. For example, in the simplified cladogram that follows you can see that Mollusca and Arthropoda are the most closely related phylums, as they shared the most recent common ancestor. However, the phylums Porifera, Cnidaria, Echinodermata, Arthropoda, and Mollusca all share a single common ancestor.

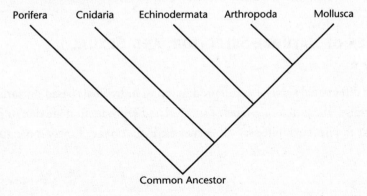

> ## ANALOGOUS STRUCTURES
>
> The wings of a bat and the wings of a butterfly are analogous.

DID YOU KNOW?

A domain is the broadest classification of organisms; therefore, organisms in different domains will be the most distantly related.

MOLECULAR BIOLOGY

Siblings share more similar DNA with each other than they do with other members of the same species. Similar species have more similar DNA than do more distantly related species (in different genera, for example). Related genera share more similar DNA with one another than they do with genera in another family.

RATES OF EVOLUTION ★★★

Prior to Darwin and even for scientists today, there is an underlying assumption that evolution takes a long time. The reasoning behind this theory is that because mutation is the ultimate source of variation and mutations that allow viable offspring are extremely rare, the probability of accumulating enough mutations to cause considerable change in organismal form requires a lot of time. Due to this assumption, prior to chemical dating techniques, geological layers were considered to be very old. Chemical dating has allowed modern scientists to assess the age of geological layers accurately.

Scientists have proposed two more hypotheses related to the rate of evolution. The first hypothesis is **punctuated equilibrium**. The hypothesis of punctuated equilibrium suggests that changes in organismal form did not take millions of years. Instead, very large changes in form happened relatively quickly (i.e., over thousands or tens of thousands of years) and were maintained thereafter over long periods of time. There is some evidence to suggest this hypothesis may be true, such as the Cambrian explosion in the fossil record, the discovery of cascading developmental genes, and the observation of large changes in phenotypic expression caused by single base-pair mutations. The second, the **molecular clock hypothesis**, is the notion that genetic mutations occur in a genome at a linear rate. Assuming the molecular clock hypothesis is true, one could extrapolate the age of divergence of two organisms by counting the number of genetic differences in their genomes. This latter hypothesis assumes that mutation rates are constant over time and between species; a likely incorrect pair of assumptions.

VARIATION, MODES OF NATURAL SELECTION, AND SEXUAL SELECTION ★★★★

Natural selection is the differential survival and reproduction of individuals based on variation in genetically controlled traits. These differing rates of survival and reproduction are due to forces in the environment and/or to forces exhibited by other species. Evolutionary fitness is measured by

the reproductive success of a species. To understand natural selection, it is necessary
to understand variation.

Ultimately all variation originates in the mutation of DNA in an individual's
genome. For a mutation to have an impact on evolution, it must occur in a gamete
and be passed on to offspring. If mutation occurs in a gamete that forms a zygote,
the offspring will inherit the allele for that trait and pass it on to its offspring.
Genetic variation can also occur during recombination in meiosis I. Most mutations
are harmful, but occasionally a mutation exists in viable offspring. These offspring
may then exhibit a phenotype that differs from the rest of the population.

The three types of variation that can occur in a population are stabilizing selection,
directional selection, and disruptive (diversifying) selection. If a population
is subject to **stabilizing selection**, extremes at both ends of a phenotype are
eliminated, resulting in less genetic variability. For example, if the variation in color
of a bird species ranges from dark gray to white and the population is subject to
stabilizing selection, the medium-gray phenotype will be most common. If the
population is under **directional selection**, one extreme is selected against but not
the other (e.g., dark gray but not white) so that the average in the population moves
in one direction. **Disruptive (diversifying) selection** favors both extremes but
selects against the average(s), which would mean dark gray and white are selected
over medium-gray in the case of the birds.

> ## VARIATION
>
> Variation that occurs
> in a population will
> have a distribution
> based upon the kind
> of natural selection
> that is taking place in
> the population.

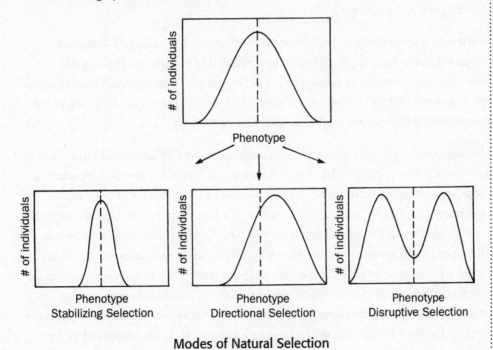

Modes of Natural Selection

Sexual selection is a force exhibited by a member of the same species of the opposite sex. This is most commonly seen when a female selects particular mating characteristics in a male. Many male birds have bright mating plumage that makes them conspicuous in a forest and more prone to attack from predators. Male ungulates often have huge antlers that grow every year and make it more difficult to travel through dense woods. The bright feathers on the male bird and the large antlers on the male deer are characteristics that are **selective disadvantages** in the animals' natural environment, but are **selective advantages** when it comes to courtship and mating.

KEY POINT

The balance between survival and the ability to mate dictates how a species evolves in terms of natural selection.

SPECIATION ★★★★

Over a dozen different concepts of speciation have been promoted in scientific literature, but the most prevalent by far is the **biological species concept (BSC)**. This concept states that a species is defined by a naturally interbreeding population of organisms that produces viable, fertile offspring. In other words, different species are organisms that can't breed with each other or don't naturally breed with each other due to certain barriers. There are two kinds of barriers to interbreeding: prezygotic and postzygotic.

Prezygotic barriers to interbreeding include isolation of species due to ecological, temporal, behavioral, or mechanical factors, or physiological incompatibility of gametes. **Postzygotic barriers** include ultimate inviability or sterility of hybrid organisms from the interbreeding of two species. Hybrid organisms may not die off in one generation, but ultimately the offspring of the mating of two species will die without producing offspring of their own.

Geographic isolation is not the only factor that can cause speciation, so additional terms have been defined to describe how species evolve due to isolating mechanisms. **Allopatric speciation** is when one population is separated into two distinct populations by some geographic barrier such as the movement of a tectonic plate or the elevation of a mountain range. After the original population is no longer able to share its alleles, it evolves into distinct populations that have a high probability of acquiring distinctive traits. In contrast, **sympatric speciation** occurs when individuals within a population acquire distinctively different traits while in the same geographic area. Sympatric speciation requires some other form of reproductive isolation, such as those previously mentioned. **Parapatric speciation** is less definitive. This occurs when two populations are able to interbreed along a border, but the exchange of alleles is negligible compared to the amount of genetic exchange occurring within each population. A narrow zone of hybridization exists at the meeting of the two populations, but the two populations never coalesce into one.

AP BIOLOGY LAB 1: ARTIFICIAL SELECTION
INVESTIGATION ★★★★

In this inquiry investigation, you will explore real-time natural selection using Wisconsin Fast Plants (Brassica) just as Darwin did with pea plants. This is a long-term laboratory investigation in which you will select, plant, grow, and tend to a population of plants to observe how natural selection acts on phenotypic variations. This is a pretty simple investigation in terms of content and equipment. However, throughout the course of this seven-week investigation, you must keep detailed records in which you quantify variation, record images of that variation, and evaluate and explain your results. These skills will be helpful for you to cultivate for your responses on the free-response questions on the AP Biology exam.

Natural selection is the mechanism that describes the reproductive success or fitness of certain traits. Scientists and farmers use artificial selection to grow preferred agricultural crops. We also see natural selection occur inadvertently when diseases and pests grow resistance to the use of antibiotics or pesticides. In this lab, as a class and independently, you will select for a trait and investigate its reproductive success in two generations. Remember that directional selection tends to increase or decrease the trait in the next population. After growing your plants and recording your observations of the two different generations, you have many options for how to analyze your data. Deciding what to do with your data is the tricky part of this investigation, but these types of analyses will help you with the grid-in and free-response questions on your exam. Let's look at two ways we can analyze sets of data.

You could use different descriptive statistics such as mean, median, range, and standard deviation to describe your populations of study. Let's take a look at a couple of these descriptive statistics. The equation for solving for standard deviation s is:

$$s = \sqrt{\frac{\sum(x_i - \bar{x})^2}{n-1}}$$

where x_i is each data point, $\bar{x}$ is the mean of the data points, and n is equal to the size of the sample. The standard deviation calculates the difference between each of your data points and the mean value of your data, allowing you to see how much variability there is in the population.

For each set of data, you will construct a histogram. A histogram is a graphical representation of the distribution of the data. Using your histograms and other statistical analyses such as a t-test or chi-squared test, you can determine the validity of the differences in the populations. The formula for a chi-squared test is:

$$\chi^2 = \sum \frac{(o - e)^2}{e}$$

where o is the number of individuals with an observed genotype and e is the number of expected individuals with an observed genotype.

All of this mathematical problem solving might seem intimidating, but it is essential for real-world biology practice and for the AP Biology exam. You can take a deep breath, though; all of these formulas will be provided to you on your exam. You will not be required to memorize these equations and formulas.

AP BIOLOGY LAB 2: MATHEMATICAL MODELING: HARDY-WEINBERG INVESTIGATION ★★★★

This lab is easy to perform, but it causes confusion for some students because it involves quantitative and analytical skills. In this investigation, you will develop a mathematical model to investigate the relationship between allele frequencies in populations of organisms. You will use a spreadsheet to build a model based on the Hardy-Weinberg relationship to determine how allele frequencies change from one generation to the next. This model will help you to see how selection, mutation, and migration can affect these inheritance patterns.

Now comes the hard part: putting information into equations. The way to determine the frequencies of the alleles in your breeding population is to assume the population is in **Hardy-Weinberg equilibrium**. If your population is in Hardy-Weinberg equilibrium, two things will be true. First, the addition of the frequencies of the alleles will equal one. The simple formula is:

$$p + q = 1$$

where the frequency of the dominant allele is indicated by p and the frequency of the recessive allele is indicated by q. For this experiment, $p = T$ and $q = t$. The frequencies of the phenotypes in a Hardy-Weinberg population follow the equation:

$$p^2 + 2pq + q^2 = 1$$

The genotypes in a Hardy-Weinberg population are indicated by each term on the left side of the previous equation. The frequency of the homozygous dominant phenotype in the population above is:

$$T \times T = T^2 = p^2$$

The frequency of the heterozygote phenotype is:

$$2 \times T \times t = 2Tt = 2pq$$

and the frequency of the homozygous recessive phenotype is:

$$t \times t = t^2 = q^2$$

Normally, success in mastering Hardy-Weinberg problems lies in the reading of the question. You usually have to figure out p and q in order to plug in the terms. If the question states "frequency of alleles in a population," then you should start with $p + q = 1$ and solve for p or q. If, on the other hand, the problem says "frequencies of organisms that express the trait (dominant or recessive)," then you start with $p^2 + 2pq + q^2 = 1$ and figure your frequencies of p and q from this data. Once you have calculated p and q, you can plug those values into the equation where appropriate to figure out frequencies of homozygotes, heterozygotes, or carriers for a trait.

HARDY-WEINBERG EQUILIBRIUM—FIVE ASSUMPTIONS

There are five assumptions that must be met before assuming a population is under Hardy-Weinberg equilibrium:

1. The population is very large and not subject to small perturbations in the frequencies of alleles. There are no bottleneck effects.

2. The population is isolated from both immigration and emigration. There is no gene flow.

3. There is no mutation.

4. There is no selective breeding, and mating is random between individuals.

5. There is no genetic drift. All genotypes code for phenotypes that have equal chance of viability and reproduction. There is no selection of phenotypes.

Keep in mind that the theoretical Hardy-Weinberg population is the benchmark to which all naturally occurring populations are compared. There are probably few if any populations in complete equilibrium. Comparing the expected frequencies under the previously given assumptions to what actually occurs in a population gives insight into which of these assumptions is being violated. In this way, a scientist can determine which evolutionary or environmental forces prevent a population from maintaining equilibrium.

For this experiment, we can make assumptions about the theoretical alleles of our model population. Let's say that 5 percent of the population is homozygous recessive. This means that the frequency of the recessive allele can be determined by letting $q^2 = 0.05$, which means the frequency of the dominant allele can be determined with the equation:

$$p = 1 - q = 1 - 0.22 = 0.78$$

The frequencies of the homozygous dominant genotype and the heterozygous genotype, respectively, are:

$$p^2 = 0.78 \times 0.78 = 0.61$$
$$2pq = 2 \times 0.78 \times 0.22 = 0.34$$

KEY POINT

There is a good likelihood that you will see some sort of Hardy-Weinberg question in either the grid-in or free-response section of the exam.

You can check your work by adding all the genotypic frequencies and making sure that they total 1:

$$0.05 + 0.61 + 0.34 = 1$$

After developing your Hardy-Weinberg model and exploring how random events can affect allele frequencies over generations, you will identify and then test different factors that affect evolution of allele frequencies. In addition to measuring allele frequencies, your model should track changes in population size, number of generations, selection (fitness), mutation, migration, and genetic drift.

AP BIOLOGY LAB 3: COMPARING DNA SEQUENCES TO UNDERSTAND EVOLUTIONARY RELATIONSHIPS WITH BLAST INVESTIGATION ★★★★

Since 1990, scientists have been working to develop a comprehensive library of genes from several species, including humans, mice, fruit flies, and *E. coli*. BLAST (Basic Local Alignment Search Tool) is a powerful bioinformatics program that helps scientists to compare genes from different organisms catalogued in this library. Information derived from BLAST can, therefore, show scientists the evolutionary relationships between different organisms.

In this investigation, you will use BLAST to compare several genes from different organisms, and then construct a cladogram to represent the evolutionary relationships among these different species. While cladograms can be constructed using many different factors, including the presence of different morphological traits (wings, gills, etc.), in this investigation you will use DNA evidence to establish the similarity among different organisms.

Locating and sequencing genes in different organisms not only helps us to better understand evolutionary relationships among organisms, but it can also provide important insights into genetic diseases. Species with smaller genomes, such as the fruit fly and mouse, are easier for scientists to study than humans. When scientists locate a disease-causing gene in a fruit fly or mouse, they can then use BLAST to see if there is a similar sequence in the human genome.

Using BLAST requires several specific steps and an abundance of information, none of which you need to memorize for the AP Biology exam. Rather, you should be able to apply similarities and differences in morphology and genetics to determine the evolutionary relationships between different species. Drawing and analyzing cladograms (also known as phylogenetic trees) is a skill you should practice and be able to perform successfully.

Let's use some data from an example in this investigation to construct a cladogram based on genetic data. In humans, the GAPDH gene produces a protein that catalyzes a step in glycolysis. The following table shows the percentage similarity between the GAPDH gene and protein in humans compared with four different species.

Species	Gene Percentage Similarity	Protein Percentage Similarity
Chimpanzee (*Pan troglodytes*)	99.6%	100%
Dog (*Canis lupus familiaris*)	91.3%	95.2%
Fruit fly (*Drosophila melanogaster*)	72.4%	76.7%
Roundworm (*Caenorhabditis elegans*)	68.2%	74.3%

Just as with the cladogram shown earlier in this chapter, we can use this data to construct a cladogram that tells us how these species are related to each other based on GADPH. There are five species being compared here, so our cladogram will have five branches.

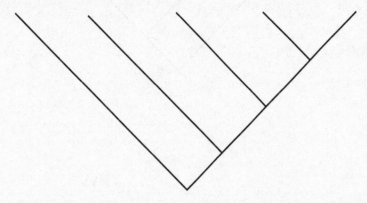

Reading the table, we see that humans and chimpanzees have the most similarities in their GADPH genes and proteins. Therefore, humans and chimpanzees would be placed closest together on our cladogram as shown.

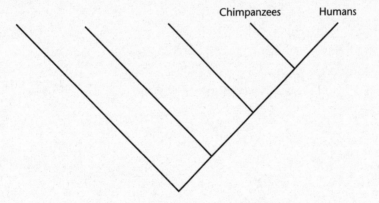

We can make deductions for the remaining three species based on their similarity to humans. Roundworms' GADPH genes have the least in common with humans, so roundworms should be placed farthest from humans on the cladogram. Dogs are closer to humans than fruit flies but farther away than chimpanzees.

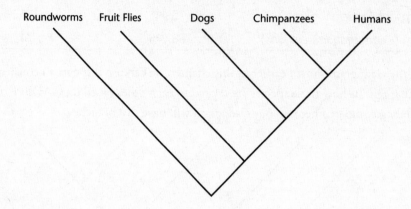

CHAPTER 4: DIVERSITY OF ORGANISMS

BIG IDEA 1: The process of evolution drives the diversity and unity of life.

IF YOU ONLY LEARN FOUR THINGS IN THIS CHAPTER . . .

1. Scientists use the Linnaean system of binomial nomenclature to identify organisms based on their relationships to one another.

2. The divisions of classification are, from largest to smallest: domain, kingdom, phylum, class, order, family, genus, and species. Organisms that share a phylum are more closely related than those sharing only a kingdom, and so on. Commonly, organisms are identified by genus and species (e.g., *Felix domesticus*).

3. Domains include Archaea, Bacteria, and Eukarya (eukaryotes). Kingdoms within the Eukarya include Plantae (plants), Animalia (animals), Fungi, and Protoctista.

4. A phylogeny is a diagram indicating evolutionary relationships between specimens. It depicts an ancestral species with other species branching off from it. More closely related taxa are nearer to one another in the diagram. The diagram shows anagenetic evolutionary changes (which do not lead to speciation), as well as cladogenetic evolutionary changes (which do).

INTRODUCTION

This chapter covers a variety of Biology topics, including binomial nomenclature, the divisions of classification, and the diversity of life and distinguishing features. The concepts presented in this chapter should acquaint you with the difficulty of creating a language with which scientists from all over the world can communicate about the forms of life they observe.

CATALOGING BIODIVERSITY: THE LINNAEAN SYSTEM

In 1758, the Swedish naturalist Carl von Linné (a.k.a. Carolus Linnaeus, 1707–1778) published a version of the *Systema Naturae*, which used a consistent method of naming organisms called **binomial nomenclature** in an attempt to construct a "natural classification" that would reveal order in the universe. The study of biology has never been the same since.

CLASSIFICATION ★★

Scientists need a way not only to identify the kinds of organisms they observe, but a way to communicate to other scientists which animal(s) they study. If, for example, a scientist refers to an organism by a regional name in a native language, it is difficult for a scientist on the other side of the world to identify the organism in question or to correlate any associative information by relationship with other organisms. The way the scientific community has addressed this need has been to create a hierarchical system of naming called the **Linnaean system of classification**. The following are the Linnaean classifications of the common housefly, the vinegar fly, the human, and cultivated corn.

LINNAEAN CLASSIFICATION SYSTEM

Linnaean Hierarchy	Housefly	Vinegar Fly	Human	Corn
Kingdom	Animalia	Animalia	Animalia	Plantae
Phylum (Division)	Arthropoda	Arthropoda	Chordata	Anthophyta
Class	Insecta	Insecta	Mammalia	Liliopsida
Order	Diptera	Diptera	Primates	Poales
Family	Muscidae	Drosophilidae	Hominidae	Poaceae
Genus	*Musca*	*Drosophila*	*Homo*	*Zea*
Species	*domestica*	*melanogaster*	*sapiens*	*mays*

The Latin binomial for an organism, which includes the genus and the species (e.g., *Musca domestica, Homo sapiens*), provides an internationally applied standard to which all people can refer. In addition, the collective group names to which the organism belongs imply an evolutionary relationship if the classification system follows phylogeny (evolutionary events involved with the origin of a group of organisms). The two species of flies given in the table belong in the same order, humans belong in the same kingdom, and corn is in a completely different kingdom. The point of common origin of these organisms can be extrapolated by simply knowing the binomial.

THE DIVERSITY OF LIFE AND DISTINGUISHING FEATURES ★★★

In addition to naming organisms, scientists are always trying to simplify things. These simplifications usually take the form of generalities. In general, there are few generalities that apply to biology (irony intentionally applied here). Classification, however, has always relied on the similarities and differences that occur between species. You cannot review all of microbiology, cell biology, zoology, and botany in one chapter, but this section presents a summary of some of the major groups and their characteristics.

> ## IMPORTANT
>
> Until recently, the first level of classification of organisms was the **kingdom**. There were five recognized kingdoms (**Monera, Protoctista, Fungi, Plantae,** and **Animalia**) because too many organisms defied the simple classification of being either a plant or animal. Now all of life is first divided into three domains. You should be familiar with the domains for the AP Biology exam.

The first two domains, **Archaea** and **Bacteria**, include all of the prokaryotes. Remember that **prokaryotes** are unicellular organisms with cell walls and no membranes on any of their subcellular structures. The Archaea are special organisms that produce methane or possess adaptations for extreme environmental conditions, such as high temperatures or high salt concentrations. The Archaea are separated into three groups: **Methanogens** (methane makers), **Halophiles** (salt loving), and **Thermophiles** (heat loving). New species of Archaea are often discovered in deep-sea thermal vents or hot springs. Bacteria compose the rest of the prokaryotes. There are distinct chemical differences between the Archaea and Bacteria, but you should know that the Archaea have adapted to extreme conditions. The last domain is the **Eukarya**, which encompasses the rest of life on Earth and the groups that you should spend the majority of your time reviewing.

The Eukarya are split into four groups: Protoctista, Fungi, Plantae, and Animalia. Simply put, the Eukarya are all of the **eukaryotes**. They can be divided into four kingdoms by the information in the next table. Keep in mind that the table contains general information and there are always exceptions in biology.

THE KINGDOMS OF THE EUKARYA

Kingdom	Structure	Reproduction	Examples
Protoctista	Microscopic; mostly unicellular; 9 + 2 cilia or flagella	Asexual; no embryo	Amoeba, paramecia, diatoms, slime molds
Fungi	Multicellular; no cilia or flagella; chitinous cell wall	Haploid or dikaryotic; haploid spores	Yeasts, molds, mushrooms
Plantae	Large, sessile, multicellular; cellulose cell wall	Fertilization of female by male gamete; alternation of generations	Mosses, gymnosperms, angiosperms
Animalia	Large, motile, multicellular; use muscles for movement; no cell walls	Sperm and egg form zygote	Sea stars, worms, insects, mammals

Be familiar with the distinguishing features and some representative members of the major divisions of Plantae. There are at least ten divisions of Plantae. To illustrate the diversity, check out: **Mosses**, **Ferns**, **Conifers**, and **Angiosperms**. The following table summarizes some of the major differences between each division of Plantae.

THE MAJOR PLANT DIVISIONS

Division	Structure	Reproduction
Mosses	Nonvascular; produce a mat consisting of many plants and rootlike structures called rhizoids	Water necessary for swimming sperm; haploid gametophyte is dominant generation; homosporous
Ferns	Vascular; compound leaves called fronds develop from coiled fiddlehead; true roots	Water necessary for swimming sperm; diploid sporophyte is dominant generation; homosporous
Conifers	Vascular; produce cones with internal gametophyte; evergreen, needlelike leaves	Wind pollinated; diploid sporophyte is dominant generation; heterosporous
Angiosperms	Vascular; produce flowers and fruit	Wind or animal pollinated; diploid sporophyte is dominant generation; heterosporous

Be familiar with the hypothesis that terrestrial plants evolved from green algae (Protoctista) because they share several features:

1. Chlorophyll *a* and accessory pigments

2. Thylakoid membranes stacked locally into grana

3. Cell wall made of cellulose

4. Carbohydrate stores made of starch

5. Cell plate dividing cytoplasm in cytokinesis formed from Golgi complex vesicles

Whereas the major determinant of plant divisions is mode of reproduction, the major determinants in animal phyla are body plans and development. The first branch in the animal tree has to do with body symmetry. The most primitive animals have either no body symmetry, as with the sponges that grow in amorphous masses, or **radial symmetry**. The rest of the animals have **bilateral symmetry**. Animals with radial symmetry have a side with a mouth (oral) and one without (aboral) but no front or back end or left or right side. Animals with bilateral symmetry have a head and tail and a right and left side in addition to a front and back or top and bottom.

> ### CHLOROPHYLL
>
> You do not need to know the molecular structure of chlorophyll *a* for the AP Biology exam. Focus on the function of chlorophyll instead.

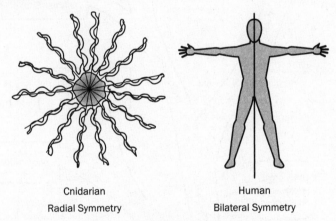

Cnidarian
Radial Symmetry

Human
Bilateral Symmetry

Radial and Bilateral Symmetry

Sponges are sessile filter feeders. They actively draw water through their bodies with specialized cells that have flagella. Different sponges are characterized by the composition of their internal **spicules**, which can be made of calcium carbonate, silica, or a soft protein called spongin.

The animals with radial symmetry include the Cnidarians, which encompass jellies, corals, and sea anemones. Simply put, Cnidaria are bags with a mouth. Their mouths are lined with stinging tentacles containing cnidocytes (*cnide* means "nettle"). The remaining animals are further subdivided by the way they develop internally. There are two characteristics to consider. One is whether the animal has a true **coelom** (a mesoderm-lined cavity in the body between the gut and the outer body wall).

This separates animals into **acoelomates** (no coelom), **pseudocoelomates** (possessing a body cavity not completely lined by mesoderm), and **coelomates**. The other characteristic to consider is whether during development the zygote has determinate or indeterminate cleavage. Animals with determinate cleavage have early developmental cells with a predetermined fate. If you separate a cell after initial cleavage, it will not develop into a complete animal—it will be missing parts. Animals with indeterminate cleavage have early developmental cells that can go on to produce a whole animal. This is what makes identical twins possible. Animals with determinate cleavage are called protostomes, and animals with indeterminate cleavage are called deuterostomes. Like that of most groups of organisms, the current classification of animals is constantly changing, as new characteristics are discovered and analyzed. For example, the use of DNA analysis has led to recent advances in genetic research of animal relationships.

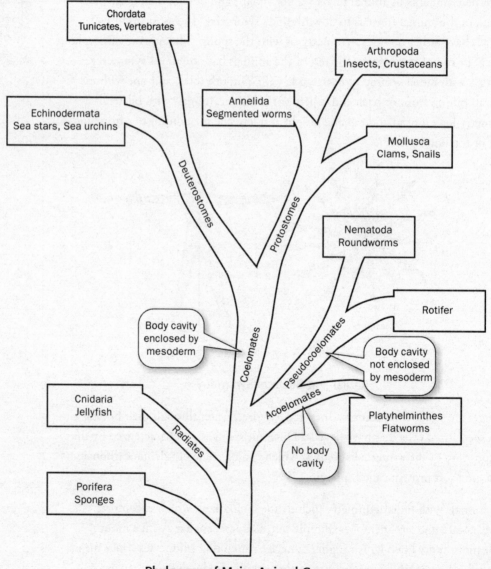

Phylogeny of Major Animal Groups

Use the next table to review some of the diagnostic features of some of the more important animal groups.

ANIMAL GROUPS OF IMPORTANCE

Phylum	Diagnostic Feature(s)	Examples
Mollusca	Muscular foot, visceral mass containing gut, mantle (shell)	Octopus, snail, clam
Annelida (segmented worms)	Repeating body segments, ventral nerve cord with segmented ganglia	Earthworm, leech, sea mouse
Arthropoda	Segmented exoskeleton and appendages, compound eyes	Lobster, horseshoe crab, spider, insects
Echinodermata	Radial symmetry, water vascular system, endoskeleton of calciferous plates	Sea star, sea urchin, sea cucumber
Chordata	Notochord, dorsal hollow nerve cord, pharyngeal slits	Fish, amphibians, reptiles, mammals

KEY POINT

On the AP Biology exam, you will likely be given data and asked to choose which taxa (organisms) are most closely related among several pairs of taxa.

RELATEDNESS AND PHYLOGENETIC SYSTEMATICS ★★★★

Systematists are scientists who study and formulate classifications as well as the relationships among taxa. These relationships are illustrated in the form of a **phylogeny** (branching evolutionary tree) that is usually obtained by analyzing data matrices composed of characteristics that are shared among the taxa in the analysis. These trees show how taxa are related based on the common evolved characteristic (or several characteristics) that they share. The organisms on the same branches of the phylogenetic tree are more closely related than those on different branches. The following tree illustrates some misconceptions about phylogenies.

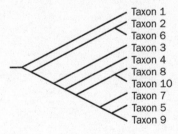

Taxon 1
Taxon 2
Taxon 6
Taxon 3
Taxon 4
Taxon 8
Taxon 10
Taxon 7
Taxon 5
Taxon 9

Sample Phylogeny

This hypothetical phylogeny shows the evolutionary relationships among ten taxa. A phylogenetic tree does not show a time scale. Taxon 7 shares a common ancestor, which isn't shown on the tree, with taxa 5 and 9. Taxon 7 is not the ancestor of taxa 5 and 9, nor does it necessarily have more characteristics in common with the ancestral taxon of taxa 7, 5, and 9 than taxa 5 or 9 do. It could have just as many **anagenetic** (evolutionary changes without speciation) changes along its branch as **cladogenetic** (evolutionary events that lead to speciation) changes that lead to taxon 10. Taxon 4 is more closely related to taxa 8, 10, 7, 5, and 9 than it is to taxon 3. Even though taxon 3 is close to taxon 4 on the tree, it is not on the same branch that includes taxa 4, 8, 10, 7, 5, and 9. Likewise, taxon 2 is more closely related to taxon 6 than it is to taxon 1. Taxa 2 and 6 belong to the same branch that does not include taxon 1.

APPLYING THE CONCEPTS: BIG IDEA ONE REVIEW QUESTIONS

EVOLUTION

This Review Quiz is designed to assess your content knowledge, and is not necessarily representative of the actual AP Biology exam. So don't panic if you see a new question type or a question asking about a relatively low-yield detail from the chapter.

1. Which of the following is the likely source of energy for the synthesis of the small organic molecules that presumably predated the first forms of life on Earth?

    (A) Fermentation by bacteria

    (B) Photosynthesis by microscopic algae

    (C) Lightning from constant storms

    (D) Shifts in ocean currents

2. The earliest forms of life on Earth were recorded from fossil deposits dating from about 3.5 billion years ago. Which of the following were likely found in these deposits?

    (A) Microscopic fungi

    (B) Microscopic protozoa

    (C) Bacteria

    (D) Viruses

3. A population of birds on the coast range in size from 9 cm to 15 cm, with the majority of the birds being 12 cm. Another population of birds exists on a distant island off the same coastline, but all of these birds are 15 cm. What is the most likely explanation for the difference in sizes of the members of these populations?

    (A) Sexual selection

    (B) Mutation

    (C) Disruptive selection

    (D) Genetic bottleneck (founder effect)

4. Which of the following is not an assumption made about a Hardy-Weinberg population in equilibrium?

    (A) There are an equal number of males and females in the population.

    (B) The population is large.

    (C) There is no immigration into the population.

    (D) There is no mutation.

5. Which of the following would be a statement most likely supported by Lamarck?

    (A) Four out of five hyena pups die because of a lack of resources in the environment.

    (B) The occurrence of new forms is a random event produced by a change in heritable factors.

    (C) The wings of a butterfly and the wings of a bat are analogous structures.

    (D) The long neck of a giraffe is a result of its ancestors continually trying to reach higher branches.

6. Which of the following is the ultimate source of all variation within and among populations?

    (A) Natural selection in different environments

    (B) Viable genetic mutations in gametes

    (C) Nonrandom mating between individuals with different traits

    (D) The diversity of habitats that exist on Earth

7. All of the following provide evidence for evolution EXCEPT which?

   (A) The fossil record

   (B) Comparative anatomy

   (C) Embryology

   (D) Food webs

8. The assumption that genetic mutations occur at a regular rate over time is part of which hypothesis?

   (A) Recombination

   (B) Natural selection

   (C) Molecular clock

   (D) Speciation

9. All of the following pairs of structures are homologous EXCEPT which?

   (A) The wing of a bat and the wing of a bird

   (B) The leg of a human and the leg of a centipede

   (C) The scales on a bird's legs and the scales on a lizard

   (D) The shell of a clam and the shell of a snail

Questions 10–13 refer to the following list.

   (A) Allopatric speciation

   (B) Sympatric speciation

   (C) Stabilizing selection

   (D) Sexual selection

10. This type of evolution favors the average members of a population, while eliminating the extremes.

11. A male peacock has such outrageously colored plumage during mating season that its survival is reduced.

12. The Isthmus of Panama separates two species of fish that are more closely related to each other than they are to the other species of fish in the bodies of water in which they exist.

13. Two closely related but reproductively isolated species of bees live in the same forest; one bee gathers nectar from a flower that peaks in early spring and the other gathers nectar from a flower that peaks in early summer.

Questions 14–16 refer to the following list.

(A) Cambrian

(B) Devonian

(C) Jurassic

(D) Cretaceous

14. Dinosaurs diversified during this geologic period.

15. Angiosperms developed during this geologic period.

16. Chordates developed during this geologic period.

ANSWERS AND EXPLANATIONS

1. C

It should be obvious that the correct answer will not include any energy that comes from an organism. The energy to create the first forms of life could not come from the first forms of life. This rules out (A) and (B) right away. Ocean currents were never part of the discussion for the origins of early life, so (D) can be ruled out. Early Earth was a place of continual thunderstorms, which correlates with (C).

2. C

The first thing that stands out in this question is the age of the fossil deposits, dated 3.5 billion years ago, which means the fossils should be very, very "primitive." All of the possible answers are microscopic organisms and *small* often means less complex, so you might pick a choice based on this key word, but the answer is not based on the organism being small. Also, the smallest organisms among the choices are viruses, but there are no fossil records for viruses. The oldest fossil records are from stromatolite deposits containing prokaryotes.

3. D

There is no apparent genetic difference between mainland and island populations. The birds on the island don't exhibit the same range of variation. The birds on the island could be an example of a very strong stabilizing selection, but this isn't one of the choices and is not likely. The occurrence of only large birds on the island could be an example of sexual selection, (A), in which only large birds mate successfully, but this is much less likely than the hypothesis that only large birds can make it to the distant island. This question is tricky because when discussing evolutionary scenarios, one can imagine all sorts of possibilities and all of the answers become possible, if not probable. When a new population is formed from a narrow range of the phenotypes in an original population, this is called a genetic bottleneck. The analogy is that the range of phenotypes is narrowed to a select few as if they were being forced through the narrow neck of a bottle.

4. A

This is another tricky question because the wording you learned in association with the assumptions of the Hardy-Weinberg population is most likely different from the wording given in the answer choices. Choices (B), (C), and (D) can be ruled out because they contain key words about assumptions made about a Hardy-Weinberg population at equilibrium. Fitness is a measure of an organism's ability to reproduce. The correct choice is (A), but it still isn't completely obvious after the other choices are ruled out. The assumption of Hardy-Weinberg is that all individuals have an equal chance of mating and that mating is random. If the number of males and females in a population were heavily skewed to only a few individuals of either sex, there would be nonrandom mating, but in most cases the population is large enough to prevent this. Random mating is still likely to occur if there are 10,010 males and 9,990 females, for example.

5. D

Lamarck believed that acquired traits are passed on to offspring. Some of the statements given are part of the developing history of the perceptions of evolution, but are not attributable to Lamarck. A giraffe striving to reach higher branches during its lifetime is an act that is not preserved in the genome. Should this action lead any given individual to develop a longer neck, increased neck size would be an acquired trait that cannot be passed on to offspring.

6. B

All evolutionary change ultimately comes from genetic mutation. Genetic mutations in gametes are passed on from generation to generation.

7. D

This is one of those knowledge-based questions that are hard to reason through without the facts. If you don't immediately know the answer, the best possible approach is to consider all of the choices within the context of the question. Fossils, (A), can be associated with evolution. Comparative anatomy, (B), and embryology, (C), are part of form and function, another topic in AP Biology. Food webs, (D), are part of another topic as well, ecology. However, you might remember that Darwin used comparative anatomy and embryology in his studies and that these techniques are still relied upon today. The correct answer is (D).

8. C

Even if you don't know the answer right off, you can still figure out the answer to this question. A rate is always a change in a quantity over a given period of time. The correct answer is (C), molecular clock, a clock being a tool to measure time.

9. B

Okay, quick review: Analogous structures usually share a similar function but do not have a common ancestral origin. Homologous structures might have completely different functions, but are composed of the same building blocks. There is only one obvious answer choice. The leg of a human and the leg of a centipede both allow the organism to walk, but they are completely different structures. A human leg has an internal skeleton and a completely different developmental origin than the leg of any arthropod.

10. C

Always remember stabilizing selection as a bell curve in which the middle trait is most prevalent.

11. D

Sexual selection affects factors that help an organism obtain a mate and often acts against the survival of the organism.

12. A

This is an example of geographic isolation. An initial population of fish was separated by the isthmus and acquired new traits after being isolated.

13. B

This is an example of prezygotic isolation due to ecological and, perhaps, behavioral mechanisms. Because the two bees exist in the same space, they are sympatric.

14. C

Dinosaurs diversified during the Jurassic period, although they originated in the Triassic period.

15. D

Angiosperms developed during the Cretaceous period.

16. A

Chordates developed during the Cambrian period.

DIVERSITY OF ORGANISMS

This Review Quiz is designed to assess your content knowledge, and is not necessarily representative of the actual AP Biology exam. So don't panic if you see a new question type or a question asking about a relatively low-yield detail from the chapter.

1. The Cnidaria are unique among animals because they possess which of the following characteristics?

   (A) Radial symmetry

   (B) Tentacles with stinging cells

   (C) Water vascular system

   (D) Muscular foot

2. Which group of organisms is the most prolific of all the organisms that have ever existed on Earth, with over a million existing species known and perhaps several million more to discover?

   (A) Bacteria

   (B) Vertebrates

   (C) Fungi

   (D) Arthropods

3. A phylogenetic tree constructed from data about characteristics of taxa provides all of the following information EXCEPT what?

   (A) The time of divergence between two taxa

   (B) A framework with which to classify organisms

   (C) A hypothesis of the way characters evolved among the taxa

   (D) A hypothesis of whether similar characteristics have a common origin (homologous) or are convergent (analogous)

4. Based on the current scientific consensus of evolutionary relationships among living organisms, which of the following pairs of taxa share the closest relationship with each other?

   (A) Mushroom and moss

   (B) Sea star and human

   (C) Bacterium and amoeba

   (D) Snail and jelly

5. The prokaryotic group Archaea differs from the other prokaryotes by being able to withstand extreme environmental conditions such as

   (A) high pressure.

   (B) high N_2 concentration.

   (C) high O_2 concentration.

   (D) high temperature.

6. This heterotrophic group of organisms can be microscopic and has chitinous cell walls.

   (A) Actinopoda (radiolarians)

   (B) Ciliophora (ciliates)

   (C) Chlorophyta (green algae)

   (D) Fungi

7. A scientist returns from an expedition in the Sargasso Sea with a peculiar organism that is apparently unknown to science, but has a water vascular system. The scientist can begin to classify the organism by placing it in which of the following groups?

   (A) Annelida

   (B) Arthropoda

   (C) Echinodermata

   (D) Cnidaria

8. Linnaeus provided the modern synthesis of classification with his consistent application of a system of

   (A) binomial nomenclature.

   (B) cladistics.

   (C) phylogenetic systematics.

   (D) natural selection.

9. Which of the following terrestrial plant groups fulfills a life cycle in which a haploid gametophyte is the dominant generation?

   (A) Conifers

   (B) Mosses

   (C) Cycads

   (D) Ferns

10. All of the following groups of organisms are members of the Protoctista EXCEPT

    (A) diatoms.

    (B) amoebas.

    (C) rotifers.

    (D) paramecia.

11. Organisms classified in which category show radial cleavage during early development and have a dorsal nerve cord?

    (A) Protostomes

    (B) Deuterostomes

    (C) Parazoans

    (D) Acoelomates

12. Which organism has radial symmetry?

    (A) Frog

    (B) Coral polyp

    (C) Rabbit

    (D) Kangaroo

Questions 13–16 refer to the following list.

   (A) Endosperm

   (B) Angiosperm

   (C) Gymnosperm

   (D) Tracheids

13. This group of plants includes the pine.

14. This tissue nurtures the developing plant zygote.

15. These cells produce tubes and fibers that function in fluid transport in plants.

16. This is the most diverse group of land plants.

ANSWERS AND EXPLANATIONS

1. B

Your first impulse might be to choose (A) because one of the features that separate taxa at the base of the tree of Animalia is general body plan, but remember that echinoderms also have radial symmetry and aren't closely related to Cnidaria. Scientists often use a diagnostic feature to name a group of organisms, which is true in this case. Cnidaria are named for their *cnide* (Greek for "nettle"), the stinging tentacles they possess.

2. D

Both bacteria and fungi have many species with many new ones to discover, but whenever you encounter a question asking about the group with "over a million species," "the most diversity," "the most diverse group," and so on, think either arthropods or insects. There are more species of arthropods than all the rest of the species combined. The more familiar group (vertebrates) does not compare in terms of numbers of species.

3. A

The question says that the phylogenetic tree was constructed from character data, which is important for developing hypotheses of character evolution, (C). It also means that the tree is constructed based on more objective methodologies and provides an objective way to compare different individual hypotheses of characteristics from total evidence, (D). Not all scientists use phylogenetic information to classify organisms, however. A more pragmatic approach based on similarity (e.g., most people refuse to call birds reptiles) is sometimes used, but in most cases, a phylogenetic tree is a good basis for classification, (B). The correct answer is (A). A tree can show relative evolutionary positions of taxa, but does not identify exactly when the evolutionary events happened. Even a strong fossil record can only suggest minimum and maximum ages of divergence, not exact dates.

4. B

According to evolutionary theory, all of these taxa are related at some point on the Tree of Life. To answer this question correctly you need to choose the pair of taxa that are highest on the same branch of the tree. Another way to look at this question is to find the two taxa that are in the most inclusive taxonomic group. Two species in the same genus would be very closely related. The correct answer is (B). Chordates and echinoderms both share the characteristic of being the only deuterostomes among all of Animalia. Mushrooms are not even in the same kingdom as mosses, so (A) is incorrect. Bacteria and amoebas, (C), are both microscopic, but are not in the same domain. Snails are protostomal coelomates with bilateral symmetry, and jellies are acoelomates with radial symmetry; therefore, (D) is incorrect.

5. D

There are three groups of Archaea: methanogens (they produce methane), halophiles (they live in environments with very high concentrations of salt, or NaCl), and thermophiles (they live in environments with very high temperatures). The characteristic that all three groups share is the ability to exist in extreme environments. From the list of possible answers, the only one that is applicable is high temperature, (D).

6. D

The term *heterotrophic* refers to an organism's need to obtain energy from sources other than itself. In other words, the organism isn't photosynthetic. Animals are heterotrophs because they eat other organisms. The correct answer is not the photosynthetic organism in the list, (C), because these organisms are autotrophic. Actinopoda and Ciliophora are heterotrophic, but Fungi are organisms that are characterized as having cells with chitinous cell walls, (D).

7. C

This question is rather straightforward if you remember which group of organisms has a water vascular system. Unfortunately, of the possible answers, all can be found in marine ecosystems so you cannot rule out possibilities based on ecology. One way to remember this system is to think of the movement of sea stars. Sea stars move by pumping fluid in and out of their tubelike feet. The correct answer is (C), because sea stars belong in the group Echinodermata.

8. A

All of the possible answers are actual terms used in classification, but (B), (C), and (D) are methodologies and theories that postdate Linnaeus. Although the concept of changing forms was around at the time of Linnaeus, he did not believe in evolution and thought that species were immutable. Linnaeus is credited for providing a framework of consistent classification that was based on a binomen composed of a genus and species name.

9. B

To get this answer correct, think of the most primitive plants. In most terrestrial plants, it is the sporophyte that is the dominant form and this is considered an "advanced" characteristic for survival on land. The primitive terrestrial plants are the ones that still retain the gametophyte as the dominant form. Of the possible answers, mosses are the most primitive.

10. C

With a question like this, if you don't know the answer right away, you are left with little recourse but to eliminate the possible answers you are sure are incorrect and then guess which answer is correct. The obvious incorrect answers are amoebas and paramecia because these are common protoctists examined under the microscope in pond samples. Diatoms are very small protoctists, often with beautiful shells that look sculpted from glass. Rotifers are multicellular pseudocoelomates closely related to roundworms, so they do not belong in kingdom Protoctista.

11. B

Organisms classified as deuterostomes show radial cleavage during early development and have a dorsal nerve cord.

12. B

The coral polyp has radial symmetry, while the frog, rabbit, and kangaroo have bilateral symmetry. In radial symmetry, lines of symmetry meet at a point at the center of the organism. In bilateral symmetry, an organism has two equal parts; a plane divides the body into two halves that are mirror images of each other.

13. C

Gymnosperms is the group of plants that includes the pine.

14. A

The endosperm is the tissue that nurtures the developing plant zygote.

15. D

Tracheid cells in plants produce tubes and fibers that function in fluid transport.

16. B

Angiosperms, or flowering plants, are the most diverse group of land plants.

CHAPTER 5: BIOLOGICAL MOLECULES

BIG IDEA 2: Biological systems utilize free energy and molecular building blocks to grow, reproduce, and maintain dynamic homeostasis.

IF YOU ONLY LEARN SIX THINGS IN THIS CHAPTER . . .

1. Organic compounds are molecules that contain carbon. All living matter is made up of nitrogen, carbon, hydrogen, oxygen, phosphorus, and sulfur (N'CHOPS).

2. Types of chemical bonds include ionic, hydrogen, and covalent, in which atoms share electrons with other atoms. Polar compounds are held together by ionic bonds and dissolve in water, while nonpolar compounds do not mix with water.

3. In hydrolysis reactions, a compound often dissolves in water. Dehydration reactions release water. Oxidation-reduction reactions involve the gain (reduction) or loss (oxidation) of electrons. Anabolic reactions create larger molecules from smaller ones, and catabolic reactions break down larger molecules. Exergonic reactions release energy, while endergonic reactions require energy.

4. Living systems require free energy and matter from the environment to grow, reproduce, and maintain homeostasis. Organisms survive by coupling chemical reactions that increase entropy with those that decrease entropy.

5. The water molecule's polar nature leads to surface tension (enabling capillary action) and makes it an effective solvent. It expands rather than contracts when it freezes. These properties make water essential to life on Earth.

6. Enzymes are proteins that facilitate reactions by binding to the substrate and reducing the activation energy.

STARTING SMALL: ATOMS, MOLECULES, AND REACTIONS CENTRAL TO BIOLOGY

The world around us follows a hierarchy of organization. All life on Earth is connected, from the smallest individual units of matter (atoms and molecules) to complex organisms. In this chapter, we will discuss the basic building blocks that compose the living world. Although some organisms are more complex than others, different levels of organization do not correlate with levels of complexity. An individual cell with its multitude of chemical reactions is just as dynamic and complex as an entire community of species. This chapter will review the most crucial concepts of the building blocks of life before concluding with sample questions.

ELEMENTS ESSENTIAL TO LIFE AND ORGANIC COMPOUNDS ★

Ninety-nine percent of all living matter is made up of only four elements. These elements are nitrogen (N), carbon (C), hydrogen (H), and oxygen (O). Phosphorus (P) and sulfur (S) account for almost all of the remaining 1 percent of living matter. All six elements are important in biochemistry, especially carbon (C).

HELPFUL MNEMONIC DEVICE

The mnemonic device N'CHOPS (for nice chops!) can be used to remember the main elements of life.

MACRO-MOLECULES

Macromolecules such as carbohydrates, lipids, proteins, and nucleic acids are examples of organic compounds that serve as building blocks in all living organisms.

Even though macromolecules are not covered in this guide, it is important that you familiarize yourself with the *general* structures and functions of the carbohydrates, lipids, proteins, and nucleic acids.

A molecule containing carbon (C) is called an **organic** molecule or organic compound. Life wouldn't exist without organic compounds. Carbon earned its place as a staple in biology because it can bond to other atoms or to itself in four different equally spaced directions, allowing complex molecules of almost unlimited size and shape to be formed. The structural properties of the carbon atom have allowed for the formation of molecular compounds—such as DNA and enzymes—that have unique chemical identities and functions. The message to take home is that there are only a few elements important to biology (N'CHOPS) and that carbon is the main element of life because of the variety of organic compounds it can form.

CHEMICAL BONDS ★

The carbon atom shares its electrons fairly easily with other atoms, forming **covalent bonds**. In contrast, atoms like heavy metals—calcium (Ca), magnesium (Mg), potassium (K), and sodium (Na)—don't share electrons well. These metals form bonds that have charges between the atoms called **ionic bonds**. **Hydrogen bonds** are a third type of bond that occurs when a hydrogen atom in one molecule, such as water, is attracted to an electronegative atom in another molecule, such as nitrogen, oxygen, or fluorine. Unlike covalent and ionic bonds, which are intramolecular bonds (within a single molecule), hydrogen bonds are intermolecular bonds (between two or more molecules).

A compound that has mostly covalent bonds, with electrons evenly distributed around the molecules that are sharing electrons, is called **nonpolar**. If the electrons within the compound are unevenly shared, it is said to be **polar**.

Compounds that easily dissolve in and mix with water are called **hydrophilic**, or water loving. **Hydrophobic** compounds do not dissolve in water; they repel it. These compounds do not combine with water at all. An example of a hydrophobic compound would be oil, which sticks together and forms beads when dropped in water. A hydrophilic compound would be table salt, NaCl.

KEY POINTS

The key points to remember about chemical bonds are as follows:

- Polar compounds have unevenly shared electrons, resulting in a negative charge.

- Nonpolar compounds share electrons equally so that the molecules within the compound have no positive or negative poles.

- Nonpolar molecules do not dissolve in water, but polar molecules do dissolve in/ mix with water.

CHEMICAL REACTIONS ★

Several basic types of chemical reactions are presented throughout the AP Biology course, and the College Board assumes you are familiar with them. Chemical reactions fit into four categories.

1. HYDROLYSIS, DEHYDRATION, AND IONIC REACTIONS

Hydrolysis is the decomposition of something in the presence of water. It can involve creating ions in solution.

$$NaCl(s) \rightarrow Na^+(aq) + Cl^-(aq)$$

The (s) indicates that the substance is in a solid state and the (aq) indicates an aqueous state.

Hydrolysis can also involve donating a water molecule by splitting the molecule into H^+ and OH^- and attaching these to different molecules in a compound. Here is the hydrolysis of sucrose into glucose and fructose.

Sucrose + H_2O → Glucose + Fructose

Dehydration is the exact opposite reaction to hydrolysis, as it involves the release of water. Ionic reactions occur when one atom donates an electron to another atom, thereby becoming a positive ion, while the other atom attains a negative charge. The difference in charge between the two atoms holds them together.

$$NaCl(aq) + AgNO_3(aq) \rightarrow AgCl(s) + NaNO_3(aq)$$

2. Oxidation-Reduction Reactions

Oxidation-reduction reactions involve the gain (reduction) or loss (oxidation) of electrons, usually as a result of the action of an acid or a base on another substance. Because oxidation and reduction occur simultaneously during a reaction, the terms are coupled. Most importantly, oxidation-reduction reactions regularly take place between organic and inorganic compounds in living cells and provide the mechanism for energy transfer in biology. The loss or gain of electrons can be viewed by creating half reactions for the compounds involved in the full reaction. Creating half reactions is a concept from chemistry.

Here is the half reaction converting ethanol to acetic acid:

$$CH_3CH_2OH + H_2O \rightarrow CH_3CO_2H + 4H^+ + 4e^-$$

The four electrons given up on the right side of the equation indicate that ethanol is being oxidized during the reaction. If electrons were being added on the left side of the equation, we would know that the reaction was a reduction.

3. Anabolism and Catabolism Reactions

Anabolism and catabolism are different ways of describing oxidation-reduction reactions. Anabolism is the process of synthesizing complex materials from simple substances (like reduction). You can remember this by recalling that athletes take anabolic steroids to BUILD muscle. Catabolism is the opposite, breaking down complex material into simple parts (like oxidation). This is what happens during digestion when complex food particles are broken down into molecules like simple sugars, fatty acids, and amino acids.

4. Exergonic and Endergonic Reactions

Exergonic and endergonic reactions are used to explain the use or generation of energy. A reaction that releases energy is said to be exergonic. A reaction that requires energy is called endergonic.

FREE ENERGY IN LIVING SYSTEMS ★★★★

All living things require the capture or harvest of free energy from the environment to grow, reproduce, and maintain dynamic homeostasis. However, living systems must also follow the laws of thermodynamics, which means that entropy or disorder is always increasing. To balance the movement toward disorder, organisms must take in more energy than is used by the organism.

Though you do not need to memorize the specific steps of the reactions that help living things to capture and use free energy, it is important to remember that endergonic and exergonic reactions are typically connected together. For example, the catabolic breakdown of ATP to ADP is an energetically favorable exergonic reaction. This release in free energy is coupled to endergonic reactions such as the synthesis of glutamine, an essential amino acid.

THE IMPORTANCE OF WATER IN BIOLOGY ★★★

A water molecule is formed between two H atoms bonded covalently to a single O atom. The oxygen molecule tends to control all of the electrons by keeping them away from the hydrogen atoms, giving the oxygen a slightly negative charge, which is balanced by slightly positive charges on the hydrogen atoms. This is an example of a **polar covalent bond**. The water molecule looks like Mickey Mouse because it has a big head (the oxygen atom) and two big ears (the hydrogen atoms).

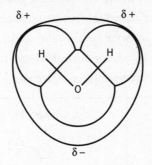

The head has a slightly negative charge and the ears have a slightly positive charge, which makes water a polar compound. The shape and charge of this molecule give it unique properties.

Water is the only substance on Earth that commonly exists in all three physical states (gas, liquid, and solid). The substance has a high specific heat so it serves as a temperature stabilizer for other compounds and vaporizes at a relatively high temperature.

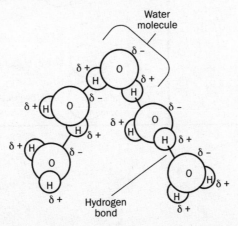

Water plays a key role in hydrolysis, condensation, and other chemical reactions that are essential to life. It is also fundamental to the biological activity of nucleotides, carbohydrates, and proteins. There are two specific characteristics of water that are particularly important to remember. First, the polar nature of water makes it "sticky." The positive charges on the hydrogen atoms cling to the negative charges on the oxygen atoms between molecules; i.e., water molecules attract one another, causing water to have high surface tension. This is what makes water bead on windshields and form round raindrops. Even matter with greater density can float on top of water if it doesn't break the surface tension.

This surface tension is also the force behind **capillary action**, by which water (and anything dissolved in it) will climb up a thin tube or move through the spaces of a porous material until it is overcome by gravity. It is as if each water molecule drags along the one behind it, as well as any nutrients dissolved in the water. Capillary action plays an important role in moving nutrients and other metabolites through living things. Plants' roots, for example, take in water from the soil, full of minerals and dissolved nutrients; capillary action draws the water and its load through the plant.

Second, the polar nature of water makes it a good solvent. Capillary action would be a lot less useful if water lacked this property. Fortunately, water molecules are happy to surround positive and negative ions and readily dissolve other polar compounds. In addition, hydrogen bonds can form between the hydrogen and oxygen atoms of the water molecule and surrounding molecules.

DID YOU KNOW?

Water expands rather than contracts when it freezes.

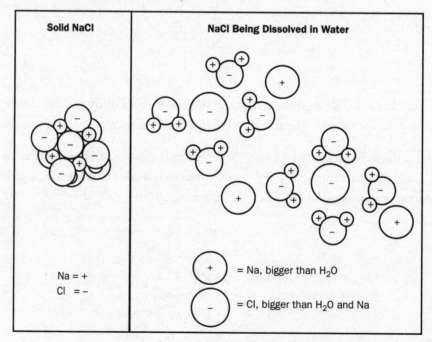

Water's role as a solvent is another reason it is so important for living things. Chemicals that are dissolved by the water in blood can be carried around the body rapidly and easily; this is how sugar gets to muscles, for example. Each living cell is, in large part, a membrane surrounding chemicals dissolved in water. It is only by being dissolved in water that these chemicals can participate in many of the biological reactions that keep living things alive.

DON'T FORGET!

Remember that many organic compounds are nonpolar or have nonpolar regions and won't mix with water. This is an important aspect of segregating biological activity. All of life survives only because water exists on Earth. The chemistry that supports life occurs almost exclusively in water and relies on its presence.

STRUCTURE AND FUNCTION OF BIOLOGICALLY SIGNIFICANT FUNCTIONAL GROUPS AND MACROMOLECULES ★

Carbon forms bonds with various other elements to create characteristic functional groups. These functional groups are important for two reasons:

1. The functional group(s) in a compound are the source of the compound's name.

2. Functional groups are the sites of chemical reactivity for the compound.

Following is an easy reference table for the biologically important functional groups. The functional groups listed are all moderately to very polar and important in biological reactions.

FUNCTIONAL GROUPS, MOLECULAR STRUCTURES, AND THEIR PROPERTIES

Functional Group	Structure	Name of Compound It Forms	Example	Some Important Properties
Amino	$-N\begin{smallmatrix}H\\\\H\end{smallmatrix}$	Amine	CH_3NH_2 Methylamine	1. Acts as a base, which affects electronegativity of amino acids 2. Formation of peptide bond between amino acids

(continued on next page)

Carbonyl	(aldehyde structure)	Aldehyde	CH_3CHO Acetaldehyde	1. Highly reactive carbonyl group 2. Highly reactive carbon near carbonyl group 3. Important intermediaries in several biological reactions
	(ketone structure)	Ketone	$(CH_3)_2CO$ Acetone	
Carboxyl	(carboxylic acid structure)	Carboxylic acid	$CH_3(CH_2)_2$ CO_2H Butyric acid	1. Weak acids able to donate a H^+ ion to several biological reactions
Hydroxyl	$-O-H$	Alcohol	CH_3CH_2OH Ethanol	1. Makes compounds soluble in water
Phosphate	(organic phosphate structure)	Organic phosphate	$C_{10}H_8N_4O_2NH_2$ $(OH)_2(PO_3H)_3H$ Adenosine triphosphate (ATP)	1. Storage and transfer of energy
Sulfhydryl	$-SH$	Thiol	$CH_3H_7NO_2S$ Cysteine	1. Stabilizes protein structure

GRAPHING FOR LABS

For labs requiring graph construction and calculations, make sure that your graph is labeled thoroughly and accurately. Make sure you identify appropriate axes, show units, plot points accurately, and assign your graph a title.

AP BIOLOGY LAB 13: ENZYME ACTIVITY INVESTIGATION ★★★★

The enzyme catalysis lab is designed to allow you to explore structural proteins and environmental conditions that affect chemical activity in the natural world. In the lab, you will measure the chemical activity of the enzyme catalase as it breaks down hydrogen peroxide (H_2O_2) into water and oxygen gas. Specific experimental conditions—temperature, substrate and enzyme concentrations, pH, and so on—will be altered to help you understand the effects of environmental conditions on a specific chemical reaction. This experiment helps you to learn the experimental method through manipulation of experimental treatments compared with a control treatment.

Understanding the experiment itself can be encapsulated in just a few graphs. One or more of these graphs may be on the exam. The first graph is about free energy and enzyme activity.

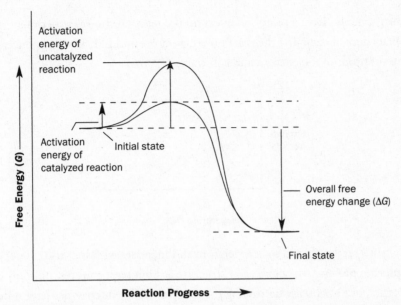

Before a reaction can take place, it must reach a point called its **activation energy** by receiving enough energy from the environment, termed **free energy**, in the form of heat or kinetic energy. An enzyme catalyzes a reaction by lowering the activation energy needed (it requires less free energy) to allow the reaction to take place. Hydrogen peroxide breaks down into water and oxygen gas on its own, but at an incredibly slow rate. The enzyme catalase lowers the activation energy of the reaction and the reaction happens very quickly.

Notice on the graph that the amount of free energy in the system changes over the progress of the reaction. Reactants of the reaction have an initial amount of energy, receive energy from the

ENZYMES

Many reactions that occur in biology can be sped up through the addition of enzymes that lower activation energy. Natural compounds readily metabolize and synthesize in the presence of certain enzymes, and biological systems control chemical activity largely through synchronizing the presence or absence of enzymes. Enzymes also control the release of energy so that cells don't combust during certain reactions from a sudden influx of energy.

environment, and then energy is released with the products. Remember that the energy of the reactants and the energy of the products is the same, whether the reaction is enzyme mediated or not. All the enzyme does is lower the activation energy for the molecule to do "what comes naturally" and makes the reaction happen. When the reaction is complete, the enzyme is free and helps to speed up the reaction for another H_2O_2 molecule. This lab experiment demonstrates firsthand what occurs in all living systems on a regular basis and points out the importance of these specialized proteins.

The environment can also have a profound effect on the reaction. In this experiment, you measure the experimental effect of changing the reaction temperature and pH, as well as the effect of changing concentrations of the enzyme and substrate.

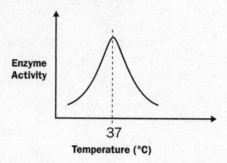

Catalytic activity is greatly affected by temperature and increases with increasing temperature. Because enzymes are proteins, they lose their structure at high temperatures, not only eliminating catalytic properties, but essentially destroying the protein. Different enzymes have different specific temperatures. The activity of animal catalase (catalase occurs in plants, too) peaks at about normal body temperature, or 35–40°C. Once the temperature increases beyond this temperature range, the catalase proteins "die."

How does temperature change the reactivity? Remember that the reaction depends on free energy, so increasing temperature will also increase the amount of free energy in the form of both heat and kinetic energy (all the molecules will be moving faster). Because all of the molecules in the system will be moving faster at a higher temperature, there is an increased rate of collision between enzyme and substrate molecules. In essence, there are more chance meetings.

The pH of the environment affects the reactivity of enzymes in the same way as temperature in that there is a level of peak activity.

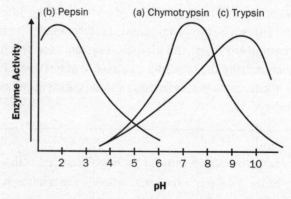

In this graph, you can see the peak activity for different common digestive enzymes. Pepsin, for example, has a peak activity at a very low pH because it catalyzes reactions in the stomach, which is very acidic. The peak activity for catalase occurs at a pH of 7 to 8, which is fairly neutral. The pH of a system affects protein activity by altering the structure of the protein. Proteins have a tertiary structure based on the electrochemical properties of the amino acids in the chain. The charges on these amino acids can change as the pH changes because of the available H^+ and OH^- ions.

The last things you need to understand from this lab are the effects of changing the amount of enzyme and substrate available.

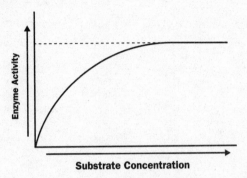

You may think that as you add more H_2O_2, you get more reaction. You would be right . . . to a point. Eventually, no matter how much substrate you add, the reaction maxes out. This is because there are a limited number of enzyme molecules in the system. These enzymes can handle only one molecule of H_2O_2 at a time. Think of enzymes as cashiers at the grocery store. If a bunch of customers want to check out at the same time, they can only be helped one at a time by any single cashier. Once all of the cashiers are busy, a customer has to wait for the next available cashier. Thus, H_2O_2 has to wait for the next available enzyme to catalyze the reaction.

Changing the enzyme concentration affects the reaction in a similar way, except the reaction becomes almost instantaneous as the number of enzyme molecules becomes equal to the number of substrate molecules. An increase in enzyme concentration is like every customer in the grocery store having his own cashier. A customer would move through the line as quickly as the cashier could process her order. In the world of enzymes, orders are processed very quickly. Catalase can convert almost 6 million molecules of H_2O_2 to H_2O and O_2 each minute.

By completing this simple experiment with one enzyme catalyzing one reaction, you not only learn about catalase and H_2O_2, you also learn about enzyme properties and the effect the environment has on chemical reactions. The type of reaction completed in the experiment was oxidation-reduction, where the active site of the enzyme catalase is a heme group like in hemoglobin, so it has a similar tertiary and quaternary structure. The reason the catalase enzyme exists in animal and plant tissue is to remove H_2O_2, an oxidizer that is dangerous to living cells and must be removed when it is created as a byproduct of metabolism. It should be easy for you to see now how biology is so intricately connected with chemistry.

CHAPTER 6: CELLS

BIG IDEA 2: Biological systems utilize free energy and molecular building blocks to grow, reproduce, and maintain homeostasis.

IF YOU ONLY LEARN FOUR THINGS IN THIS CHAPTER . . .

1. Cells are the basic structural and functional unit of all known living organisms.
2. There are two major cell types: prokaryotic cells and eukaryotic cells.
3. Prokaryotic cells have no real nucleus, have circular DNA, and reproduce using binary fission. Prokaryotes include archaea and bacteria. These cells are small and move via flagella.
4. Eukaryotic cells have a nucleus, linear DNA, highly structured cell membranes, and organelles, and they reproduce via mitosis and meiosis. Eukaryotes include protists, fungi, plants, and animals. These cells are large and move via a variety of methods including flagella and cilia.

THE CELL AND CELLULAR STRUCTURES

There are two main types of cells—**prokaryotes** and **eukaryotes**. Prokaryotes are much simpler than eukaryotes; they are composed of a plasma membrane, cytoplasm, cell wall, DNA, ribosomes, and simple microtubules. Bacteria are an example of prokaryotic cells. Eukaryotes include both plant and animal cells. These cells are much more complex than prokaryotes and contain numerous organelles.

PROKARYOTES ★★★★

Prokaryotes are the more basic, simpler of the two major cell types. These cells are considered to be the older form of cells. There are three major regions of prokaryotic cells. The inside region is called the **cytoplasmic region**. It contains the circular DNA that makes up the genetic material of the cell. Because prokaryotes do not contain a nucleus, DNA is condensed inside a nucleoid. Prokaryotes are useful in science because they carry extrachromosomal DNA elements called **plasmids**. Plasmids are circular bits of DNA that can be added or changed to allow for the addition or suppression of certain functions based on the coding of inserted DNA sections.

The **cell envelope** usually consists of a cell wall that covers a plasma membrane and may sometimes also include another protective layer called the **capsule**. The envelope provides structure as well as a protective filter for the cell. Most prokaryotes contain a cell wall that acts as yet another protective barrier from the cell's external environment.

The outside of most prokaryotes has projections called **flagella**, or **pili**. Flagella are long projections or appendages that protrude from the cell body. The primary role of flagella is locomotion, but they can also function as a sensory structure.

EUKARYOTIC CELLS AND ORGANELLES ★★★★

All multicellular organisms (such as you, a tree, or a mushroom) and all protoctists (such as amoebas and paramecia) are eukaryotic. The eukaryotes include the protoctists (protists), fungi, animals, and plants. Eukaryotic cells are enclosed within a lipid bilayer cell membrane, as are prokaryotic cells. Unlike prokaryotes, eukaryotic cells contain membrane-bound organelles (see figure). An organelle is a structure within the cell with a specific function that is separated from the rest of the cell by a membrane. The presence of membrane-bound organelles in eukaryotes allows eukaryotic cells to compartmentalize activities in different parts of the cell, making them more efficient. Compartments within a cell allow the cell to carry out activities such as ATP production and consumption within the same cell and control each independently.

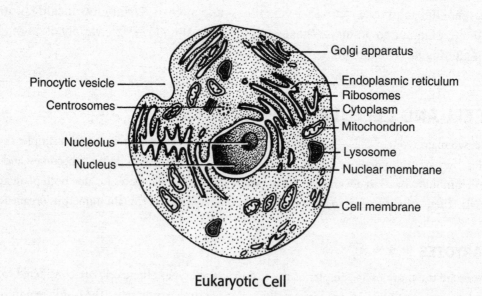

Eukaryotic Cell

ORGANELLE STRUCTURE AND FUNCTION ★★★★

NUCLEUS

The genetic material, the DNA genome, is found in the largest organelle of the animal cell, the nucleus. The nucleus is separated from the rest of the cell by the nuclear envelope, a double membrane that has a large number of nuclear pores for communication of material between the interior and exterior of the nucleus. The pores are large enough to allow proteins to pass through but are also selective in the proteins that are transported into the nucleus or excluded from the nucleus. Special sequences in proteins signal a protein to be imported into the nucleus. While the prokaryotic genome is generally found in a single circular piece of DNA, the eukaryotic genome in each cell is split into chromosomes. Chromosomes contain the DNA genome complexed with structural proteins, called histones, that help package the large strands of DNA in each chromosome within the limited space of the nucleus. Genes in the DNA genome are read (transcribed) to make RNA, which is processed in the nucleus before it is exported to the cytoplasm, where the RNA is read in turn (translated) to make proteins. The basic information flow of the cell is DNA to RNA to protein. The DNA genome is replicated in the nucleus when the cell divides. Other metabolic activities such as energy production are excluded from the nucleus. The structure and function of the eukaryotic genome will be presented later in more detail.

A dense structure within the nucleus in which ribosomal RNA (rRNA) synthesis occurs is known as the nucleolus. The nucleolus is not surrounded by a membrane but is the site of assembly of ribosomal subunits from RNA and protein components. After assembly, the ribosomal subunits are exported from the nucleus to the cytoplasm to carry out protein synthesis.

RIBOSOMES

Ribosomes are not organelles but are large, complex structures in the cytoplasm that are involved in protein production (translation) and are synthesized in the nucleolus. They consist of two subunits, one large and one small. Each ribosomal subunit is composed of ribosomal RNA (rRNA) and many proteins. Free ribosomes are found in the cytoplasm, while bound ribosomes line the outer membrane of the endoplasmic reticulum. Proteins that are destined for the cytoplasm are synthesized by ribosomes free in the cytoplasm, while proteins that are bound for one of the several membranes or that are to be secreted from the cell are translated on ribosomes bound to the rough endoplasmic reticulum. Prokaryotic ribosomes are similar to those of eukaryotes, composed of rRNA and proteins that form two different size subunits that come together to perform DNA synthesis. Prokaryotic ribosomes are, however, smaller and simpler than eukaryotic ribosomes. Mitochondria and chloroplasts also have their own ribosomes of these organelles, which are distinct from those of the eukaryotic cytoplasm and more closely resemble prokaryotic ribosomes.

ENDOPLASMIC RETICULUM

The **endoplasmic reticulum (ER)** is an extensive network of membrane-enclosed spaces in the cytoplasm. The interior of the ER between membrane layers is called the lumen, and at points in the ER, the lumen is continuous with the nuclear envelope. If a region of the ER has ribosomes lining its outer surface, it is termed rough endoplasmic reticulum (rough ER); without ribosomes, it is known as smooth endoplasmic reticulum (smooth ER). Smooth ER is involved in lipid synthesis and the detoxification of drugs and poisons, and has the appearance of a network of tubes, while rough ER is involved in protein synthesis and is a series of stacked plates. Proteins that are secreted or found in the cell membrane, the ER, or the Golgi are made by ribosomes on the rough ER. Proteins synthesized on the rough ER cross into the lumen of the rough ER during synthesis. A hydrophobic sequence of amino acids at the amino terminus of proteins as they are synthesized determines whether the protein will be sorted into the secretory pathway starting at the rough ER or synthesized in the cytoplasm. Proteins that are secreted will have only one hydrophobic signal sequence, the signal peptide, and will be inserted into the ER lumen when they are synthesized, then released from the cell later. Proteins that are destined to be membrane bound have hydrophobic transmembrane domains that are threaded through the rough ER membrane as the protein is synthesized. When the protein reaches the correct membrane destination along the secretory pathway, additional signals in the protein sequence and structure will cause the protein to stay localized at the current location.

Small regions of ER membrane bud off to form small round membrane-bound vesicles that contain newly synthesized proteins. These cytoplasmic vesicles are then transported to the Golgi apparatus, which is the next stop along the secretory pathway.

Golgi Apparatus

The **Golgi** is a stack of membrane-enclosed sacs, usually located in the cell between the ER and the plasma membrane (see figure). The stacks closest to the ER are called the cis Golgi and the stacks farthest from the ER, closer to the plasma membrane, are called the trans Golgi. Vesicles containing newly synthesized proteins bud off of the ER and fuse with the cis Golgi. In the Golgi, these proteins are modified and then repackaged for delivery to other destinations in the cell. For example, the Golgi carries out post-translational modification of proteins through glycosylation, the process of adding sugar groups to the proteins to form glycoproteins. Many proteins destined for the plasma membrane have carbohydrate groups added to the surface of the protein facing the exterior of the cell.

After processing in the cis Golgi, proteins are packaged in vesicles that move to the next layer in the stack, where they fuse and release their contents. Proteins proceed in this manner from one stack to the next until they reach the trans Golgi. In the trans Golgi, proteins are sorted into vesicles based on signals in different proteins that indicate their final destination. The nature of the signal varies but includes the protein primary sequence, structure, and post-translational modifications. Once packaged into vesicles, the vesicles move on to their final destination. The final destination for a protein may include the lysosome, the plasma membrane, or the exterior of the cell. Some proteins are retained in the Golgi or the ER. Proteins that are destined for the plasma membrane as transmembrane proteins are inserted in the membrane in the ER as they are synthesized and maintain their orientation in the membrane as they move from the ER to the Golgi to the vesicle to the plasma membrane. Proteins that are secreted from the cell are inserted in the ER lumen during protein synthesis and remain in the lumen of the ER to the Golgi, where they form secretory vesicles. The last step in secretion is the fusion of the secretory vesicle with the plasma membrane, releasing the contents of the vesicle to the cellular exterior.

LYSOSOMES

Lysosomes contain hydrolytic enzymes involved in intracellular digestion that break down proteins, carbohydrates, and nucleic acids. For white blood cells, the lysosome may degrade bacteria or damaged cells. For a protist, lysosomes may provide food for the cell. They also aid in renewing a cell's own components by breaking them down and releasing their molecular building blocks into the cytosol for reuse. A cell in injured or dying tissue may rupture the lysosome membrane and release its hydrolytic enzymes to digest its own cellular contents.

The lysosome maintains a slightly acidic pH of 5 in its interior, a pH at which lysosomal enzymes are maximally active. The contents of the lysosome are isolated from the cytoplasm by the lysosomal membrane, keeping the pH distinct from the neutral pH of the cytoplasm. The optimal pH and compartmentalization of lysosomal enzymes prevent the rest of the cellular contents from degrading.

PEROXISOMES

Peroxisomes contain oxidative enzymes that catalyze reactions in which hydrogen peroxide is produced and degraded. Peroxisomes break fats down into small molecules that can be used for fuel; they are also used in the liver to detoxify compounds, such as alcohol, that may be harmful to the body. The peroxides produced in the peroxisome would be hazardous to the cell if present in the cytoplasm, because these molecules are highly reactive and could covalently alter molecules such as DNA. Compartmentalization of these activities within the peroxisome reduces this risk.

MITOCHONDRIA

Mitochondria are the source of most energy in the eukaryotic cell as the site of aerobic respiration. Mitochondria are bound by an outer and inner phospholipid bilayer membrane (see figure). The outer membrane has many pores and acts as a sieve, allowing molecules through on the basis of their size. The area between the inner and outer membranes is known as the intermembrane space. The inner membrane has many convolutions called cristae, as well as a high protein content that includes the proteins of the electron transport chain. The area bounded by the inner membrane is known as the mitochondrial matrix and is the site of many of the reactions in cell respiration, including electron transport, the Krebs cycle, and ATP production.

Mitochondria are somewhat unusual in that they are semiautonomous within the cell. They contain their own circular DNA and ribosomes, which enable them to produce some of their own proteins. The genome and ribosomes of mitochondria resemble those of prokaryotes more than eukaryotes. In addition, they are able to self-replicate through binary fission. Mitochondria are believed to have developed from early prokaryotic cells that began a symbiotic relationship with the ancestors of eukaryotes, with the mitochondria providing energy and the host cell providing nutrients and protection from the exterior environment. This theory of the origin of mitochondria and the modern eukaryotic cell is called the endosymbiotic hypothesis.

SPECIALIZED PLANT ORGANELLES

Plants also have some organelles that are not found in animal cells. **Chloroplasts** are found only in plant cells and some protists. With the help of one of their primary components, chlorophyll, chloroplasts function at the site of photosynthesis, using the energy of the sun to produce glucose. Chloroplasts have two membranes, an inner and an outer membrane. Additional membrane sacs called **thylakoids** inside the chloroplast are derived from the inner membrane and form stacks called **grana**. The fluid inside the chloroplast surrounding the grana is the **stroma**. The thylakoid membranes contain the chlorophyll of the cell.

Like mitochondria, chloroplasts contain their own DNA and ribosomes and exhibit the same semi-autonomy. They are also believed to have evolved via symbiosis of a photosynthetic early prokaryote that invaded the precursor of the eukaryotic cell. In this arrangement, the chloroplast precursor cell provided food and received protection. Photosynthetic prokaryotes today carry out photosynthesis in a manner similar to the chloroplast.

Vacuoles are membrane-enclosed sacs within the cell. Many types of cells have vacuoles, but plant vacuoles are particularly large, taking up 90 percent of the cell volume in some cases. Plants use the vacuole to store waste products, and the pressure of liquid and solutes in the vacuole helps the plant to maintain stiffness and structure as well.

All plant cells have a cellulose cell wall that distinguishes them from animal cells, which lack a cell wall. The cell wall of plants is also distinct from the peptidoglycan cell wall of bacteria and the chitin cell wall of fungi. The cell wall provides structure and strength to plants.

CILIA AND FLAGELLA ★★

Cilia and flagella are both anchored into the cell membrane by arrangements of microtubule triplets, which are called basal bodies. Because the microtubules in cilia and flagella must be rebuilt often, tubulin dimers use these basal bodies as the foundation to make new microtubules, which are used to maintain cilia and flagella.

As you can see in the following figure, cilia and flagella are composed of long stabilized microtubules arranged in a **"9 + 2" structure** (nine pairs of microtubules surrounding two central microtubules for added stability). These nine doublets slide past each other as dynein proteins grab neighboring tubules and pull them. This rapid sliding generates the force needed for the cilia or flagella to quickly beat back and forth and cause movement.

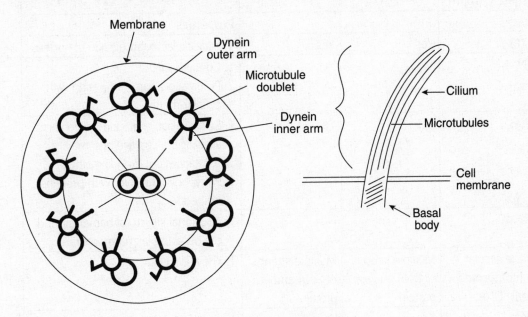

Cilium Cross-Section

CHARACTERISTICS OF CELLS

	Prokaryotes	Eukaryotes		
		Plant Cells	Animal Cells	
Size	0.2–500µm, most 1–10µm	Most 30–50µm	Most 10–20µm	
Structure				**Properties**
Cytoplasm	Yes	Yes	Yes	1. Intracellular matrix outside of nucleus
Nucleus	No	Yes	Yes	1. Contains DNA 2. Pores allow communication with cellular matrix
Plasma membrane	Yes	Yes	Yes	1. Selective barrier around cell contents allowing the passage of some substances but excluding others 2. Phospholipid bilayer with proteins embedded
Cell wall	Most	Yes	No	1. Additional structural barrier around cell outside plasma membrane
Chromosomes	One circular chromosome, only DNA	Multiple strands of DNA and protein	Multiple strands of DNA and protein	1. The cell's DNA
Ribosomes	Yes	Yes	Yes	1. Site of protein synthesis (translation)
Endoplasmic reticulum (ER)	No	Yes	Yes	1. Site of attachment for ribosomes 2. Protein and membrane synthesis 3. Formation of vesicles for transport
Golgi complex	No	Yes	Yes	1. Synthesis, accumulation, storage, and transport of products
Lysosomes	No	Some vacuoles function as lysosomes	Usually	1. Vesicle containing hydrolytic enzymes
Vacuoles or vesicles	No	Yes	Some	1. Membrane-bound sacs in the cytoplasm
Mitochondria	No	Yes	Yes	1. Site of cellular respiration
Plastids	No	Yes	No	1. Group of plant organelles that includes chloroplasts 2. Site of photosynthesis 3. Carbohydrate storage
Microtubules (Cilia or Flagella)	Simple	On some sperm	Complex (9 + 2 arrangement)	1. Tubes of globular protein, tubulins 2. Provides structural framework for cell 3. Provides motility
Centrioles	No	No	Yes	1. Cell center for microtubule formation

CHAPTER 7: MEMBRANE TRAFFIC

BIG IDEA 2: Biological systems utilize free energy and molecular building blocks to grow, reproduce, and maintain dynamic homeostasis.

IF YOU ONLY LEARN THREE THINGS IN THIS CHAPTER . . .

1. All cells are membrane-enclosed bodies of cytoplasm.

2. Cell membranes are made up of a lipid bilayer that includes a hydrophobic and a hydrophilic region. Specific structures embedded within the membrane help to facilitate transport.

3. The cell membrane is selectively permeable, meaning it allows certain things through while keeping others out. Water diffuses across the membrane from areas of lesser to greater solute concentration (osmosis). While certain things can cross the membrane in the processes of diffusion or facilitated diffusion, which do not require energy, others require the expenditure of energy for active transport against the concentration gradient.

INTRODUCTION

One way the living world stays compartmentalized is with **membranes**. Cells and cell **organelles** are surrounded by a membrane, a selectively permeable barrier that segregates cell contents from the outside world. In this chapter, you'll review some basics about the cell cycle. In the investigation exercise, you'll learn how the membrane allows transport of certain materials between compartments. Cells and cell organelles come in many different sizes to form simple or complex organisms.

MEMBRANES ★★★★

All cells are surrounded by a **plasma membrane**. In eukaryotic cells, the **nucleus** and most of the organelles are surrounded by plasma membranes. Membranes are composed mostly of lipids, which is another word for fats. **Lipids** are full of nonpolar covalent bonds, so they are hydrophobic and do not dissolve in water. Because membranes are composed primarily of lipids, most of the material in membranes will not mix well with water. Most of the lipids in membranes have a phosphate group attached to one end, so they are called **phospholipids**. This charged end is polar and happy to be in water, which is why it is termed hydrophilic. The other end of the phospholipid is a tail that is nonpolar, and it turns in toward the center of the membrane. Tails on phospholipids are hydrophobic, because the tails do not mix well with water. The attraction between the nonpolar regions of these phospholipids creates the foundation for the bilayer of the membrane. The lipid ends group together like the insides of a sandwich surrounded by polar barriers.

Cytoplasm

Aqueous environment

— Polar phosphate bead
— Nonpolar lipid chains
— Polar phosphate bead

Aqueous environment

Extracellular environment

Phosphate and Lipid Chains

Embedded among all of these membrane lipids are **proteins**, **carbohydrates**, and **sterols** (cholesterol). Some proteins are embedded on the outer surface, some on the inner surface, and some span the entire width of the plasma membrane (these usually function as transport proteins). Some surface proteins have sugar groups attached to them, called glycoproteins. Within the cell membrane, each component performs its own function. For example, some proteins that span the width of the cellular membrane act as channels for ion transport. Nerve cells have a higher density of these proteins than other cells. Each component of a cell membrane contributes to how the membrane functions. Proteins act as transport molecules, receptor sites, attachments to the cytoskeleton, and surface enzymes. Carbohydrates on the surface of the cell and glycoproteins contribute to cell recognition, particularly in the immune response. Cholesterol contributes fluidity to the membrane.

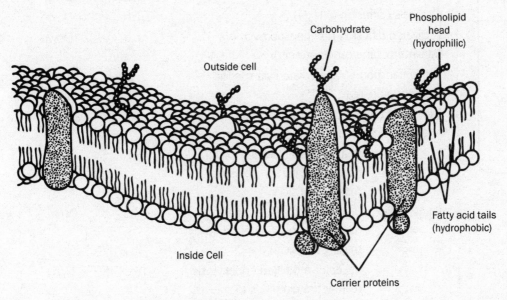

Lipid Bilayer Plasma Membrane

Membranes are **selectively permeable**, which means that they allow some things to pass through, but not others. Nonpolar molecules and small uncharged polar molecules (e.g., H_2O and CO_2) easily pass through the lipid matrix of the membrane. Charged ions and large uncharged polar molecules such as glucose do not easily pass through the membrane. **Diffusion**, the movement of molecules from an area of higher concentration to an area of lower concentration, will be reviewed in the lab section of this chapter, but other kinds of transport are summarized in the table that follows. Osmosis is the diffusion of water across a semipermeable membrane.

Recall that molecules are in constant motion and tend to spread out if there is nothing in the way. These molecules move from areas of high concentration to areas of low concentration. This is called moving down or with the **concentration gradient**. Often, a cell expends energy to create or maintain a gradient between the cytoplasm and the extracellular environment, causing molecules to move against the concentration gradient (active transport).

TYPES OF TRANSPORT ACROSS THE CELL MEMBRANE

Type of Transport		Requires Energy?	Concentration Gradient
Passive (diffusion)		No	Down
Osmosis (diffusion of H_2O)		No	Down
Facilitated diffusion (ion transport via transmembrane carrier protein)		No	Down
Active transport (ATP mediated via specific receptor proteins)		Yes	Against
Exocytosis (vesicle fuses with cell membrane, releasing molecules to extracellular environment)		Yes	N/A
Endocytosis	Phagocytosis ("cellular eating," membrane invaginates solid, making a vesicle)	Yes	N/A
	Pinocytosis ("cellular drinking," membrane invaginates liquid, making a vesicle)	Yes	N/A
	Receptor-mediated endocytosis (carrier protein binds to specific substances for invagination)	Yes	N/A

MEMBRANE PROTEINS ★★★★

Membrane proteins, like the membrane phospholipids, usually have carbohydrate groups attached to them so that the outside surface of the plasma membrane is extremely sugar rich.

Membrane-spanning proteins have regions that are hydrophobic as well as regions that are hydrophilic, with the nonpolar (hydrophobic) regions passing through the nonpolar interior of the membrane and the polar areas sticking both into the cytoplasm and out into the extracellular space. Other proteins can be located completely intracellularly or extracellularly, anchored to the cell membrane by a variety of special lipids. These proteins are made in the cytosol and bind into the cell membrane only because they subsequently have a lipid molecule attached to their structure. Following are some of the ways in which proteins regularly associate with the lipid bilayer.

Transmembrane proteins are involved not only in carrying materials across the membrane, but also in cell recognition, cell adhesion, cell signaling, and enzymatic reactions. Recall that most proteins sticking up from the surface of the membrane are covered in carbohydrates on the extracellular surface. The term used to describe the protein- and carbohydrate-rich coating on the cell surface is **glycocalyx**. Keep in mind that these sugars reside exclusively on the exterior of the membrane.

TRANSPORT MECHANISMS ★★★★

The main limiting factors that determine whether a molecule will be able to pass through a cell's membrane are the *size of the particle* and *its charge* (polarity). Simply stated, the molecules quickest to pass through the lipid bilayer are those that are *small* and *nonpolar*, because the interior of the membrane is far too hydrophobic for others to make it through without assistance. This assistance can come in the form of membrane-spanning proteins, which either can bind to extracellular molecules and bring them inside the cell via a conformational change or can open up a temporary tunnel through the membrane lipids so that the molecules can pass through.

Simple diffusion refers to the movement of particles down their concentration gradient from a region of higher concentration to a region of lower concentration. This form of transport takes place directly through the cell membrane lipid bilayer without using any form of energy or membrane proteins in order to move particles. Again, small nonpolar molecules move most freely by simple diffusion. Examples include water (small but polar), carbon dioxide (nonpolar), and oxygen (nonpolar). The simple diffusion of water is referred to as osmosis and occurs from a region of higher water concentration to a region of lower water concentration. For water to be in high concentration, the amount of dissolved solute (salts, sugars, etc.) must be low, and vice versa for water in low concentration. So, although water diffusion works like any other passive diffusion in terms of movement from high → low concentration, it is generally stated that water moves from an area of low *solute* concentration to one of higher *solute* concentration.

Solutions low in solute concentration relative to other solutions are said to be **hypotonic**, whereas solutions higher in solute than others are **hypertonic**. When two solutions have the same solute concentration as each other, they are said to be

ISOTONIC SOLUTIONS

In isotonic solutions, water and solutes can and do move across the membrane, yet movement inward is always balanced by reciprocal movement outward. This means that overall concentrations of water and solute on opposite sides of the membrane do not change.

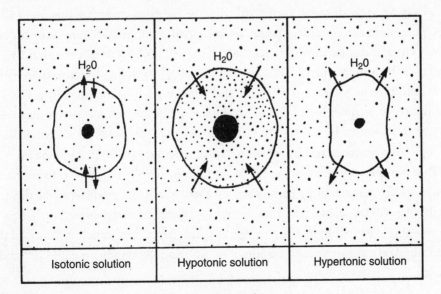

| Isotonic solution | Hypotonic solution | Hypertonic solution |

Osmosis

isotonic. The cell membrane effectively separates two distinct solutions: One is the extracellular environment and the other is the cytoplasm. If the outside of a cell is hypertonic, higher in solute concentration than the inside (e.g., a cell has just been moved from fresh water to salt water), water will move out of the cell into the high solute solution. If possible, some of those solutes will also move into the cell until a balance has been established so that both areas are equivalent in solute concentration. In hypotonic solutions, cells generally take on water, sometimes until they burst.

Facilitated diffusion, otherwise known as passive transport, involves the use of channel or carrier proteins embedded in the membrane to allow molecules to diffuse down a gradient. The structures of the proteins involved in this type of transport are very similar and amino acid sequences are highly conserved across many species. In some cases, these proteins act as pores for ions; in other cases, they may open and close in response to external signals. Keep in mind that, because cells naturally have a negatively charged cytoplasm, the opening of ion channels favors the movement of positively charged ions into the cytoplasm. This combination of solute concentration and an electrical gradient is called the **electrochemical gradient** and is the key determinant of what moves into and out of membranes when passive transport channels are open.

Active transport: Some membrane proteins use the energy of ATP to change the protein's conformation in the membrane so that molecules can be brought into and out of the cell against their concentration gradients. These **ATPase pumps** are found in every membrane of cells, and they are extremely important in the maintenance of unequal concentrations of certain ions across the lipid bilayer—something that we will later see is *essential for processes like nerve signal conduction.*

A commonly cited example of active transport is the Na^+-K^+ ATPase membrane pump, whose conformational change uses the energy of ATP breakdown to pull 2 K^+ (potassium) ions into a cell while kicking out 3 Na^+ (sodium) ions at the same time. This transport of molecules in opposite directions is known as **antiport**, and it can be contrasted with pumps that pull two different molecules in the same direction (**symport**). Because the pump, which is present in all cell membranes, pumps out three positive charges for every two it brings in, the inside of the cell remains negatively charged compared with the outside of the cell under normal conditions. Yet, the more important role the pump plays is that it helps to control the solute concentration within the cytoplasm of cells, thereby preventing cells from shrinking or swelling too much when the extracellular environment becomes too hypertonic or hypotonic. Another ion that you may see on the exam is Ca^{2+} (calcium), which is kept in extremely low concentrations in cell cytoplasm, yet is stored in high concentrations within the endoplasmic reticulum (ER). This is done by using Ca^{2+}-**ATPase pumps**, embedded in the ER membrane, to actively transport calcium from the cytoplasm into the ER lumen. This naturally sets up a strong calcium gradient across the ER membrane that is used, for example, by muscle cells to regulate muscle contraction. When depolarized by an action potential from a nerve cell, the specialized ER of muscle cells (called the sarcoplasmic reticulum) releases its store of calcium ions, flooding the cytoplasm with Ca^{2+} and leading to rapid contraction of the cell. Because only one ion moves through these channels, they are known as **uniport** pumps.

As a last example, those ATPases that manufacture ATP in the mitochondria and chloroplast as part of the **electron transport chain** are simply ATPase membrane-transport proteins working in a reverse manner from how they usually work. Rather than ATP hydrolysis driving changes in protein structure so that ions can pass through the membrane, it seems that these pumps are driven by the flow of H^+ ions moving through them.

Endocytosis and **exocytosis** are two mechanisms of transport that can move large molecules and even entire cells through the cell membrane. To accomplish this, the cell membrane actually invaginates, or pinches inward, to form a pocket in which the material to be transported can fall. In the case of endocytosis, this invagination pinches off completely, forming a vesicle that contains the transported material and can move freely within the cytoplasm. In exocytosis, a vesicle containing material to be expelled simply merges with the lipid bilayer, and the material is pushed off into the extracellular space.

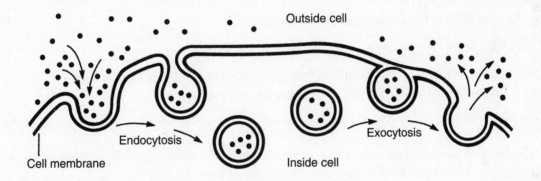

Endocytosis and Exocytosis

AP BIOLOGY LAB 4: DIFFUSION AND OSMOSIS INVESTIGATION ★★★★

The properties of diffusion and osmosis, discussed at the beginning of this chapter, will be explored in this investigation. Recall that selectively permeable membranes let some molecules through but not others. For this investigation, it is important to know that:

- Isotonic solutions are two solutions that have the same concentration of solute.

- A hypotonic solution has a lower solute concentration than another solution.

- A hypertonic solution has a higher solute concentration than another solution (think *hypo-* "below" and *hyper-* "above").

Keep in mind that water potential is the measure of force a solution has for pulling or drawing water into it. The more negative the water potential, the stronger its pulling force.

Dialysis bags allow for the movement of water but not ions. A common osmosis experiment is to fill dialysis bags with different solutions. These bags are tied at each end and put into hypertonic, hypotonic, and isotonic solutions. The water will move in the direction of the hypertonic solution. Suppose you tied your dialysis bags and put them in solutions to complete the experiment but didn't get the results you expected. Skewed results can be obtained by not tying the knot on your bag tight enough, not getting all the solution washed off the outside of your bag, or by some other slight oversight. Often, one group gets usable results for one part of the lab and another group gets another part right, so everyone shares the "good" results. Occasionally, classes have really good luck and all the groups get "good" results. Either way, you need to know what happened in order to score well on the exam. You can expect to see an osmosis question in one form or another on the exam. It may be in the multiple-choice section or it may be a free-response question.

KEY SKILLS

Two key skills of the diffusion and osmosis lab are:

- Measuring the effects (e.g., weight change) of osmosis
- Determination of the osmotic concentration/water potential of an unknown tissue or solution using solutions of known concentrations

The first skill uses experimental methods to obtain results. Even if you know how osmosis works, it is important to know how to measure it. You can't really watch water molecules or sugar and starch molecules move because they're too small. In this investigation, you learn how to measure things you don't see by observing things you can. To obtain results, the weight change of a dialysis bag or a piece of potato can be measured. An indicator color will appear if glucose or starch occurs in a solution. If a bag or piece of potato increased in weight, it gained water and had a more negative water potential than (was hypertonic to) the solution it was placed into, and vice versa. If a solution produced an indicator color, the semipermeable membrane allowed the passage of molecules that the dye is an indicator for.

The second key skill deals with using a standard to figure out the "identity" of an unknown and using observed data to interpolate expected data. Part two of the experiment begins with a dialysis bag full of solution (the concentration of which is unknown). Questions to ask during the experiment include: What is the osmolarity/water potential of the unknown, and how do you measure it? The first parts of the investigation teach you to observe the effects of osmosis by measuring the change in weight. Samples are weighed and placed into solutions with known concentrations, usually ranging from distilled water to a high concentration like 1 M. Percentage weight change can be plotted on a graph like this:

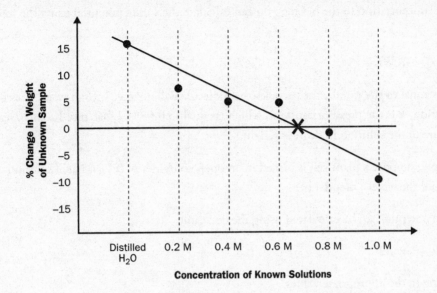

An imaginary line can be drawn that nearly bisects all data points. Note where the line crosses the x-axis. At this point, there is no change in weight of the unknown sample, which represents the point at which the osmolarity/water potential of the unknown is the same as a known solution equal to that point on the x-axis (on the graph, this is about 0.675 M). By assuming change in weight is linear with respect to change in concentration, the expected concentration of a theoretical solution can be determined. The concentration of the unknown can be estimated by comparing it to the concentration of the theoretical solution (as we saw in the graph above, this is approximately 0.675 M). This type of experimental design and interpolation is very common in biological research. The College Board expects you to be able to design simple experiments like this to test simple hypotheses.

Water potential (represented by the Greek letter psi (ψ)) predicts which way water diffuses. Water potential is calculated from the solute potential (ψ_S), which is dependent on solute concentration, and the pressure potential (ψ_P), which results from the exertion of pressure on a solution. When a solution is open to the atmosphere, the pressure potential is equal to 0 because there is no tension on that solution.

$$\psi = \psi_P + \psi_S$$
water potential = pressure potential + solute potential

In an open beaker of pure water, the water potential is equal to zero. There is no solute and no tension on the solution, so the solute and pressure potentials are zero. If you add solute in a measured concentration to the beaker, you can calculate the solute potential using the following formula:

$$\psi_S = -iCRT$$

Where i is equal to the number of particles the molecule will dissolve into in water, C is the molar concentration, R is the pressure constant, which is equal to 0.0831 L bar/mol K, and T is the temperature of the solution in degrees Kelvin.

For example, suppose a plant cell is placed in an open container of 0.1 M NaCl solution at 25°C. What would the water potential be?

The solute potential can be calculated using the equation:

$$\psi_S = -iCRT$$

Substituting in the appropriate values,

$$\psi_S = -(2)(0.1 \text{ mol/L})(0.0831 \text{ L bar/mol K})(273 \text{ K})$$

$$\psi_S = -4.5 \text{ bars}$$

Because it is an open container, the pressure potential is equal to 0. Therefore, the water potential is:

$$\psi = \psi_P + \psi_S$$

$$\psi = 0 + -4.5 \text{ bars}$$

$$\psi = -4.5 \text{ bars}$$

A negative water potential means that water is likely to diffuse out of the cell from a place of high water potential to a place of low water potential.

CHAPTER 8: PHOTOSYNTHESIS

BIG IDEA 2: Biological systems utilize free energy and molecular building blocks to grow, reproduce, and maintain homeostasis.

IF YOU ONLY LEARN THREE THINGS IN THIS CHAPTER . . .

1. Photosynthesis is the energy foundation for almost all living systems. In addition, photosynthesis provides almost all of the oxygen present in the Earth's atmosphere.

2. All photosynthetic organisms use chloroplasts and mitochondria to perform photosynthesis and cellular respiration. Photosynthesis and cellular respiration are essentially reverse operations of each other.

3. Photosynthesis has two main parts—the light cycle and the dark cycle (the latter is usually called the Calvin or Calvin-Benson cycle). Light reactions produce energy and dark reactions make sugars. The light reactions occur in the interior of the thylakoid, while the Calvin-Benson cycle occurs in the stroma.

INTRODUCTION ★★★★

To survive, all organisms need energy. ATP is an energy intermediary used to drive biosynthesis and other processes. ATP is generated in mitochondria using the chemical energy of glucose and other nutrients. The energy foundation of almost all ecosystems is photosynthesis. Plants are autotrophs, or self-feeders, that generate their own chemical energy from the energy of the sun through photosynthesis. The chemical energy that plants get from the sun is used to produce glucose. This glucose can then be burned in plant mitochondria to make ATP, which is used to drive all of the energy-requiring processes in the plant, including the production of proteins, lipids, carbohydrates, and nucleic acids. Similarly, animals can eat plants to extract the energy for their own metabolic needs. In this way, photosynthesis is the energy foundation of most living systems.

PHOTOSYNTHESIS ★★★★

To begin, remember that although all organisms containing photosynthetic pigments can perform photosynthesis, cellular respiration still takes place in the cells of these organisms. Plant cells have both chloroplasts and mitochondria to perform photosynthesis and cellular respiration. Photosynthesis is actually the reverse reaction of cellular respiration.

$$6CO_2 + 12H_2O + \text{light energy} \rightarrow C_6H_{12}O_6 + 6O_2 + 6H_2O$$

The most amazing aspect of photosynthesis is the ability of photosynthetic proteins to split water molecules (H_2O). Once the water molecules have been split, the oxygen atoms are immediately released as O_2 and the hydrogen (H) atoms donate their electrons, which are used to form ATP and NADPH. The two H atoms combine with the C and O atoms from CO_2 to form carbohydrates and more water. Sugars are used for energy storage, and O_2 is a waste product that other organisms use for respiration.

Photosynthesis takes place in two stages: the **light reactions** and the **dark (Calvin)**, or **light-independent**, reactions. In the light reactions, light energy is harnessed to produce chemical energy in the form of ATP and NADPH in a process called photophosphorylation. The dark reactions complete carbon fixation, the process by which CO_2 from the environment is incorporated into sugars with the help of energy from the reduction of ATP and NADPH. Light reactions produce energy and dark reactions make sugars.

Photosynthesis will be explored further in the lab section of this chapter. Comparisons of C3, C4, and CAM plants may show up within the data provided for a question. Most plants are C3. This means that the initial products of C fixation are two three-carbon molecules (phosphoglycerate, or PGA), synthesized through the intermediate enzyme **rubisco**. In C4 plants, CO_2 is initially fixed into a four-carbon molecule (oxaloacetate, also found in the Krebs cycle) by the intermediate enzyme phosphoenol pyruvic acid (PEP) in a "mesophyl" cell. This four-carbon molecule later releases a CO_2 molecule when it enters a bundle sheath cell. The enzyme PEP is much more likely to bind to CO_2 because it has a higher affinity for CO_2 than rubisco. C4 plants have a physiological advantage in hot, arid environments where they often have to limit the opening of stomata during the day.

PHOTOSYNTHETIC ORGANISMS

Photosynthetic organisms are primary **producers**, providing food that supplies the rest of the food web. This "food" starts out in the form of glucose.

DID YOU KNOW?

If the process of photosynthesis had not evolved and produced atmospheric O_2, you wouldn't see most of the organisms that now live on Earth.

Plants that go through C4 photosynthesis are grasses, which include semi-arid to arid crops like corn and sorghum. Many succulent plants, such as cacti, use an alternative method of limiting water loss in arid environments and are called **CAM (crassulacean acid metabolism)** plants because they collect CO_2 at night when it is cooler. CO_2 is then stored in the form of organic acids. C3, C4, and CAM plants all carry out the dark reactions in the Calvin cycle. However, C4 plants complete a carbon fixation step in separate *parts* of the plant, and CAM plants complete a carbon fixation step at separate *times*.

Photophosphorylation is driven by light energy absorbed by pigments in chloroplasts. White light is composed of many different wavelengths. Plants in particular have developed ways to use light of more than one wavelength. Chloroplasts can only use light energy if the energy is absorbed. Visible color is caused by reflected light. Plants reflect green light, so green light is not very useful in photosynthesis.

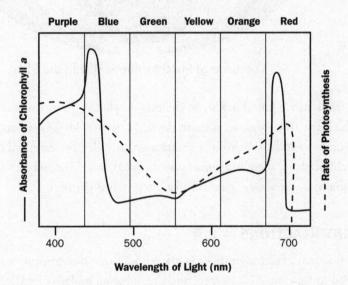

This image shows action spectra for the rate of photosynthesis (dotted line) and the absorbance of chlorophyll *a*. Note that the two action spectra do not correlate identically, indicating that the rate of photosynthesis depends on the presence of other photopigments as well. Chlorophyll *a* is medium green, chlorophyll *b* is yellow-green, and the carotenoids range in color from yellow to orange. Chlorophyll *a* is the only photopigment that participates in light reactions, but chlorophyll *b* and the carotenoids indirectly supplement photosynthesis by providing energy to chlorophyll *a*. (The carotenoids absorb light the chlorophyll cannot and transfer the energy to the chlorophyll.) The point to keep in mind is that many different photopigments absorb light energy from different wavelengths during photosynthesis. If you want plants to be healthy, give them blue and red light, not green.

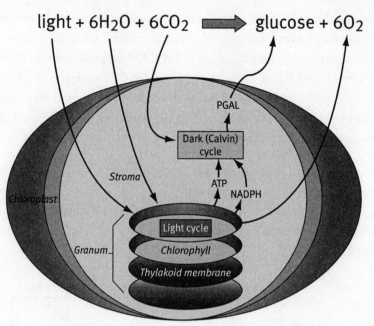

$$light + 6H_2O + 6CO_2 \Rightarrow glucose + 6O_2$$

Locations of Photosynthesis Within the Plant

Chloroplasts are found mainly in the mesophyll, the green tissue in the interior of the plant. The stomata, pores in the leaf's surface, let CO_2 in and O_2 out. The opening and closing of the stomata is controlled by the guard cells. Inside the chloroplasts, there is stroma (a dense fluid) and thylakoid sacs (arranged into chlorophyll-containing, pancake-like stacks called grana).

LIGHT REACTIONS ★★★

The first part of photosynthesis is made up of light reactions, in which light energy is used to generate ATP, oxygen, and the reducing molecule NADPH. The molecule that captures light energy to start photosynthesis is a pigment called chlorophyll, found in the thylakoid membranes of the chloroplast. Chlorophylls absorb most wavelengths of visible light, with the exception of green. Chlorophyll does not absorb green, rather it reflects green light, which makes plants appear green. Chlorophyll is used by two complex systems in the thylakoid membrane, called photosystems I and II. Each photosystem is a complex assembly of protein and pigments in the membrane. When photons strike chlorophyll, electrons are excited and transferred through the photosystems to a **reaction center**. When electrons reach the reaction center, the reaction center gives up excited electrons that enter an electron transport chain where they are used to generate chemical energy as either reduced NADPH or ATP.

Two different processes occur in the photosystems, **cyclic photophosphorylation** and **noncyclic photophosphorylation**. Both are used to generate ATP but in different ways. The ATP, in turn, is used to generate glucose in the dark reactions. Cyclic photophosphorylation occurs in photosystem I to produce ATP. In the cyclic method, electrons move from the reaction center, through an electron transport chain, then back to the same reaction center again (see figure). The reaction center in photosystem I includes a chlorophyll called P700 because its maximal light absorbance occurs at 700 nm. This process does not produce oxygen and does not produce NADPH.

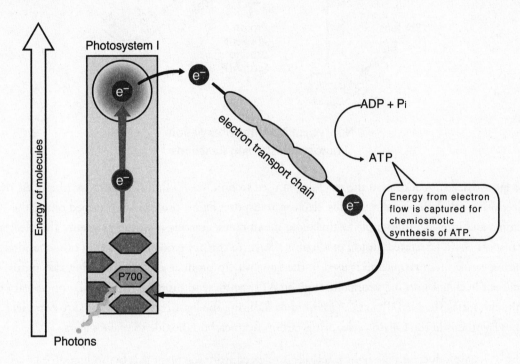

Cyclic Photophosphorylation

Noncyclic photophosphorylation starts in photosystem II (see figure below). In noncyclic photophosphorylation, chlorophyll pigment absorbs light and passes excited electrons to a reaction center, a process equivalent to cyclic photophosphorylation. The photosystem II reaction center contains a P680 chlorophyll, distinct from photosystem I. From the photosystem II reaction center, the electrons are passed to an electron transport chain. In this case, however, the electrons are not returned to the reaction center at the end of the electron transport chain but are passed to photosystem I. Photosystem II replaces the electrons it lost by getting them from water, producing oxygen in the process. In this case, the electrons that enter photosystem I are used to produce NADPH.

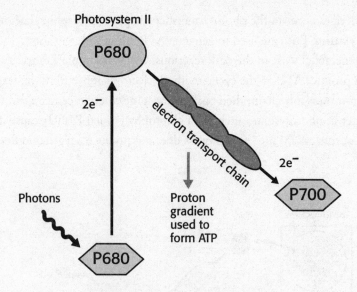

Noncyclic Photophosphorylation:
Photosynthesis Light Reactions

So far, we have not addressed the mechanism used to produce ATP during photosynthesis. As the electrons work their way through the electron transport chains, protons are pumped out of the stroma and into the interior of the thylakoid membranes, creating a proton gradient. This electron transport chain-generated proton gradient is similar to the pH gradient created in mitochondria during aerobic respiration and is used in the same way to produce ATP. Protons flow down this gradient back out into the stroma through an ATP synthase to produce ATP, similar once again to mitochondria. The NADPH and ATP produced during the light reactions are used to complete photosynthesis in the Calvin cycle, using carbon from carbon dioxide to make sugars.

The oxygen produced in the light reactions is released from the plant as a byproduct of photosynthesis. Starting about 1.5 billion years ago, photosynthesis helped to create the oxygen-rich atmosphere found on Earth today, which allowed the evolution of animals requiring the efficient energy metabolism provided by aerobic respiration. The oxygen produced through photosynthesis today maintains Earth's oxygen and is a key to the continued functioning of the biosphere.

CALVIN-BENSON CYCLE ★★

The first portion of photosynthesis, in which ATP, oxygen, and NADPH are produced, is sometimes called the light reactions because light energy is required to energize electrons and drive the reactions forward. In the remaining part of photosynthesis, the energy captured in the light reactions as ATP and NADPH is used to drive carbohydrate synthesis. This cycle, often called the Calvin cycle but also known as the **Calvin-Benson cycle**, fixes CO_2 into carbohydrates, reducing the fixed carbon to carbohydrates through the addition of electrons. The NADPH provides the reducing power for the reduction of CO_2 to carbohydrates, and air provides the carbon dioxide. CO_2 first combines with, or is fixed to, ribulose bisphosphate (RuBP), a five-carbon sugar with two phosphate groups. The enzyme that catalyzes this reaction is called rubisco and is the most abundant enzyme on Earth. The resulting six-carbon compound is promptly split, resulting in the formation of two molecules of 3-phosphoglycerate, a three-carbon compound. The 3-phosphoglycerate is then phosphorylated by ATP and reduced by NADPH, which leads to the formation of phosphoglyceraldehyde (PGAL). This can then be utilized as a starting point for the synthesis of glucose, starch, proteins, and fats. (Note: The steps of the Calvin cycle and the structure of the molecules involved are not required knowledge for the AP Biology exam.)

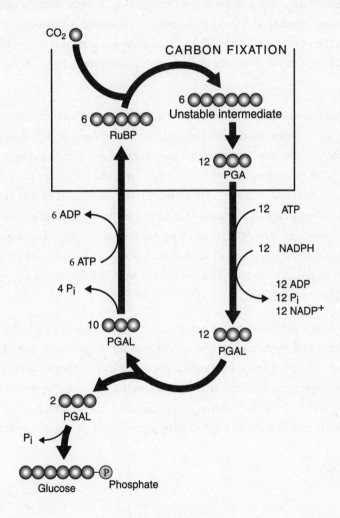

Calvin-Benson Cycle

AP BIOLOGY LAB 5: PHOTOSYNTHESIS INVESTIGATION ★★★★

In this investigation, you will learn how to measure the rate of photosynthesis by determining oxygen production. To review, photosynthesis is the process by which autotrophs capture free energy (in the form of sunlight) to build carbohydrates. The process is summarized by the following reaction:

$$2H_2O + CO_2 + \text{light} \rightarrow \text{carbohydrate } (CH_2O) + O_2 + H_2O$$

To determine the rate of photosynthesis by a plant cell, you can measure the production of O_2 or the consumption of CO_2. The process is not that simple, though, because cellular respiration is coupled with aerobic respiration in which the oxygen produced is simultaneously consumed. Because measuring the consumption of carbon dioxide typically requires expensive equipment and complex procedures, you will use the floating disk procedure to measure the production of oxygen.

In this procedure, you use a vacuum to remove all air and then add a bicarbonate solution to plant (leaf) disk samples. These leaves sink until placed in sufficient light, when photosynthesis produces enough oxygen bubbles to change the buoyancy of the disk, causing it to float to the surface. Many different factors can affect the rate of photosynthesis in the real world (i.e., intensity of light, color of light, leaf size, type of plant), but the results of this experiment can also be influenced by different procedural factors (i.e., depth of your solution, method of cutting disks, size of leaf disks). You will effectively be measuring net photosynthesis. The standard measurement to use after determining how long it takes each disk to float to the surface is ET_{50}, the estimated time for 50 percent of the disks to rise. This measurement will help you to aggregate your data for the second half of the investigation.

After learning how to use the floating disk procedure, you will choose one factor that affects the rate of photosynthesis, and then develop and conduct an investigation of that variable. When you compare the ET_{50} for different levels of your chosen variable, you should observe that ET_{50} goes down as rate of photosynthesis goes up. This creates a nontraditional graph that is not the best display of your data (below left). Alternatively, you can use $1/ET_{50}$, which will show increasing rates of photosynthesis and, therefore, a graph with a positive slope (below right).

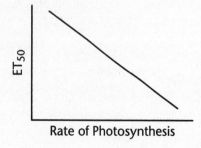

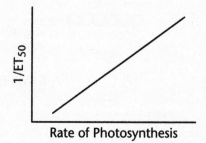

CHAPTER 9: CELLULAR RESPIRATION

BIG IDEA 2: Biological systems utilize free energy and molecular building blocks to grow, reproduce, and maintain homeostasis.

IF YOU ONLY LEARN THREE THINGS FROM THIS CHAPTER . . .

1. Cellular respiration is an efficient catabolic pathway and yields ATP. Cellular respiration is an aerobic process. The metabolic reactions of cell respiration occur in the eukaryotic mitochondria and are catalyzed by reaction-specific enzymes.

2. Cellular respiration can be divided into several stages: glycolysis, pyruvate decarboxylation, the citric acid cycle, and the electron transport chain.

3. Cellular respiration is a complex process that requires many different products and specialized molecules. Focus on the requirements and overall net production for the major steps.

INTRODUCTION ★★★★

Cells combine enzymatic regulation with biomolecules to make essential pathways, through which they are able to create from scratch certain necessary molecules, build energy, communicate with other cells, and break down nutrients, wastes, and toxins. These pathways include glycolysis, fermentation, and cellular respiration, as well as photosynthesis and other biosynthetic reactions. Earlier, you learned about photosynthesis; now we will discuss the complement of photosynthesis: cellular respiration. But first, we will review the major aspects of the flow of energy.

Cellular metabolism is the sum total of all chemical reactions that take place in a cell. These reactions can be generally categorized as either anabolic or catabolic. Anabolic processes are energy-requiring, involving the biosynthesis of complex organic compounds from simpler molecules. Catabolic processes release energy as they break down complex organic compounds into smaller molecules. The metabolic reactions of cells are coupled so that energy released from catabolic reactions can be harnessed to fuel anabolic reactions.

TRANSFER OF ENERGY ★★★★

The ultimate energy source for living organisms is the sun. Autotrophic organisms, such as green plants, convert sunlight into energy stored in the bonds of organic compounds (chiefly glucose) during the anabolic process of photosynthesis. Autotrophs do not need an exogenous supply of organic compounds. Heterotrophic organisms obtain their energy catabolically, via the breakdown of organic nutrients that must be ingested. Note in the following energy flow diagram that some energy is dissipated as heat at every stage.

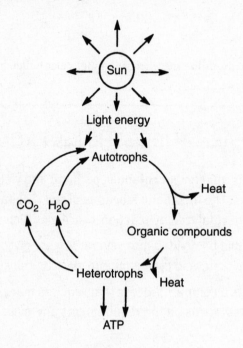

Energy Flow

CELLULAR RESPIRATION ★★★

Cellular respiration is the most efficient catabolic pathway used by organisms to harvest the energy stored in glucose. Whereas glycolysis yields only 2 ATP per molecule of glucose, cellular respiration can yield 36–38 ATP. Cellular respiration is an aerobic process; oxygen acts as the final acceptor of electrons that are passed from carrier to carrier during the final stage of glucose oxidation. The metabolic reactions of cellular respiration occur in the eukaryotic mitochondria and are catalyzed by reaction-specific enzymes.

Cellular respiration can be divided into five stages: glycolysis, fermentation, pyruvate decarboxylation, the citric acid cycle, and the electron transport chain.

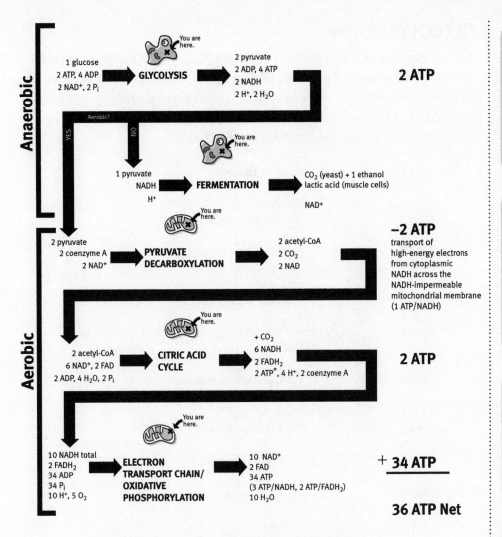

*The citric acid cycle yields a direct product of 2 GTP. The 2 GTP subsequently donate their phosphate to 2 ADP to form 2 ATP and regenerate the original 2 GDP.

General Overview of Cellular Respiration

REMEMBER

Fermentation ends with the conversion of pyruvate to lactic acid (muscle cells) or ethanol (yeast).

DON'T FORGET

Anaerobic respiration occurs in the absence of O_2 and results in the *net* of 2 ATP per molecule of glucose.

REMEMBER

In the Krebs cycle, the breakdown of acetyl-CoA generates NADH, $FADH_2$, and ATP.

REMEMBER

In the ETC, O_2 is the final electron acceptor.

GLYCOLYSIS ★★

The first stage of glucose catabolism is glycolysis. Glycolysis is a series of reactions that lead to the oxidative breakdown of glucose into two molecules of pyruvate (the ionized form of pyruvic acid), the production of ATP, and the reduction of NAD^+ into NADH. All of these reactions occur in the cytoplasm and are mediated by specific enzymes. The glycolytic pathway is as follows:

REMEMBER

You do not need to know all these steps or the intermediates.

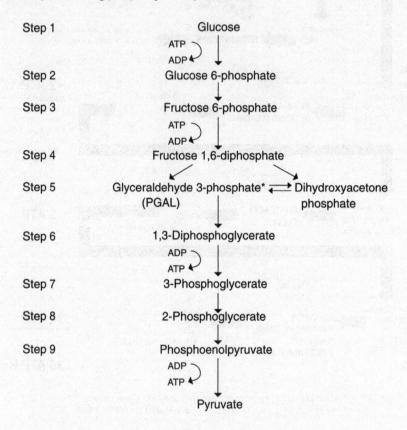

Step 1   Glucose

Step 2   Glucose 6-phosphate

Step 3   Fructose 6-phosphate

Step 4   Fructose 1,6-diphosphate

Step 5   Glyceraldehyde 3-phosphate* ⇌ Dihydroxyacetone phosphate
          (PGAL)

Step 6   1,3-Diphosphoglycerate

Step 7   3-Phosphoglycerate

Step 8   2-Phosphoglycerate

Step 9   Phosphoenolpyruvate

          Pyruvate

* NOTE: Steps 5–9 occur twice per molecule of glucose (see text).

Glycolysis

From one molecule of glucose (a six-carbon molecule), two molecules of pyruvate (a three-carbon molecule) are obtained. During this sequence of reactions, 2 ATP are used (in steps 1 and 3) and 4 ATP are generated (2 in step 6 and 2 in step 9). Thus, there is a net production of 2 ATP per glucose molecule. This type of phosphorylation is called substrate level phosphorylation, because ATP synthesis is directly coupled with the degradation of glucose without the participation of an intermediate molecule such as NAD^+. One NADH is produced per PGAL, for a total of 2 NADH per glucose.

The net reaction for glucose is:

Glucose + 2ADP + $2P_i$ + $2NAD^+$ → 2 Pyruvate + 2ATP + 2NADH + $2H^+$ + $2H_2O$

This series of reactions occurs in both prokaryotic and eukaryotic cells. However, at this stage, much of the initial energy stored in the glucose molecule has not been released and is still present in the chemical bonds of pyruvate. Depending on the capabilities of the organism, pyruvate degradation can proceed in one of two directions. Under anaerobic conditions (in the absence of oxygen), pyruvate is reduced during the process of **fermentation**. Under aerobic conditions (in the presence of oxygen), pyruvate is further oxidized during **cell respiration** in the mitochondria.

FERMENTATION ★★★★

NAD^+ must be regenerated for glycolysis to continue in the absence of O_2. This is accomplished by reducing pyruvate into ethanol or lactic acid. Fermentation refers to all of the reactions involved in this process—glycolysis and the additional steps leading to the formation of ethanol or lactic acid. Fermentation produces only 2 ATP per glucose molecule.

Alcohol fermentation commonly occurs only in yeast and some bacteria. The pyruvate produced in glycolysis is decarboxylated to become acetaldehyde, which is then reduced by the NADH to yield ethanol. In this way, NAD^+ is regenerated and glycolysis can continue.

$$CO_2$$

Pyruvate (3C) $\longrightarrow$ Acetaldehyde (2C) $\qquad$ NADH $\longrightarrow$ NAD^+ H^+ $\longrightarrow$ Ethanol (2C)

Lactic acid fermentation occurs in certain fungi and bacteria and in human muscle cells during strenuous activity. When the oxygen supply to muscle cells lags behind the rate of glucose catabolism, the pyruvate generated is reduced to lactic acid. As in alcohol fermentation, the NAD^+ used in step 5 of glycolysis is regenerated when pyruvate is reduced. In humans, lactic acid may accumulate in the muscles during exercise, causing a decrease in blood pH that leads to muscle fatigue. Once the oxygen supply has been replenished, the lactic acid is oxidized back to pyruvate and enters cellular respiration. The amount of oxygen needed for this conversion is known as the oxygen debt.

PYRUVATE DECARBOXYLATION ★★★★

The pyruvate formed during glycolysis is transported from the cytoplasm into the mitochondrial matrix where it is decarboxylated; i.e., it loses a CO_2, and the acetyl group that remains is transferred to coenzyme A to form acetyl-CoA. In the process, NAD^+ is reduced to NADH.

Pyruvate (3C) + Coenzyme A $\qquad$ NAD^+ $\longrightarrow$ $NADH + H^+$ $\longrightarrow$ Acetyl-CoA (2C)

CITRIC ACID CYCLE ★★

The citric acid cycle is also known as the Krebs cycle or the **tricarboxylic acid cycle** (TCA cycle). The cycle begins when the two-carbon acetyl group from acetyl-CoA combines with oxaloacetate, a four-carbon molecule, to form the six-carbon **citrate**. Through a complicated series of reactions, 2 CO_2 are released, and oxaloacetate is regenerated for use in another turn of the cycle.

For each turn of the citric acid cycle, 1 ATP is produced by substrate level phosphorylation via a GTP intermediate. In addition, electrons are transferred to NAD$^+$ and FAD, generating NADH and FADH$_2$, respectively. These coenzymes then transport the electrons to the electron transport chain, where more ATP is produced via oxidative phosphorylation (see diagram). Studying the cycle, we can do some bookkeeping; keep in mind that for each molecule of glucose, 2 pyruvates are decarboxylated and channeled into the citric acid cycle. (Note: Memorization of the steps and intermediates of the citric acid cycle is not necessary for the AP Biology exam.)

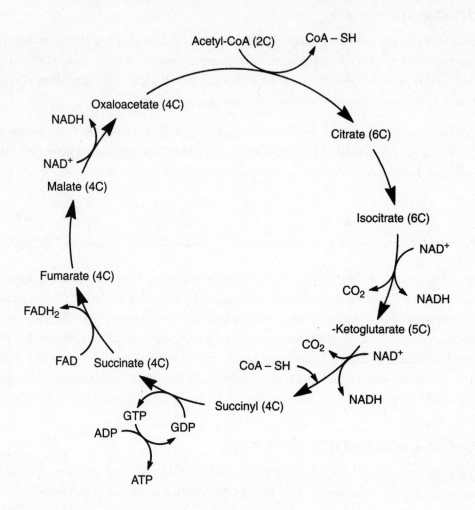

The Citric Acid Cycle

The *net reaction* of the citric acid cycle per glucose molecule is:

$$2 \text{ Acetyl-CoA} + 6\text{NAD}^+ + 2\text{FAD} + 2\text{GDP} + 2\text{P}_i + 4\text{H}_2\text{O} \rightarrow 4\text{CO}_2 + 6\text{NADH} + 2\text{FADH}_2 + 2\text{ATP} + 4\text{H}^+ + 2\text{CoA}$$

ELECTRON TRANSPORT CHAIN ★★★

The **electron transport chain (ETC)** is a complex carrier mechanism located on the inside of the inner mitochondrial membrane. During oxidative phosphorylation, ATP is produced when high-energy-potential electrons are transferred from NADH and $FADH_2$ to oxygen by a series of carrier molecules located in the inner mitochondrial membrane. As the electrons are transferred from carrier to carrier, free energy is released, which is then used to form ATP. Most of the molecules of the ETC are cytochromes, electron carriers that resemble hemoglobin in the structure of their active site. The functional unit contains a central iron atom, which is capable of undergoing a reversible redox reaction; that is, it can be alternatively reduced and oxidized.

FMN (flavin mononucleotide) is the first molecule of the ETC. It is reduced when it accepts electrons from NADH, thereby oxidizing NADH to NAD^+. Sequential redox reactions continue to occur as the electrons are transferred from one carrier to the next; each carrier is reduced as it accepts an electron and is then oxidized when it passes it on to the next carrier.

The last carrier of the ETC, cytochrome a_3, passes its electron to the final electron acceptor, O_2. In addition to the electrons, O_2 picks up a pair of hydrogen ions from the surrounding medium, forming water. (Note: The names of the specific electron carriers in the electron transport chain are not required knowledge for the AP Biology exam.)

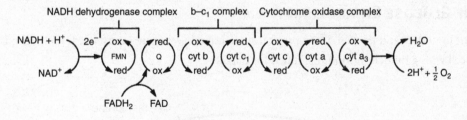

Electron Transport Chain

$$3 \times 6 \text{ NADH} \longrightarrow 18 \text{ ATP}$$

$$2 \times 2 \text{ FADH}_2 \longrightarrow 4 \text{ ATP}$$

$$1 \times 2 \text{ GTP (ATP)} \longrightarrow 2 \text{ ATP}$$

Without oxygen, the ETC becomes backlogged with electrons. As a result, NAD^+ cannot be regenerated and glycolysis cannot continue unless lactic acid fermentation occurs. Likewise, ATP synthesis comes to a halt if respiratory poisons such as **cyanide** or **dinitrophenol** enter the cell. Cyanide blocks the transfer of electrons from cytochrome a_3 to O_2. Dinitrophenol uncouples the electron transport chain from the proton gradient established across the inner mitochondrial membrane.

ATP GENERATION AND THE PROTON PUMP ★★

The electron carriers are categorized into three large protein complexes, NADH dehydrogenase, the cytochrome b–c_1 complex, and cytochrome oxidase, as well as

PLEASE NOTE

Everything the human body does to deliver inhaled oxygen to tissues (discussed in later chapters) comes down to the role oxygen plays as the final electron acceptor in the electron transport chain. Without oxygen, ATP production is not adequate to sustain human life. Similarly, the CO_2 generated in the citric acid cycle is the same carbon dioxide we exhale.

NOTE

Notice that the electron transport chain (ETC) is like an assembly line where the majority of ATP is generated.

the carrier molecule Q. There are energy losses as the electrons are transferred from one complex to the next; this energy is then used to synthesize 1 ATP per complex. Thus, an electron passing through the entire ETC supplies enough to generate 3 ATP. NADH delivers its electrons to the NADH dehydrogenase complex so that for each NADH, 3 ATP are produced. However, $FADH_2$ bypasses the NADH dehydrogenase complex and delivers its electrons directly to carrier Q (ubiquinone), which lies between the NADH dehydrogenase and cytochrome $b-c_1$ complexes. Therefore, for each $FADH_2$, there are only two energy drops, and only 2 ATP are produced.

The operating mechanism in this type of ATP production involves coupling the oxidation of NADH to the phosphorylation of ADP. The coupling agent for these two processes is a proton gradient across the inner mitochondrial membrane, maintained by the ETC. As NADH passes its electrons to the ETC, hydrogen ions (H^+) are pumped out of the **matrix**, across the inner mitochondrial membrane, and into the **intermembrane space** at each of the three protein complexes. The continuous translocation of H^+ creates a positively charged acidic environment in the intermembrane space. This electrochemical gradient generates a proton-motive force, which drives H^+ back across the inner membrane and into the matrix. However, to pass through the membrane (which is impermeable to ions), the H^+ must flow through specialized channels provided by enzyme complexes called ATP synthetases. As the H^+ pass through the ATP synthetases, enough energy is released to allow for the phosphorylation of ADP to ATP. The coupling of the oxidation of NADH with the phosphorylation of ADP is called **oxidative phosphorylation**.

REVIEW OF GLUCOSE CATABOLISM ★★★

It is important to understand how all of the previously described events are interrelated. Following is a eukaryotic cell with a mitochondrion magnified for detail.

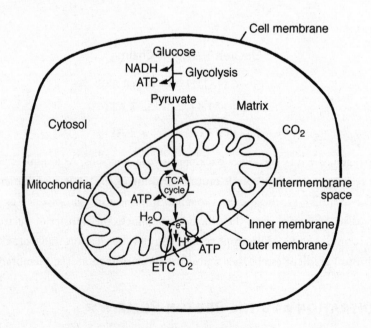

Locations of Glucose Catabolism

To calculate the net amount of ATP produced per molecule of glucose, we need to tally the number of ATP produced by substrate level phosphorylation and the number of ATP produced by oxidative phosphorylation.

SUBSTRATE LEVEL PHOSPHORYLATION

Degradation of one glucose molecule yields a net of 2 ATP from glycolysis and 1 ATP for each turn of the citric acid cycle with two turns per glucose. Thus, a total of 4 ATP are produced by substrate level phosphorylation.

OXIDATIVE PHOSPHORYLATION

Two pyruvate decarboxylations yield 1 NADH each for a total of 2 NADH per glucose. Each turn of the citric acid cycle yields 3 NADH and 1 $FADH_2$, for a total of 6 NADH and 2 $FADH_2$ per glucose molecule. Each $FADH_2$ generates 2 ATP, as previously discussed. Each NADH generates 3 ATP except for the 2 NADH that were reduced during glycolysis; these NADH cannot cross the inner mitochondrial membrane and must transfer their electrons to an intermediate carrier molecule, which delivers the electrons to the second carrier protein complex, Q. Therefore, these NADH generate only 2 ATP per glucose. Therefore, the 2 NADH of glycolysis yield 4 ATP, the other 8 NADH yield 24 ATP, and the 2 $FADH_2$ produce 4 ATP, for a total of 32 ATP by oxidative phosphorylation.

The total amount of ATP produced during eukaryotic glucose catabolism is 4 via substrate level phosphorylation plus 32 via oxidative phosphorylation, for a total of 36 ATP. (For prokaryotes the yield is 38 ATP, because the 2 NADH of glycolysis don't have any mitochondrial membranes to cross and therefore don't lose energy.) See the following table for a summary of eukaryotic ATP production.

A QUICK REFERENCE TO ENERGY PRODUCTION IN CELLULAR RESPIRATION

Glycolysis

2 ATP invested (steps 1 and 3)	−2 ATP
4 ATP generated (steps 6 and 9)	+4 ATP (substrate)
2 NADH × 2 ATP/NADH (step 5)	+4 ATP (oxidative)

Pyruvate Decarboxylation

2 NADH × 3 ATP/NADH	+ 6 ATP (oxidative)

Citric Acid Cycle

6 NADH × 3 ATP/NADH	+18 ATP (oxidative)
2 $FADH_2$ × 2 ATP/$FADH_2$	+4 ATP (oxidative)
2 GTP × 1 ATP/GTP	+2 ATP (substrate)
	Total + 36 ATP

AP BIOLOGY LAB 6: CELLULAR RESPIRATION INVESTIGATION ★★★★

This investigation is much more about the experimental method than cellular respiration. In this investigation, you learn how to set up and use a respirometer to measure the change in volume of a gas, which can be assumed to be O_2, from the germinating seeds placed in water. The effect of increasing temperature on the rate of gas volume change (O_2 utilization) is also measured during the lab. One of the problems with this lab is that gas expands when it is heated. Sometimes the volume increases so much it blows the dye out of the end of the respirometer tube. You will also have the opportunity to ask your own questions and conduct your own investigations about cellular respiration.

This experiment is an excellent example of using control specimens to isolate experimental variables. Put quite simply, the dependent variable being measured is the change in gas volume in the respirometer. The hypothesis is that the change in gas volume is being caused by the utilization of O_2 by the germinating seeds. Other factors can contribute to a change in gas volume (i.e., temperature and pressure) so these variables need to be isolated from the variables you are interested in (i.e., rate of respiration of the seeds). To isolate variables you are interested in from other variables, glass beads should be subjected to the same experimental treatment (change in temperature) that the target specimens, the live seeds, are subjected to. The glass bead sample is the control group. Any measurable change in the dependent variable in the control group must be removed from the dependent variable of the target group.

Let's say that when you heated your seeds to 35°C, the volume of gas in the respirometer of the target group (live seeds) decreased by 0.3 mL, but the volume of gas in the respirometer of the control group (glass beads) increased by 0.1 mL. Something occurred in the respirometer of the control group that caused an increase in gas volume. Perhaps the change in volume in the respirometer was created by the expansion of heated gas, or expansion was caused by CO_2 not being absorbed by the soda lime or other CO_2 absorbent. Either way, the dependent variable of temperature must be taken away from the target group because a change in volume should have occurred in the respirometer of the target group as well as the respirometer of the control group. As a result, it can be assumed that an additional 0.1 mL (for a total of 0.4 mL) of O_2 was likely used by the germinating seeds. In the experiment, 0.4 mL of gas were not measured because 0.1 mL of gas was obscured by the expansion of gas from some unknown factor, likely increased temperature. An accurate measure of the target group can never be obtained if the control group isn't included in the experiment.

Though this investigation is typically conducted during the study of Big Idea Two (energy and cellular processes), it also connects to content and processes described under Big Idea One (evolution) and Big Idea Four (ecology). Think about how cellular respiration is a conserved evolutionary process or how different ecological habitats have modified the capture and use of free energy by different organisms. By thinking outside the box, you will be better able to apply your knowledge across the entire AP Biology exam. This is particularly helpful when addressing free-response questions, which often ask you to make connections between content and process.

CHAPTER 10: THERMODYNAMICS AND HOMEOSTASIS

Big Idea 2: Biological systems utilize free energy and molecular building blocks to grow, reproduce, and maintain homeostasis.

IF YOU ONLY LEARN FIVE THINGS IN THIS CHAPTER . . .

1. There are three laws of thermodynamics. Together they discuss the conservation of energy, entropy, and absolute zero.

2. Bioenergetics, or biological thermodynamics, studies how chemical energy is broken down and converted to usable energy within the biological system. This can be at the ecosystem, organismal, or cellular level.

3. Chemiosmosis produces energy from the movement of H^+ ions across a membrane against a concentration gradient in both photosynthesis and respiration. Catabolism is the breaking down of complex substances into simple substances, making energy available in the process.

4. Homeostasis is the process by which a stable internal environment is maintained within an organism. Our primary homeostatic organs are the kidneys, the liver, the large intestine, and the skin.

5. Thermoregulation is all the physiological processes that come together to maintain a stable body temperature in warm-blooded animals.

IT'S ALL ABOUT ENERGY: WHERE IT COMES FROM, WHERE IT GOES, AND HOW IT'S NOT LOST

The processes that follow are important to all life on Earth and warrant a quick review of the major laws of thermodynamics. Energy is expensive to create metabolically and thus we need to make sure that it is preserved in the best ways possible. We will begin with a quick review of the major laws of thermodynamics.

THERMODYNAMICS ★★★★

Thermodynamics deals with heat, energy, and work and can be discussed in terms of temperature, internal energy, entropy, and pressure. For your AP Biology exam, you need to be familiar with the three laws of thermodynamics.

THE FIRST LAW OF THERMODYNAMICS

The increase in internal energy of a closed system is equal to the difference of the heat supplied to the system and the work done by the system. This is a variation of the law of the conservation of energy that states that energy cannot be created or destroyed, it can only change form. The classic example of this is how kinetic energy can be converted to potential energy or (within the body) chemical energy can be converted to kinetic energy.

THE SECOND LAW OF THERMODYNAMICS

The second law states that heat cannot spontaneously flow from a colder location to a hotter location. Over time, differences in temperature, pressure, and chemical potential tend to even out in a physical system that is isolated from the outside world. For example, a watch that is driven by a watch spring will run as the potential energy in the spring is converted into kinetic energy. When the potential energy runs out, the watch will stop unless new energy is reapplied to the spring to rewind it. In a biological system, potential energy inside carbohydrates is converted to kinetic energy. However, this process of energy transfer will also release some energy in the form of heat. This is also called entropy—as energy is transferred from one form to another, some is lost as heat and the energy decreases.

THE THIRD LAW OF THERMODYNAMICS

The entropy of a system approaches a constant value as the temperature approaches zero. Therefore, at absolute zero (the coldest possible temperature), entropy reaches its minimum value. At absolute zero, nothing can be colder and no heat energy remains in a substance.

BIOENERGETICS

Biological thermodynamics, also known as bioenergetics, is the study of energy transformation in biology. This involves looking at energy transformations and transductions in and between living things, including their major functions down to the cellular level, and understanding the function of the chemical processes underlying these transductions. Bioenergetics and biochemistry are very large and complex topics. This chapter will help to simplify these concepts into the major things you will need for the AP exam.

Exam questions place an emphasis on integration and synthesis of concepts in this chapter and how the processes discussed relate to ecology, physiological systems, and form following function. Information is broadly applied in synthesis questions. Photosynthesis and cellular respirations are covered in detail in earlier chapters, but this chapter will use energy examples from them. Therefore, if you are reviewing these chapters out of order, you may want to come back to this chapter after you have familiarized yourself more with these processes.

In some biology textbooks, extensive diagrams of chains of chemical reactions leading from small building blocks such as CO_2 and H_2O to complex products such as glucose are shown. Sometimes a textbook shows the reverse, illustrating a diagram of cells breaking down complex molecules to release chemical energy. Diagrams containing such information will not be shown in this chapter. (See the Additional Resources listed earlier for materials you can use to view extensive diagrams of these reactions.) This chapter presents concepts of thermodynamics and homeostasis, and you will find out how your knowledge of that information will be assessed on the exam. It is important to remember that all of the processes discussed in this chapter revolve around the production or utilization of adenosine triphosphate (ATP), the molecule that powers all living systems.

IMPORTANT

As you study, be sure to know the structure of ATP and how it releases and absorbs energy.

COUPLED REACTIONS AND CHEMIOSMOSIS ★★★

A coupled reaction is one in which transport across a membrane is coupled with a chemical reaction. Of all the things to remember from this chapter, there is one that is most important: Chemiosmosis is used by cells to generate ATP by moving H^+ ions across a membrane, down a concentration gradient. Special membrane proteins called ATPases create proton channels to convert ADP to ATP when a proton passes through. Aerobic respiration and photosynthesis utilize chemiosmosis, whereas glycolysis and other forms of ATP creation do not.

The mitochondrion moves H^+ into the **intermembrane space** via the **electron transport chain**, which creates a proton gradient across the inner membrane. The energy to do this comes from the breakdown of food. The chloroplast moves H^+ into the **thylakoid space** in a way very similar to the mitochondrion, but it drives the oxidative phosphorylation process with light energy. This process is called **photophosphorylation**.

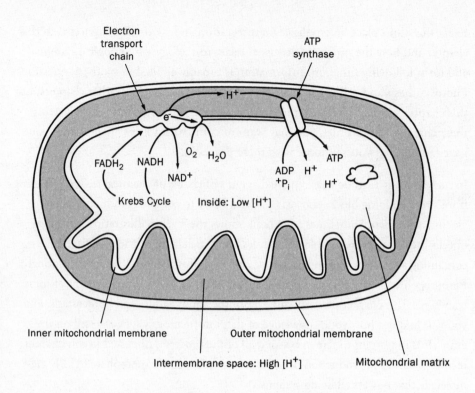

Chemiosmosis in the Mitochondrion

The structures that perform chemiosmosis are examples of form specialized for function. The invaginations of the inner membrane (cristae) inside the mitochondrion provide an increased surface area. Even the proteins in the membrane are aligned in a way that spans the membrane width, and their juxtaposition provides the proper architecture to allow chemical reactions to take place. The most important thing to remember about fine protein structures is that their location and form are what allow chemiosmosis to happen.

ANAEROBIC AND AEROBIC CATABOLISM ★★

Catabolism is the breaking down of complex substances into simple substances, making energy available in the process. In **anaerobic catabolism**, there is no electron transport chain or oxygen (O_2) available to carry out a reaction. Anaerobic reactions require almost as much energy to carry out as they yield. From an evolutionary perspective, it is much more advantageous to have the ability to utilize the oxidative properties of oxygen (O_2), as reactions in **aerobic catabolism** do, because more energy can be generated.

The following table includes a few comparisons between the three different components of cellular respiration. In the presence of oxygen (aerobic catabolism, or **aerobic respiration**), all three processes can occur. However, in the absence of oxygen, only glycolysis can occur; this is anaerobic catabolism (or **anaerobic respiration**).

ANAEROBIC AND AEROBIC CATABOLISM

Reaction	Location	Original Substance	Reaction Product(s)	O_2	Energy Source	ATP Cost	ATP Yield	Net ATP
Glycolysis (yielding ATP directly from NADH via fermentation)	Cytoplasm	Glucose	Pyruvic acid	No	2 NADH	2	4	2
Citric acid (Krebs) cycle	Mitochondrial matrix	Citric acid (from pyruvate/ acetyl-CoA)	Oxaloacetic acid (with many intermediates)	Yes	6 NADH 2 $FADH_2$	2	2	0
Electron transport chain (oxidative phosphorylation)	ATPases across inner mitochondrial membrane and inner mitochondrial matrix	NADH $FADH_2$	NAD^+ FAD	Yes	Proton motive force	0	34	34

Total yield from one glucose molecule via both anaerobic and aerobic processes	36

> **REMEMBER**
>
> You do not need to memorize the details of glycolysis, Krebs cycle, or the electron transport chain. Focus instead on the overall processes and the ATP costs and yields of each.

Keep in mind that not all energy comes from sugar in the form of glucose. Organisms use food in the form of carbohydrates (starch), protein, and fat. Carbohydrates ultimately break down or are transformed into glucose. The amino acids from proteins ultimately enter the citric acid cycle as pyruvate and acetyl-CoA. Fats are broken down into glycerols that can be converted to pyruvate and fatty acids that are converted to acetyl-CoA. All of the pyruvate and acetyl-CoA from the breakdown of food enter the citric acid cycle and the electron transport chain.

THERMOREGULATION AND HOMEOSTASIS ★★★★

Homeostasis is the process by which a stable internal environment is maintained within an organism. Important homeostatic mechanisms include the maintenance of a water and solute balance (**osmoregulation**), which will be covered later in this book, regulation of blood glucose levels, and the maintenance of a constant body temperature. In mammals, the primary homeostatic organs are the kidneys, the liver, the large intestine, and the skin.

THERMOREGULATION

The hypothalamus is the brain region that acts to control the body temperature of an organism that is able to set its internal temperature. Mammals fall into this category. Hormones such as **epinephrine** or thyroid hormones, released by the adrenal glands, can increase metabolic rate and therefore heat production. Muscles can generate heat by contracting rapidly (**shivering**). Heat loss is regulated through the contraction or relaxation of precapillary sphincters, as explained in chapter 22. Alternative mechanisms are used by some mammals to regulate body temperature. Panting is a cooling mechanism that results in the evaporation of water from the respiratory passages. Sweating is also important in increasing evaporative heat loss so that the body cools. Fur is used to trap heat, and hibernation during the winter conserves energy by decreasing heart rate, breathing, and metabolism. Animals able to regulate their internal temperature even in the face of a changing external temperature are called endotherms, or homeotherms. Mammals and birds are capable of this type of regulation, yet reptiles, amphibians, and most other animals are not and are known to be **cold-blooded**, or **ectotherms**.

There are two laws you should know that relate heat and body size:

1. Bigger bodies produce less body heat per pound per hour.
2. Bigger bodies lose less body heat per pound per hour.

Metabolic heat production drops in a very specific manner as body size increases. Compared with an elephant, which might weigh 10,000 pounds, a small mammal weighing only 1 pound produces about 10 times more heat per pound than the elephant does. Yet of the two animals, the elephant stays warmer because it has much less overall surface area compared to internal volume than does the small mammal. In other words, the small mammal gives off much more heat to its surroundings.

DID YOU KNOW?

Heat regulation is found even among some plants, such as the skunk cabbage and the lotus. Lotus plants can maintain their internal temperature at about 85 degrees Fahrenheit, even as the temperature around them drops to near 50 degrees. This seems to play an important role in pollination and reproduction.

FEEDBACK MECHANISMS ★★★★

Input is vital to an organism's success on the organismal and cellular level. Imagine you are playing a video game and you keep losing. All you know is that the video game suddenly says "game over," but you never know why. You have no information on why you lost your game and, therefore, you have no idea how to prevent it from happening again. Biological systems are the same. They require input from the surrounding environment, whether it be from within the body or the external environment, to understand what processes need to be adapted in order to survive. This input is called feedback, and organisms all possess feedback mechanisms in order to respond to stimuli. No matter if it is on the organismal or the cellular level, self-regulating mechanisms have been around as long as life has existed. These mechanisms usually fall into two categories: positive and negative feedback.

Positive feedback results when the effects of feedback from a system result in an increase in the original factor that causes the disturbance. Positive feedback mechanisms increase or accelerate the output created by a stimulus that has already been activated. For example, if you have a bank account that is earning interest, the amount of interest grows every time the account increases. The higher the account balance, the more interest is earned and the balance increases even more. This can go on and on until some other mechanism (such as withdrawing money from your account) causes a disruption in the positive feedback cycle. If there is no mechanism to stop the positive feedback loop, then the system can lose control of the cycle. For that reason, positive feedback systems are considered unstable. For example, positive feedback can be seen with climate change. An initially small perturbation in the environment can positively feed back onto itself, growing and growing until the problem yields huge effects on the climate.

Negative feedback is the opposite of positive feedback. As more feedback is received, it causes the processes that brought about the initial change to cause a change in the opposite direction. Self-regulating systems tend to function by using negative feedback. It allows for stability within a system because it reduces the effects of fluctuations. Negative feedback loops allow a system to have the necessary amount of correction at the most important time. One of the simplest examples of negative feedback is the human body's methods of thermoregulation. An increase in core body temperature will stimulate the body to produce sweat. As the body's sweat production increases, it causes a drop in body temperature. This decrease in body temperature will turn off the mechanism for sweat production.

CHAPTER 11: IMMUNE RESPONSE

BIG IDEA 2: Biological systems utilize free energy and molecular building blocks to grow, reproduce, and maintain homeostasis.

IF YOU ONLY LEARN FOUR THINGS FROM THIS CHAPTER . . .

1. The immune system can be divided into two major divisions: nonspecific (targets general infections) and specific (attacks specific disease-causing organisms by protein-to-protein interaction). Specific immunity is how we become immune to future infections from pathogens we have already fought off.

2. There are five types of antibodies and their structure is based on the same four polypeptide chains.

3. Specialized white blood cells, lymphocytes, come in two varieties: B cells and T cells. They are produced by stem cells in the bone marrow. There are three types of T cells: helper (T_H), cytotoxic (T_C), and suppressor (T_S).

4. Antibodies work through agglutination/neutralization, precipitation, and complement activation.

INTRODUCTION

The task of fighting infections is quite demanding: The human immune system must be able to respond to disease-causing organisms that are as small as a few nanometers in diameter (e.g., the polio virus) and as large as 10 meters long (e.g., a tapeworm). Because the major cells of the immune system are all approximately 10–30 micrometers in diameter, the cells must be able to fight off a range of organisms from 1,000 times smaller than they are to a million times larger! Because of this, the immune system is made up of a network of cells and organs that perform a variety of simultaneous tasks.

A VIEW OF THE IMMUNE SYSTEM: OVERALL PHYSIOLOGY

The immune system can be divided into two major divisions: nonspecific and specific. The nonspecific immune system is composed of defenses that are used to fight off infection in general and are not targeted to specific pathogens. The specific immune system is able to attack very specific disease-causing organisms by protein-to-protein interaction and is responsible for our ability to become immune to future infections from pathogens we have already fought off.

NONSPECIFIC DEFENSE ★★★★

The skin and mucous membranes form one part of the nonspecific defenses that our body uses against foreign cells or viruses. Intact skin cannot normally be penetrated by bacteria or viruses, and oil and sweat secretions give the skin a pH that ranges from 3 to 5, which is acidic enough to discourage most microbes from being there. In addition, saliva, tears, and mucus all contain the enzyme **lysozyme**, which can destroy bacterial cell walls (causing bacteria to rupture due to osmotic pressure) and some viral capsids. Mucus is able to trap foreign particles and microbes and transport them to the stomach (through swallowing) or to the outside (by coughing or blowing the nose).

Certain white blood cells are also part of the nonspecific defenses. **Macrophages** are large white blood cells that circulate, looking for foreign material or cells to engulf, which they do through **phagocytosis**. Macrophages circulate through the blood and are able to transport themselves through capillary walls and into tissues that have been infected or wounded. Once in the tissues, macrophages use their **pseudopodia** (like amoebas) to pull in foreign particles and destroy them within lysosomes. Macrophages are called **antigen-presenting cells** (APCs) because of their ability to "display" on their own cell surface the proteins that were on the surface of the cell or viral particle they have just digested. Because macrophages and other APCs do not distinguish between "self-proteins" destined for their cell membrane and "non-self" proteins previously on another organism's membrane, both types of proteins get shipped to the macrophage's cell surface. The advantage of this is that macrophages are able to display to other more specific immune system cells the **antigens** (foreign proteins) they have just encountered. That, in turn, often results in a more intensive immune response from these more specific cells (see B and T cells later in this chapter).

Neutrophils are white blood cells that are actively phagocytic like macrophages, but are not APCs. Our bodies normally produce approximately 1 million neutrophils per second, and they can be found anywhere in the body. They usually destroy themselves as they fight off pathogens. People who have decreased numbers of neutrophils circulating through their blood are extremely susceptible to bacterial and fungal infections. Other white blood cells that secrete toxic substances without fine-tuned specificity include the **eosinophils**, **basophils**, and **mast cells**.

THE INFLAMMATORY RESPONSE ★★★★

Both basophils and mast cells release large quantities of a molecule called **histamine**, responsible for dilating the walls of capillaries nearby and making those capillaries "leaky." For this reason, histamine is considered a potent vasodilator, which can lower blood pressure across the whole body if enough is released at once (blood pressure is maintained by the integrity of the capillary walls).

Leaky capillary walls allow macrophages and neutrophils to more easily reach the site of an injury. This nonspecific defense increases overall blood flow to areas of tissue injury and is responsible for the characteristic redness and heat felt in areas of injury. Basophils and mast cells have large secretory vesicles filled with histamine molecules, but they also release **cytokines**, chemicals that cause specific immune defenses to activate. Other responses to injury that are more systemic (body-wide), rather than local (at the injury site), include fever and increased production of all types of white blood cells.

ANTIHISTAMINES

Antihistamines are compounds that can limit immunologic reactions. They bind to histamine receptors and shut down their ability to cause leakage at the site of injury. Antihistamines can be lifesaving for certain reactions, particularly allergic ones where the immune system overacts to a seemingly harmless substance and leaky capillaries in the lungs and face make breathing difficult or impossible.

SPECIFIC DEFENSE ★★★★

The major specific defense of the immune system is the use of specialized white blood cells known as **lymphocytes**. These lymphocytes come in two varieties, B cells and T cells. Both are produced by stem cells in the bone marrow after embryologic development has finished. Although T cells mature in the **thymus**, B cells do not. The thymus is essential for "educating" T cells, and T cells that recognize "self" antigens (proteins found on one's own cell surfaces) are killed off so that autoimmune reactions are less likely to occur. This so-called negative selection results in the development of T cell tolerance, a necessity of the specific immune system. Yet, a positive selection process also exists whereby T cells that do not react to a specific set of glycoproteins, called MHC (major histocompatibility complex) proteins, are killed off as well. The remaining T cells are highly capable of bonding to both self-MHC molecules and a variety of foreign antigens simultaneously, which is essential for T cells to work properly.

There are three types of T cells: helper (T_H), cytotoxic (T_C), and suppressor (T_S). While T_H cells are mediators between macrophages and B cells, T_C cells are able to kill virally infected cells directly. Because virally infected cells display some viral proteins on their cell surfaces, T_C cells can bind to self-MHC proteins and viral proteins on the cell surfaces and secrete enzymes that perforate the cell membrane and kill the cell. Cytotoxic T cells are an essential part of the body's

HELPER T CELLS

Recall the antigen-presenting macrophages that dot their surfaces with foreign proteins they have digested. A certain class of T cells, known as helper T cells, is able to bind simultaneously to self-MHC proteins on the macrophage surface and the displayed foreign proteins. This combination of signals is needed to activate the helper T cells.

defenses against viruses. T_S cells are involved in controlling the immune response so that it does not run out of control. They accomplish this by suppressing the production of antibodies by B cells. It seems likely that these T_S cells are not a separate class of T cells altogether but rather certain T_H cells that secrete inhibitory chemical messengers (cytokines).

Keep in mind that T cells cannot detect free antigens; they can only respond to displayed antigens and MHC on the surfaces of cells. When they do recognize a displayed antigen, it is always in combination with a self-MHC protein displayed along with the antigen on the host cell surface. Interactions between T cells and APCs are enhanced by certain proteins that hold the T cell to the APC as it recognizes the antigen-MHC combination. One of these "holding" proteins is the CD4 protein.

B cells make up about 30 percent of one's lymphocytes, and the average B cell lives only for days or weeks. About 1 billion are made each day in the bone marrow. Every B cell has surface receptors that are identical in structure to a certain class of antibody protein that the B cell is able to produce and secrete into the bloodstream. In other words, the receptors on B cell surfaces are essentially bound Y-shaped antibody proteins that can recognize a specific set of foreign antigens (proteins found on the surfaces of foreign cells and viruses). B cells can be "activated" in one of two ways: either they can come into contact with a foreign antigen that can bind to the B cell surface receptors, or they can engulf a pathogen, displaying its antigens on the B cell surface much as a macrophage would. Then they can get stimulated to divide by chemicals released by a helper T cell (T_H) that recognizes the foreign proteins sitting on the B cell surface.

Both B and T cells each have unique cell receptors. That means that *almost every one of the several billion B and T cells in the body is capable of responding to a slightly different foreign antigen.* When a particular B or T cell gets activated, it begins to divide rapidly to produce identical **clones**. In the case of B cells, these clones will all produce antibodies of the same structure, capable of responding to the same invading antigens. B cell clones are known as **plasma B cells** and can produce thousands of antibody molecules per second as long as they live.

ANTIBODY STRUCTURE AND PRODUCTION ★★★

Antibodies are also known as **immunoglobulins** and are *produced solely by B cells.* Each immature B cell has specific antibodies stuck to its surface so that the B cell can be activated if it comes into contact with an antigen that its antibodies are specific for. A single B cell can produce billions of antibodies in its short lifetime.

ANTIBODY STRUCTURE AND CLASS ★★★

There are five types or classes of antibody: IgM, IgG, IgD, IgE, and IgA (Ig stands for immunoglobulin). Each class serves a different purpose in the body, and the same B cell can produce each type at different times during an infection. B cells always produce IgM class antibodies first, then *class-switch* to another type, usually IgG, depending upon where the B cell is in the body and the type of antigen it is responding to.

The structure of all five classes is based on the same four polypeptide chains. Two of the chains are called **heavy chains** and two are **light chains**. Both light chains are identical to each other in amino acid sequence, as are the two heavy chains:

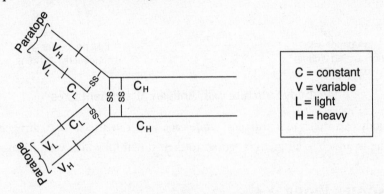

Basic Antibody Structure

Notice in the previous diagram that both the heavy chains and the light chains have **constant regions** and **variable regions**. It is the variable regions that are responsible for the specificity of a particular antibody for a particular antigen. The antigen-binding site, where antibodies can bind to foreign proteins, is also known as the **paratope**. The paratopes are each formed from a combination of variable amino acids in one heavy chain and one light chain. It should be clear from the diagram that each Y-shaped antibody has *two antigen-binding sites*.

Keep in mind that, while the B cells may have individual antibodies bound to their surface for use as receptors, the vast majority of antibodies that are produced are sent out to float freely through the bloodstream. IgG is the simplest type of antibody, with a structure approximated in the previous diagram. IgA, however, is a dimer made of two Y-shaped antibodies placed back to back, and IgM is a pentamer of five Y-shaped antibodies placed outward-facing in a circle:

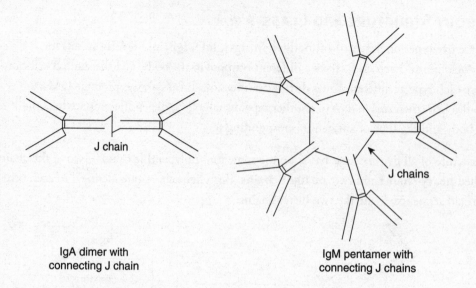

J chain

J chains

IgA dimer with
connecting J chain

IgM pentamer with
connecting J chains

Antibody Structure with Antigen Attachment Sites

Each IgA antibody can attach to 4 antigens, while each IgM can attach to 10 antigens. The ability of the antibodies to attach to antigens is central to their function, which we'll discuss later.

How Antibodies Work ★★

Agglutination/neutralization occurs when antibodies cross-link adjacent antigen molecules (on bacteria and other organisms) so that these invaders literally get stuck together by the antibodies circulating in the bloodstream. Because each Y-shaped antibody can stick to two different organisms, antibodies can cause the clumping together of many pathogens in a short time. These agglutinated bunches of bacteria or virus particles form large, insoluble masses that are no longer able to invade cells and can be easily engulfed by circulating macrophages.

Precipitation is similar to agglutination but used for soluble antigen molecules such as small bacterial toxins, which dissolve in the bloodstream. Antibody binding allows rapid phagocytosis and destruction of small proteins by macrophages.

Complement activation occurs when antibodies bound to the surfaces of foreign cells activate a system of 20 different "complement" proteins that circulate in the bloodstream. These proteins are turned on in a cascade-like fashion with each one activating the next, allowing for a great deal of control over the process.

- *Classical pathway*: This requires antibodies bound to antigens. Complement proteins bridge the gap between two adjacent antibody molecules and use a protein complex called the membrane-attack complex to lyse the cell membrane of the invader. Complement proteins also activate mast cells to release histamine, which brings more blood cells to the area.

- *Aclassical pathway*: This occurs independently of antigen-antibody binding. Cell surface molecules of many bacteria, yeasts, viruses, and protozoan parasites can cause membrane-attack complexes to form without the help of antibodies.

HYBRIDOMAS AND MONOCLONAL ANTIBODIES ★

Antibodies that arise in the natural course of fighting many pathogens are considered **polyclonal**—that is, they are produced by several different clones of plasma B cells and cover a wide range of specificity. Antibodies arising from a single clone, rapidly divided into identical B cells, are called **monoclonal**, and this single B cell has important scientific uses. The specific nature of antibody binding makes them attractive research targets for disease cures. Imagine, for example, being able to target specific cancer cells with an injection of antibodies that seek out and destroy only those cancer cells. Such antibodies can be made in the lab by fusing a myeloma cell (a cancerous, always-dividing B cell) with an antibody-producing cell from a mouse. The resulting cell, called a **hybridoma** because it is a hybrid cell from two different species, can produce almost unlimited quantities of a particular monoclonal antibody, which can be used in research.

HUMORAL VERSUS CELL-MEDIATED DEFENSES ★★★★

Within the immune system, the specific arm, which uses B and T cells to target individual microbes, can work either by secreting antibodies into the bloodstream (considered one of the body's "humors") or by directly killing cells. While antibody secretion (humoral immunity) is the job of the B cells, direct killing of cells and overall activation of the immune system (cell-mediated immunity) is the job of T cells.

THE HUMORAL RESPONSE ★★★★

When the body is exposed to an antigenic stimulus for the first time, many events must take place before a protective antibody response is produced. The period after exposure to a pathogen, but before helpful levels of antibodies have been made by B cells, is called the **lag period**. During this period, APCs such as macrophages and neutrophils process and display antigens to T_H cells, which rapidly grow and divide. These T cells activate B cells, some of which have also contacted antigens, and these B cells grow and divide as well.

After the first exposure to a microbe, the lag period lasts seven to ten days before enough antibodies are present in the blood to noticeably slow the infection. Within a week or so, the infection will have likely subsided. If the person is exposed to the same antigen a second time, there is a very quick **secondary response** (with a lag period of one to four days), and levels of antibody in the bloodstream typically reach much higher levels than they did in the **primary response**.

This is because certain antigen-specific **memory cells** remain after a primary infection. Both B and T cells form memory cells. While a typical B or T lymphocyte may live for only days or weeks, memory B and T cells can live for decades. The reasons for this are unclear, but the formation of memory cells is the basis of **immunity** and forms the concept behind **vaccination**. Small doses of antigen given as an injection or swallowed allow the body to recognize and form a primary response against the antigen so that upon actual exposure to the pathogen carrying that antigen, the body will mount a quick and sufficient response.

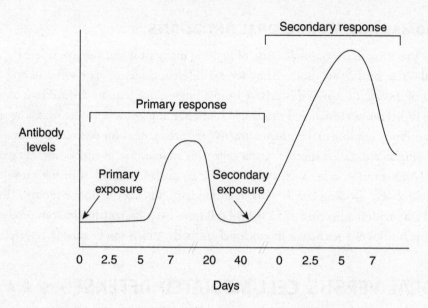

Graph of Antibody Levels During Exposure and Response

Because the lag period can be unacceptably long for some pathogens, doctors often provide preformed antibodies (usually made in horses or chickens) to people who have been exposed to a microbe or who face probable exposure. This passive immunity, to be distinguished from the active immunity that forms when B cells make their own antibodies, is temporary. Individuals who travel abroad are often given a shot of gamma globulin before leaving the country. This mixture of preformed antibodies to several tropical diseases is like a soup of temporary protection and will disintegrate within a week or two.

THE CELL-MEDIATED RESPONSE ★★★★

The timing of the T-cell response follows almost the same pattern as shown in the graph of the humoral response. After primary exposure to an antigen, specific T cells rapidly divide and form clones of identical cells. Upon secondary exposure, memory T cells left over from these clones will divide much more rapidly than they did at the first exposure. The cell-mediated response severely limits the ability of viruses to proliferate within cells, because cytotoxic (killer) T cells will destroy any cells harboring viruses. Helper T cells participate in antigen recognition and B cell activation, and natural killer cells (related to cytotoxic T cells) destroy infected and cancerous cells directly.

Altogether, nonspecific defenses such as skin, fever, and macrophages along with specific defenses such as B and T lymphocytes and the complement system make up a highly efficient and adaptable machine that serves to protect the body from a wide assortment of invaders.Recall the antigen-presenting macrophages that dot their surfaces with foreign proteins they have digested. A certain class of T cells, known as helper T cells, is able to bind simultaneously to self-MHC proteins on the macrophage surface and the displayed foreign proteins. This combination of signals is needed to activate the helper T cells.

APPLYING THE CONCEPTS: BIG IDEA TWO REVIEW QUESTIONS

BIOLOGICAL MOLECULES

This Review Quiz is designed to assess your content knowledge, and is not necessarily representative of the actual AP Biology exam. So don't panic if you see a new question type or a question asking about a relatively low-yield detail from the chapter.

1. Which of the following functional groups characterizes the structure of an alcohol?

    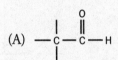

    (A)
    (B)
    (C)
    (D)

2. Organic compounds, unlike inorganic compounds, contain

    (A) hydrogen.
    (B) oxygen.
    (C) nitrogen.
    (D) carbon.

3. The bonding of two amino acid molecules to form a portion of a protein chain involves

    (A) the addition of a water molecule.
    (B) the release of a water molecule.
    (C) the addition of a nitrogen atom.
    (D) the release of a carbon dioxide molecule.

4. Starch and glycogen molecules are similar in that both are

    (A) water-soluble simple sugars.
    (B) polymers of glucose.
    (C) intermediate products in the Krebs cycle.
    (D) products of synthesis in plant cells.

5. A change in pH from 5 to 3 indicates a change in concentration of H^+ ions by a factor of

    (A) 2.
    (B) 20.
    (C) 50.
    (D) 100.

6. All of the following chemical compounds are organic EXCEPT

    (A) $Na_2S_2O_3$.
    (B) CH_4.
    (C) $(CH_2OH)_2$.
    (D) $(NH_2)_2CO$.

7. All of the following are important characteristics of water in biology EXCEPT

    (A) Cohesion provides surface tension.

    (B) It is a nonpolar solvent.

    (C) It is a medium for complex chemical reactions in organisms.

    (D) It has a high specific heat that stabilizes ambient temperature.

8. In what reaction type does water NOT play an essential role?

    (A) Condensation

    (B) Dehydration

    (C) Hydrolysis

    (D) Oxidation-reduction

9. If the free energy change of a reaction is greater than zero, then the reaction

    (A) is spontaneous.

    (B) is nonspontaneous.

    (C) is at equilibrium.

    (D) is endothermic.

10. Which is NOT a characteristic of proteins?

    (A) Can function as enzymes

    (B) Contain peptide bonds

    (C) Are important in cell signaling

    (D) Contain nitrogenous bases

Questions 11–14 refer to the following graph.
The top curve and the bottom curve represent alternate pathways for the same reaction. One pathway is enzyme-catalyzed.

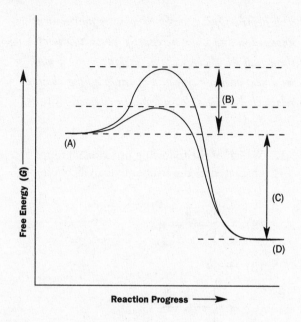

11. Represents the energy state of the reactants of the enzyme-catalyzed pathway

12. Represents the net energy change of the reaction

13. Represents the activation energy of the noncatalyzed reaction

14. Represents the energy state of the products of the enzyme-catalyzed pathway

15. What do living organisms require to grow, reproduce, and maintain homeostasis?

    (A) Enthalpy

    (B) Entropy

    (C) Free energy

    (D) Heat energy

16. What property of water allows it to exist naturally in all three states of matter?

    (A) Adhesion

    (B) Cohesion

    (C) Specific heat

    (D) Surface tension

ANSWERS AND EXPLANATIONS

1. D

There are two ways to approach this problem. First, if you remember that the functional group of an alcohol is a hydroxyl group (–OH), you know the obvious answer is (D) because it is the only choice of a functional group with a hydroxyl group. The second way to approach this problem is to think about how to construct molecular formulas, remember some examples of alcohols, and use the process of elimination. Ethanol (CH_3CH_3OH) and isopropyl alcohol or 2-propanol ($CH_3CHOHCH_3$) are common alcohols, and there is a good chance of your being familiar with them both. So what do they have in common? These compounds are composed of only C, H, and O, so (B) and (C) can be eliminated because they have an S and an N atom, respectively. Once you have eliminated these options, you may remember that (B) is the functional group for sulfhydryl and that (C) is the functional group for amines. Choice (A) can be eliminated because the molecular formula of an alcohol always denotes the hydroxyl group (–OH), and (A) shows a C bonded only to an O, not an OH group.

2. D

Organic compounds contain carbon atoms. Each carbon atom is able to form four covalent bonds, accounting for the wide variety of organic molecules.

3. B

There's some memorization involved in this question. Amino acids form long polypeptide chains called proteins. These chains get their name from the peptide bond formed between adjacent amino acids.

Amino Group   Carboxylic (Acid) Group

Amino acids have four components: a central asymmetrical α carbon, (A), bonded to an amine, (B), a carboxyl group, (C), and a side chain, (D), that dictates the chemical and physical properties of the amino acid. The peptide bond forms between the amine group of one amino acid and the carboxyl group of another.

This reaction is called dehydration synthesis, and it results in the production of a two-amino acid peptide chain and a water molecule.

4. B

Start this problem with the broad view that starch and glycogen are storage forms of sugar (carbohydrates) and are large molecules. Plants produce starch and animals produce glycogen. This rules out (D) because the correct answer can't be exclusive to either animals or plants. Only animals synthesize glycogen, and plants don't have connective tissue (not to mention that connective tissue is proteinaceous). The Krebs cycle involves the metabolism of glucose by-products, which are smaller than glucose molecules; therefore, the answer cannot be (C). Choice (A) isn't true. That leaves (B). These long chains of glucose are called polymers because they are repeating sequences of the same molecule.

5. D

Here is a test of your knowledge of the pH scale. The pH scale measures the amount of H^+ ions in a solution compared with pure water and ranges from 1 (high H^+ and very acidic) to 14 (very low H^+ and very basic). Pure water has a pH of 7. Your first response might be to pick (A) or (B) because they are obvious factors of 2, but these are not correct. The pH scale is logarithmic. Each step on the pH scale represents a tenfold change in H^+ concentration. To go from a benign pH of 7 for water to something very corrosive at a pH of 1 or 2, you need to add a lot of H^+ ions. Going two steps on the pH scale requires increasing the H^+ ion concentration by a factor of 10^2, or 100, which gives you (D).

6. A

The answer is easy to find as long as you note that the question asks for the choice that does *not* belong. Organic chemistry is about compounds containing the element carbon (represented in chemical formulas by a C). Organic compounds range from the simplest compounds such as methane (CH_4) to the most complex DNA. The only compound listed that doesn't include carbon is (A).

7. B

This is another EXCEPT question, so you have to find the choice that doesn't match. Recall that the water molecule has positive charges on the H atoms and a negative charge on the O atom. This makes it a polar molecule. Also, remember that polar and nonpolar compounds don't mix, and a nonpolar substance will not dissolve well in a polar solvent like water. Because water is polar and doesn't dissolve nonpolar compounds well, the answer is (B).

8. D

Though water can play a role in oxidation-reduction reactions, the best answer here is (D). Condensation, dehydration, and hydrolysis reactions have water as a reactant or product by definition.

9. B

If the free energy change (ΔG) of a reaction is greater than zero, then the reaction is nonspontaneous. If the free energy change of a reaction is less than zero, then the reaction is spontaneous. If the free energy change of a reaction is equal to zero, then the reaction is at equilibrium. The terms *exothermic* and *endothermic* do not refer to the free energy change of a reaction, but to the enthalpy (change in heat) of a reaction. If the change in enthalpy (ΔH) is positive, then heat is absorbed and the reaction is endothermic. If the change in enthalpy is negative, then heat is released and the reaction is exothermic.

10. D

Proteins do not contain nitrogenous bases (nucleotides do contain nitrogenous bases). The other statements are true: Proteins can function as enzymes, consist of amino acids joined by peptide bonds, and are important in cell signaling (hormones, for example).

11. A

You can actually answer three of these questions without knowing anything about enzyme activity and by using common sense. First, notice that (A) indicates state of energy while (B), (C), and (D) indicate changes in energy. Question 11 can be answered by knowing that, in general:

(1) Graphs of data move from left to right (the diagram even shows the direction of the reaction with the arrow indicating progress of the reaction).
(2) Reactants are things you start with.
(3) The *y*-axis is the measure of change in energy.

Choice (A), all the way to the *left* of the diagram, clearly indicates a stable state before energy change takes place, so it must be the answer.

12. C

This question can also be answered intuitively. The net change in anything is the amount you end with minus the amount you started with. This is the difference between the energy state of the products and the energy state of the reactants and is indicated by (C).

13. B

This is the only question in the cluster that requires a little knowledge of enzymes and activation energy: Enzymes lower the activation energy required to cause a reaction. You need to be able to recognize the activation energy on the graph, which is the difference between the energy of the reactants and the highest energy required during a reaction. There are two activation energies in the diagram, but only one is labeled. It is the one that is noncatalyzed, the one that requires more energy. The answer must be (B).

14. D

This question is at the opposite end of the spectrum from question 11. If all the way to the left is the start of the reaction, where the reactants are, then all the way to the right is the end of the reaction, where the products are. Choice (D) indicates another state of energy and occurs after both reaction pathways.

15. C

Living organisms require free energy to grow, reproduce, and maintain homeostasis. Entropy (B) and enthalpy (A) must be balanced against one another to achieve these goals. Organisms often release heat energy during exergonic reactions that increase entropy.

16. C

Water has a high specific heat, so it acts as a temperature stabilizer for many other compounds. This high specific heat capacity means that it moves through the states of matter at temperatures found throughout Earth.

CELLS

This Review Quiz is designed to assess your content knowledge, and is not necessarily representative of the actual AP Biology exam. So don't panic if you see a new question type or a question asking about a relatively low-yield detail from the chapter.

1. Plasmids (circular genetic elements present in bacteria) can

   (A) act as transposons

   (B) be found in all eukaryotic cells

   (C) replicate only when other plasmids replicate

   (D) never move from cell to cell

2. Prokaryotic and eukaryotic cells share which of the following features?

   (A) A plasma membrane

   (B) Complex flagella

   (C) Membrane-bound organelles

   (D) A membrane-bound nucleus

3. Which of the following could be identified by the presence of ribosomes, simple flagella, and a cell wall, along with the absence of membrane-bound organelles?

   (A) An amoeba

   (B) A bacterium

   (C) A muscle cell

   (D) An algae

4. Which of the following has a cell wall?

   (A) Animal cell

   (B) Plant cell

   (C) Some bacteria

   (D) B and C

5. Which of the following organelles translates the DNA code into proteins?

   (A) Chloroplast

   (B) Ribosome

   (C) Lysosome

   (D) Nucleolus

6. Which of the following is a typical component of the plasma membrane of a eukaryotic cell?

   (A) DNA

   (B) mRNA

   (C) tRNA

   (D) Cholesterol

7. A mature plant cell can be distinguished from other eukaryotic cells because it has

   (A) energy-producing mitochondria

   (B) a rough endoplasmic reticulum

   (C) chloroplasts

   (D) a large central vacuole

8. Which of the following cells would most likely have the greatest concentration of densely packed mitochondria?

   (A) A yeast cell in S phase of the cell cycle

   (B) A xylem cell in old wood of a tree

   (C) An oxygenated red blood cell

   (D) A smooth muscle cell in the diaphragm

Questions 9 and 10 refer to the plant cell figure shown. Match each term to the correct diagram label.

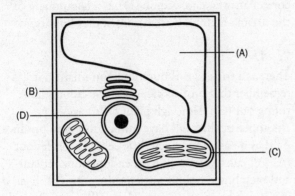

9. Chloroplast

10. Central vacuole

Questions 11–13 refer to the following list.

(A) Cilia and flagella

(B) Ribosomes

(C) Cell wall

(D) Nucleus

11. Only found in all eukaryotes

12. Found in all prokaryotes and eukaryotes

13. Found only in plants

14. Which of the following is an oxidative enzyme-containing structure that catalyzes reactions, breaks down fats for fuels, and can help detoxify compounds?

(A) Adipase

(B) Lysosomes

(C) Peroxisomes

(D) Lipase

15. Which of the following lists sizes from smallest to greatest?

(A) DNA, virus, bacteria, mitochondria

(B) virus, DNA, bacteria, mitochondria

(C) DNA, mitochondria, prokaryotes, eukaryotes

(D) virus, DNA, prokaryotes, eukaryotes

16. The 9 + 2 rule refers to

(A) ATP synthesis

(B) surface-to-volume ratio requirements for cell function

(C) cell wall thickness

(D) microtubule arrangement

ANSWERS AND EXPLANATIONS

1. A

Plasmids are able to reintegrate themselves back into the main bacterial chromosome, much as transposable elements in eukaryotic genomes can move from chromosome to chromosome using enzymes and insertion sequences. Plasmids are not found in eukaryotic cells, (B). Plasmids can replicate themselves independently of the main bacterial chromosome, and they can easily be transferred from cell to cell, so (C) and (D) are also incorrect.

2. A

The obvious answer is (A) because all cells have a plasma membrane. Because prokaryote cells are simpler than eukaryotes, they are the limiting factor for most answer choices to this question. Prokaryote cells have simple flagella composed of a single microtubule, so (B) can be ruled out. Prokaryotes don't have any membrane-bound organelles, so that rules out (C) and (D).

3. B

Only bacteria have simple flagella. With questions like this, try to find an answer choice that is very exclusive. Remember that some bacteria have cell walls, even if the walls are different in structure from plant cell walls.

4. D

Plant cells and some bacteria have a cell wall. Because both are correct, (D) has to be the answer. Animal cells have a cell membrane but do not have the rigid structure of a cell wall.

5. B

Protein translation is the same as protein synthesis, so the answer to this question should pop out at you right away. The ribosome is responsible for translating the genetic code delivered by mRNA into protein by linking together amino acids delivered by tRNA.

6. D

Some of the answers contain nucleic acids and are misleading. DNA, (A), is inside the nucleus of a eukaryotic cell, and both mRNA, (B), and tRNA, (C), are inside the nucleus or in the cytoplasm. The correct answer, cholesterol, (D), is a component of the plasma membrane that contributes fluidity.

7. D

There is a trick to this question: You might not remember that protists can also have chloroplasts, ruling out (C). Although plant cells have chloroplasts, they still have mitochondria to produce ATP, as do all other eukaryotic cells. This rules out (A). All eukaryotic cells have endoplasmic reticuli and membrane-bound nuclei, ruling out (B). Even if you didn't know the answer right off, especially with the qualifier of a "mature" plant cell, you should be able to figure out that (D) is the correct answer.

8. D

This question requires a bit of synthesis from other topics, but there is one key point to remember. If the cell has a lot of mitochondria, it's because it needs a lot of ATP to be produced to supply a lot of energy. Which of the cells seems like it needs a lot of energy? When most people think of work, they think about using muscle, so an obvious first choice would be (D). A yeast cell in S phase, (A), is duplicating DNA and producing proteins, so it needs many ribosomes. A xylem cell in old wood, (B), is probably no longer a living cell, and if it is, it's mostly transporting water to other plant cells that need it. Red blood cells, (C), are full of hemoglobin and very little else. Choice (D) is correct.

9. C

Identifying structures in diagrams is usually difficult to get around without simply knowing the structures. The question tells you that the diagram is a plant cell, but you still need to recognize structures. Identify the structures you know right away, and go back to the ones you don't. The chloroplast is identified by the granum (stacks of thylakoids) within a membrane-bound structure. Be careful not to confuse the chloroplast with the golgi apparatus that has stacks of membrane-bound structures called cisternae. The stacks of the golgi apparatus are not within a membrane.

10. A

Plant cells have a large central vacuole (membrane-enclosed sac) used for storage of organic compounds such as proteins and inorganic ions such as potassium and chloride. Water is stored here as well. As a large empty pocket in the cell, the central vacuole should be one of the easiest structures to identify in this drawing.

11. D

A nucleus is only found in eukaryotes. All eukaryotes have a nucleus. Some eukaryotes have cilia and flagella. Ribosomes are found in both prokaryotes and eukaryotes. Only plants have cell walls.

12. B

Ribosomes are found in both prokaryotes and eukaryotes.

13. C

Only plants have cell walls.

14. C

Peroxisomes contain oxidative enzymes to break down fats and are also used in the liver to detoxify the body of harmful compounds. Even though they end in *-ase*, lipase and adipase are meant to distract you from the correct answer. While they may be enzymes, they are very specific and do not work with the entire definition given.

15. A

DNA is the smallest out of everything in this list. Viruses contain DNA so they must be bigger than DNA, but viruses can infect cells so viruses must be smaller than cells. Bacteria (prokaryotes) are smaller than mitochondria. Mitochondria are located inside eukaryotic cells so they must be smaller than eukaryotes.

16. D

Cilia and flagella are composed of long microtubules arranged in a "9 + 2" structure. That means there are nine pairs of microtubules surrounding two central microtubules for added stability.

MEMBRANE TRAFFIC

This Review Quiz is designed to assess your content knowledge, and is not necessarily representative of the actual AP Biology exam. So don't panic if you see a new question type or a question asking about a relatively low-yield detail from the chapter.

1. Which of the following is not embedded in the membrane of a cell?

   (A) Cholesterol

   (B) Proteins

   (C) Nucleic acids

   (D) Carbohydrates

2. Which of the following does not require energy?

   (A) Active transport

   (B) Endocytosis

   (C) Exocytosis

   (D) Facilitated diffusion

Questions 3–5 refer to the following diagram.

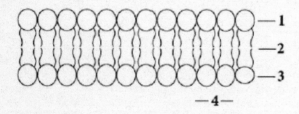

   (A) 2

   (B) 4

   (C) 1 and 3

   (D) 1 and 4

3. Which of the numbers could represent the inside of the cell?

4. Which of the numbers represents the nonpolar region?

5. Which of the numbers represents an area made of lipids?

Questions 6–9 refer to the following list.

   (A) Osmosis

   (B) Endocytosis

   (C) Exocytosis

   (D) Active transport

6. Carrier proteins bind to specific substances for invagination

7. Works against the concentration gradient

8. Specifically refers to water crossing the membrane

9. Includes phagocytosis

10. Which of the following is important for moving ions against their concentration gradients and maintaining concentration gradients across the lipid bilayer?

    (A) Transport proteins

    (B) Electron transport chain

    (C) ATPase

    (D) Membrane transport

11. Which of the following is not actively moved across the cell membrane?

    (A) Oxygen

    (B) Sodium

    (C) Potassium

    (D) Calcium

Questions 12–14 refer to the following figure.

A dialysis-tubing bag is filled with a mixture of 3 percent starch and 3 percent glucose and placed in a beaker of distilled water with a KI indicator. After one hour, the solution inside the dialysis bag has turned a dark blue, while the solution in the beaker has remained clear.

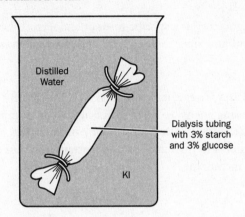

12. Which of the following is an accurate conclusion that can be made only from the observed results?

   (A) The dialysis-tubing bag weighs less.

   (B) Glucose has not diffused across the dialysis-tubing membrane.

   (C) The dialysis-tubing bag is selectively permeable.

   (D) A net movement of water into the beaker has occurred.

13. The change in solution color to blue is a positive indicator by KI for the presence of starch. Based only on the results, which of the following conclusions can be made?

   (A) The dialysis-tubing bag is permeable to starch.

   (B) The pores in the dialysis tubing are larger than KI molecules.

   (C) The dialysis-tubing bag is not permeable to glucose.

   (D) A net movement of water into the dialysis-tubing bag has occurred.

14. Which of the following best describes the system after it reaches equilibrium?

   (A) Water will have a net movement into the dialysis-tubing bag.

   (B) The osmotic pressure inside the dialysis-tubing bag will be the same as the osmotic pressure in the surrounding solution.

   (C) KI will have a net movement into the dialysis-tubing bag.

   (D) Glucose will have a net movement out of the dialysis-tubing bag.

15. Which of the following statements is true?

   (A) Active transport requires energy in the form of NADPH.

   (B) Facilitated diffusion requires a carbohydrate carrier.

   (C) Osmosis is the passive transport of nonpolar molecules.

   (D) Endocytosis and exocytosis are active transport processes that are ways to move large volumes of substances in and out of the cell.

ANSWERS AND EXPLANATIONS

1. C

Proteins, carbohydrates, and sterols like cholesterol are all embedded in all membranes. Proteins can be on the outer, inner, or across the entire membrane and can act as transport molecules (transport proteins). Carbohydrates are important for cell recognition and immunity while cholesterol helps to maintain the membrane fluidity.

2. D

The key here should be the word *diffusion*. Diffusion is the movement of molecules based on concentration gradient and thus requires no energy. Facilitated diffusion is ion transport via transmembrane carrier proteins, but it is still "powered" by the concentration gradient so it does not require any energy.

3. B

This is a diagram of the plasma membrane structure. 4 could be either the inside or the outside of a cell; however, it is the best option because none of the other options work.

4. A

1 and 3 are indicating the polar regions of the plasma membrane (the heads of the phospholipids). 2 is the nonpolar region (the tails of the phospholipids).

5. A

Two is indicating the nonpolar region of the plasma membrane, which is made up of the lipid tails of the phospholipids of the membrane.

6. B

Receptor-mediated endocytosis is when a carrier protein binds to a specific substance and envelopes it via invagination.

7. D

Active transport uses ATP to pull ions in the opposite direction of the concentration gradient. Therefore, molecules move from areas of low concentration to high concentration.

8. A

By definition, osmosis strictly refers to the movement of water from areas of high concentration to areas of low concentration.

9. B

Phagocytosis is a specific form of endocytosis when a membrane invaginates a solid and makes it a vesicle.

10. C

ATPases are enzymes that turn ATP into ADP, releasing energy to drive the chemical ion pumps that transport ions against their concentration gradients to maintain ion concentrations across a cell membrane.

11. A

Sodium, potassium, and calcium are all important ions that are kept in specific concentrations inside and outside the cell. The cell works hard to maintain the proper ionic balance so it can function properly.

12. C

This question can be answered using only analytical skills because you are expected to answer based only on the observed results. The one thing you know about the results of the experiment is that only the inside of the bag turned blue after an hour. The bag may have changed in weight, (A), but this can't be known from what is given. Glucose may have diffused into the beaker, (B), but you have no way of knowing for certain. Just as with glucose, you don't know what is going on with the water in (D). You don't know if water is moving or not.

The only correct choice is (C). The bag membrane is isolating a reaction—the change in color—because the reaction is not occurring in the surrounding solution. The only treatment applied to the bag was placing it in a solution of distilled water and KI. Either distilled water or KI is making the solution inside the bag turn blue. Why isn't the surrounding solution turning blue because the same treatments, distilled water and KI, are present there? The membrane isn't letting anything out of the bag. Because the bag is letting something go in but not letting something go out, it is selectively permeable, (C).

13. B

Now you know that KI is a positive indicator for starch, even if you didn't already know that from your laboratory experience, and that there is starch inside the dialysis-tubing bag. In this question, as in the previous question, you need to make a decision based on what you observe or don't observe. Choice (A) can't be correct because the surrounding solution would also be blue if starch were leaving the bag. Choices (C) and (D) are conclusions that can't be made based on given observations; you don't know if water or glucose is moving or what effects either is having on water potential. You know for certain that KI moved into the bag, so the membrane must have pores large enough to allow its passage, giving you (B).

14. B

When the system reaches equilibrium, it doesn't mean things stop moving. However, there is no net change in the concentration of any given molecule inside or outside the bag. Any answers that include net movement of something ((A), (C), and (D)) can be excluded, because the question states that the system is in equilibrium. The correct choice is (B). The water potential inside the bag has equilibrated with that in the surrounding solution, so there is no net movement of water.

15. D

ATP, not NADPH, is used for active transport. Facilitated diffusion requires protein carriers, not carbohydrate carriers. Osmosis is passive transport, but only of water. This statement for endocytosis and exocytosis is absolutely correct.

PHOTOSYNTHESIS

This Review Quiz is designed to assess your content knowledge, and is not necessarily representative of the actual AP Biology exam. So don't panic if you see a new question type or a question asking about a relatively low-yield detail from the chapter.

1. The energy needed to form ATP via electron transport phosphorylation comes from the net movement of

   (A) glucose into the cell by facilitated diffusion.

   (B) water molecules diffusing into an intracellular hypertonic environment.

   (C) oxygen diffusing out of the cell.

   (D) protons diffusing across a membrane with the concentration gradient.

2. Which of the following photopigments do not absorb light for photosynthesis?

   (A) Chlorophyll *a*

   (B) Chlorophyll *b*

   (C) Chlorophyll *e*

   (D) Carotenoids

3. CO_2 production can be used as a measure of metabolic rate because

   (A) the heat from metabolism increases its partial pressure.

   (B) CO_2 is a waste product of the catabolism of glucose.

   (C) a decreased cellular pH liberates gaseous CO_2.

   (D) plants need CO_2 to complete photosynthesis.

4. Plants with C4 photosynthesis are more efficient than C3 plants at

   (A) binding CO_2 and fixing carbon for light-independent reactions.

   (B) synthesizing carbon molecules into complex sugars.

   (C) utilizing energy from ATPases that perform photophosphorylation.

   (D) absorbing incident solar radiation.

5. Which of the following statements is true about both C4 plants and CAM plants?

   (A) They do not need ATP from the light-dependent reactions.

   (B) They have adapted systems for survival in cold, wet habitats.

   (C) They provide alternatives for CO_2 fixation.

   (D) They produce more O_2 than C3 plants.

6. What is the role of the H atoms in water during photosynthesis?

   (A) They stabilize pH in the thylakoid matrix.

   (B) They reduce the temperature of chloroplasts.

   (C) They magnify solar energy by refracting light waves.

   (D) They split and combine with CO_2 to make sugars and free O_2.

7. The product(s) of the light-dependent reactions of photosynthesis is/are

   (A) pyruvate.

   (B) glucose.

   (C) ATP and NADPH.

   (D) CO_2 and H_2O.

8. Which of the following compounds are both products of the light reaction in photosynthesis and reactants in the Calvin cycle?

   (A) NADPH and ATP

   (B) NADPH and O_2

   (C) $NADP^+$ and CO_2

   (D) $NADP^+$, ATP, and CO_2

Questions 9–10 refer to the following list.

   (A) Stomata

   (B) Cytoplasm

   (C) Chlorophyll

   (D) Thylakoid

9. Site of the light reactions

10. Regulates entrance of CO_2

11. Light energy is used to produce chemical energy via

   (A) carbon fixation.

   (B) photophosphorylation.

   (C) the Calvin-Benson cycle.

   (D) the electron transport chain.

12. Photosynthetic organisms are _____, meaning they provide food that supplies the rest of the food web.

   (A) plants

   (B) primary producers

   (C) bacteria and fungi

   (D) carbon fixing

13. The light reaction does not include

   (A) electron transport.

   (B) chemiosmosis.

   (C) CO_2 fixation.

   (D) charge separation.

14. Which of the following limit the dark reactions?

   (A) CO_2, water, and light

   (B) Oxygen, water, and temperature

   (C) CO_2, temperature, and light

   (D) Water, temperature, and CO_2

15. Where are the pigment molecules necessary for photosynthesis located?

   (A) Thylakoid membrane of the chloroplast

   (B) Mitochondria

   (C) Cytoplasm of the cell

   (D) Stroma of the chloroplast

16. The oxygen that is released as O_2 as a product of photosynthesis is derived from what molecules?

   (A) Chlorophyll

   (B) Glucose

   (C) Carbon dioxide

   (D) Water

ANSWERS AND EXPLANATIONS

1. D

At least one question about chemiosmosis or coupled reactions will most likely appear on the exam. The correct answer is (D) because the energy to form ATP comes from the proton motive force established by oxidative phosphorylation. Some answers can be eliminated to reduce your chance of error. Diffusion never needs or creates energy, even if it is facilitated, so (A), (B), and (C) can be eliminated.

2. C

This question tests your knowledge of the types of photopgiments. Chlorophyll *a* is the most common type of chlorophyll and absorbs many different colors of light but reflects back green and so (A) can be eliminated. Chlorophyll *b* is yellow-green and the carotenoids which are a range of colors supply energy to chlorophyll *a* and so (B) and (C) can be eliminated. Choice (D) is not a type of chlorophyll found in photosynthetic organisms.

3. B

If you don't remember that the products of the catabolism of glucose are CO_2 and H_2O, you may have more difficulty with this question. Plants need CO_2 for photosynthesis, but this has little connection to the metabolic rate of other organisms, so (D) can be eliminated. Choices (A) and (C) are tough because you can't use simple logic to eliminate them as options. You have to use a base of knowledge. To answer the question, you must either know that (B) is correct or know that the other answers are incorrect.

4. A

This is another knowledge-intensive question. Try to remember that C4 plants are grasses and that they live in dry, hot habitats. Also remember that C4 and C3 plants are more similar than dissimilar. Choice (D) is an advantage that a shade plant would capitalize on, not a desert plant. If you consider that most large plants and fruit-bearing plants are C3 plants, (B) can be eliminated. You are left with (A) and (C), which gives you a 50-50 chance at choosing the correct answer.

5. C

Because C4 plants are grasses and CAM plants are succulent plants such as cacti, (B), can be eliminated right away. C4 plants get their name from the process of creating a four-carbon molecule that has a higher affinity for CO_2, and CAM plants bind CO_2 at night when water loss is reduced, giving you (C).

6. D

This question goes back to the basics. The simplified formula for the function of water in photosynthesis is $CO_2 + 2H_2O \rightarrow CH_2O$ (sugars) $+ O_2 + H_2O$. Try to memorize this equation—you will find it useful for the exam. In photosynthesis, the water molecule is split and occurs in most of the products (sugars and H_2O).

7. C

The light-dependent reactions produce energy in the form of ATP and NADPH. The dark reactions make sugars.

8. A

The products of the light reaction in photosynthesis are ATP, NADPH, and O_2, but only ATP and NADPH are reactants in the Calvin cycle. Therefore, (A) is the correct answer.

9. D

The light reactions occur within the thylakoid.

10. A

Stomata regulate the entrance of CO_2 into the leaves of a plant.

11. B

The light reaction uses light energy to produce ATP and ADP. These can occur via cyclic photophosphorylation and noncylic photophosphorylation. The clue here should have been the word root *photo* meaning "light."

12. B

Primary producers are organisms that use photosynthesis to make glucose. These organisms form the base of the food chain. Plants is too general of a term, especially because there are some plants that use other means besides just photosynthesis to obtain nutrients. Bacteria and fungi are too specific.

13. C
The light reaction takes place in four distinct stages. Pigments capture the light photons then charge separation and electron transport occurs. Chemiosmosis occurs during electron transport. Carbon dioxide fixation occurs in the dark reactions, not in the light reactions, so the correct answer is (C).

14. C
The dark reaction is limited by temperature, light, and carbon dioxide. Remember, the dark reaction needs products from the light reaction to function. Therefore, light must be one of the limiting agents.

15. A
The thylakoid membrane of the chloroplast is where the pigments are located. The mitochondrion is the site of energy production but not the site of the pigments. The stroma and the cytoplasm do not contain pigments.

16. D
Think about the equation for photosynthesis. Glucose is a product along with oxygen, so it can't be the source of the oxygen molecules. Carbon dioxide and water are the reactants. The splitting of water to release energy also releases an oxygen molecule.

CELLULAR RESPIRATION

This Review Quiz is designed to assess your content knowledge, and is not necessarily representative of the actual AP Biology exam. So don't panic if you see a new question type or a question asking about a relatively low-yield detail from the chapter.

1. Which of the following is not one of the large protein complexes used by electron carriers

    (A) ubiquinone.

    (B) cytochrome oxidase.

    (C) NADH dehydrogenase.

    (D) ATP synthetases.

2. What is the function of O_2 in aerobic metabolism?

    (A) Oxidizes glucose, making it more soluble in water

    (B) Reduces enzymes, limiting glucose synthesis

    (C) Activates enzymes in the citric acid (Krebs) cycle

    (D) Accepts electrons through the electron transport chain

3. From an evolutionary perspective, the most primitive process in cellular energetics is most likely

    (A) fermentation.

    (B) the citric acid (Krebs) cycle.

    (C) the dark reactions in photosynthesis.

    (D) CAM reactions in photosynthesis.

4. Chemiosmosis is the process by which cells

    (A) obtain food by endocytosis.

    (B) use an electrochemical gradient to produce energy.

    (C) produce photopigments in plastids.

    (D) exchange complex molecules for water across the plasma membrane.

5. Pyruvate decarboxylation occurs in the

    (A) mitochondria inner membrane.

    (B) mitochondria matrix.

    (C) mitochondria cristae.

    (D) mitochondria intermembrane space.

6. How many ATP are produced via the chemiosmotic principle for every molecule of NADH that transfers high-energy electrons to the electron transport chain?

    (A) 1

    (B) 2

    (C) 3

    (D) 4

7. How many ATP are produced for every molecule of $FADH_2$ that transfers high-energy electrons to the electron transport chain?

    (A) 1

    (B) 2

    (C) 3

    (D) 4

Questions 8–9 refer to the following figure of a mitochondrion. Match each term with the correct letter.

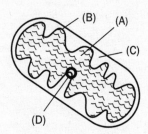

8. Site where Krebs cycle occurs

9. Contains a higher concentration of protons

10. Which of the following might interfere most directly with the process of glycolysis?

(A) A substance that closely mimics the structure of glucose but is nonmetabolic

(B) A substance that binds to oxygen and blocks it from acting as the terminal electron acceptor

(C) A compound that inactivates pyruvate by binding to it

(D) An agent that inhibits the formation of acetyl-CoA

11. Which of the following statements concerning cellular respiration is correct?

(A) The Krebs cycle directly generates ATP via substrate-level phosphorylation only.

(B) Fermentation proceeds in the absence of aerobic respiration and can work without enzymatic assistance.

(C) Eight molecules of CO_2 are produced in one turn of the citric acid cycle.

(D) Fatty acids can be β-oxidized into three-carbon units during respiration, for entry into the Krebs cycle.

12. The reaction that converts pyruvic acid into acetyl-CoA at the mitochondrial membrane is known as a(n)

(A) amination.

(B) hydrolysis.

(C) decarboxylation.

(D) phosphorylation.

13. Which of the following is true with regard to photosynthesis and respiration?

(A) NADPH produced in the dark reactions directly supplies energy for the synthesis of sugars.

(B) NADH molecules produced during glycolysis yield 3 ATP during oxidative phosphorylation.

(C) CO_2 is fixed in the light reactions of photosynthesis.

(D) Oxidative phosphorylation takes place across the inner mitochondrial membrane.

14. What is the last enzyme used during cellular respiration?

(A) Hexokinase

(B) Pyruvate decarboxylase

(C) Alcohol dehydrogenase

(D) Cytochrome C oxidase

15. The absence of which of the following substrates would not be expected to prevent oxidative phosphorylation?

(A) NAD^+

(B) O_2

(C) $FADH_2$

(D) Inorganic phosphate

16. Which one of the following is true with regard to respiration?

(A) Alcohol produced during fermentation undergoes glycolysis to produce 2 ATP.

(B) $FADH_2$ molecules yield 3 ATP each during oxidative phosphorylation.

(C) Pyruvate decarboxylation directly produces net 2 ATP.

(D) Oxidative phosphorylation takes place across the inner mitochondrial membrane.

ANSWERS AND EXPLANATIONS

1. D

The electron carriers are categorized into several large protein complexes: NADH dehydrogenase, cytochrome b-c_1 complex, cytochrome oxidase and the carrier molecule Q, also known as ubiquinone. ATP synthetases are enzyme complexes that provide specialized channels for H^+ movement. Therefore, (D) is the correct response.

2. D

This is a good question because the need for oxygen is basic to so many organisms on Earth. Recalling that an alternative name for the electron transport chain is oxidative phosphorylation may give you a hint as to the correct answer, but first let's eliminate some other answer choices. Glucose is already fairly soluble in water, so (A) can be eliminated. Oxygen is an oxidizer that doesn't reduce anything, so (B) can be eliminated. When you've shortened the list of options to (C) and (D), you have to use more of your knowledge base and a bit of logic. First of all, the question specifically asks for a function associated with metabolism. The correct choice is (D). If you can't remember that O_2 accepts electrons through the electron transport chain in aerobic respiration, you can reasonably reduce the question to a 50-50 chance. Choice (C) is associated with cellular respiration, whereas oxygen is a primary component of aerobic respiration. As such, it accepts electrons as part of aerobic respiration, giving you (D).

3. A

This question asks you to consider the world's environment over time and to synthesize this with what you know about cellular energetics. It is a good example of a question the College Board would use to test your knowledge of themes rather than concepts. Over an evolutionary timeline, the organisms that need O_2 to survive (i.e., animals) came later, so (B) can be eliminated. You may be tempted to choose either (C) or (D) because plants are more primitive, but these answers are incorrect. First of all, there was life on Earth before plants evolved (e.g., bacteria), which produced energy without photopigments. Second, two answers that pertain to the same subject can

usually be eliminated. Both (C) and (D) deal with photosynthesis. You are not likely to be expected to know the phylogeny of plant families to know whether CAM plants are primitive or derived, which eliminates (D).

4. B

Plan to see a question or two about chemiosmosis and/or coupled reactions on the exam. This is something you'll just have to memorize.

5. B

Pyruvate decarboxylation occurs in the mitochondria as all the answers suggest. However, it occurs only in the matrix and not in any other location within the mitochondria.

6. C

Three ATP are produced for every molecule of NADH that transfers high-energy electrons to the electron transport chain.

7. B

Two ATP are produced for every molecule of $FADH_2$ that transfers high-energy electrons to the electron transport chain

8. A

All cellular respiration reactions that require O_2 occur inside the mitochondrion. The Krebs (citric acid) cycle takes place in the mitochondrial matrix, indicated by (A).

9. C

The inner membrane is the site of chemiosmosis in mitochondria, so there is a proton gradient across this membrane. The electron transport chain is continually placing H^+ into the intermembrane space, so the correct answer is (C).

10. A

The enzymes that metabolize glucose in glycolysis would be stopped in their tracks by a compound that is chemically comparable to glucose yet metabolically inactive. Compounds listed in (B),

(C), and (D) would interfere in later parts of the glucose metabolism process.

11. A

It is true that the Krebs cycle generates ATP by substrate-level phosphorylation only, meaning that ADP is phosphorylated into ATP without the use of oxygen (as happens in the electron transport chain). All other statements here are incorrect regarding cellular respiration: No biosynthetic reactions, including fermentation, can occur without the help of enzymes; two molecules of CO_2 are produced in one turn of the TCA cycle; and fatty acids are oxidized into two-carbon derivatives for entry into the Krebs cycle.

12. C

Whether or not you remember that the enzyme used to convert pyruvate into acetyl-CoA at the mitochondrial membrane is called pyruvate *decarboxylase*, you should recall that this is the first reaction in cellular respiration that produces carbon dioxide as a waste product. More is produced later in the Krebs cycle. Reactions that remove carbon dioxide from a substrate are generally known as decarboxylation reactions. Aminations, (A), add amino groups ($—NH_2$); hydrolysis, (B), splits up complex biomolecules by the addition of water; and phosphorylation, (D), adds phosphorus to molecules.

13. D

Choice (A) is incorrect because NADPH is produced during the light reactions of photosynthesis. Choice (B) is tempting but is incorrect because the NADH produced during glycolysis has a net production of only 2 ATP, because 1 ATP must be expended to provide the energy for the NADH to cross the mitochondrial membranes. Choice (C) is incorrect because carbon dioxide is fixed in the dark reactions.

14. D

Oxygen is the final electron acceptor, so the enzyme must affect it directly. By thinking of things logically, you do not have to memorize every enzyme and step along the way. Even if you could not remember that the electron transport chain involved cytochrome complexes, you can still identify (D) as the only enzyme with *oxi* in the name. Additionally, all the other answers are used earlier in respiration or in side reactions.

15. A

The question is asking about which molecule is NOT necessary for the electron transport chain. NADH and $FADH_2$ are electron carriers whose electrons eventually are given to oxygen, which allows for ADP to gain another phosphate group to become ATP. Choice (A) is correct because it is the product of this reaction, not a reactant.

16. D

Choice (A) is incorrect because fermentation does not generate ATP but regenerates NAD^+. Choice (B) is incorrect because NADH is used to produce 3 ATP, and $FADH_2$ can only produce 2 ATP. Choice (C) is incorrect because pyruvate decarboxylation does not produce any ATP but serves as a link between glycolysis and the citric acid cycle.

THERMODYANMICS AND HOMEOSTASIS

This Review Quiz is designed to assess your content knowledge, and is not necessarily representative of the actual AP Biology exam. So don't panic if you see a new question type or a question asking about a relatively low-yield detail from the chapter.

1. What are the special membrane proteins that create proton channels?

    (A) Transmembrane proteins

    (B) ATPases

    (C) Sodium pumps

    (D) Calcium symporters

2. Which of the following glands contributes to thermoregulation by adjusting the resting metabolic rate?

    (A) Pancreas

    (B) Thyroid

    (C) Anterior pituitary

    (D) Adrenal

Questions 3–6 refer to the following list:

    (A) The first law of thermodynamics

    (B) The second law of thermodynamics

    (C) The third law of thermodynamics

3. This law states that heat must flow from an area of more heat to an area of less heat.

4. This law gives us the definition of absolute zero.

5. This law defines entropy.

6. This law states that energy can only change form.

Questions 7–9 refer to the following list.

    (A) Anaerobic catabolism

    (B) Aerobic catabolism

    (C) Anaerobic respiration

    (D) Aerobic respiration

7. This is the breaking down of complex substances where there is no oxygen available; also called anaerobic respiration.

8. This is glycolysis without oxygen; also called anaerobic catabolism.

9. This generates the most energy out of all of these reactions; also called aerobic respiration.

Questions 10–11 refer to the following list.

    (A) Negative feedback

    (B) Positive feedback

10. A cancer cell has uncontrolled cell division and growth. As the cell divides, it stimulates the other cells in the area to also divide. These cells then stimulate more cells to divide. What type of feedback is occurring?

11. The brain secretes the hormone ghrelin into a person's bloodstream, thereby causing an increase in appetite. The ghrelin levels in the blood increase until all the receptors are bound. The brain then responds by stopping the secretion of ghrelin. Now the person is no longer hungry. What type of feedback is occurring?

12. Which of the following is the process of maintaining a stable internal environment?

(A) Sweating

(B) Shivering

(C) Homeostasis

(D) Panting

13. Which of the following is NOT true?

(A) The larger the body, the less heat per pound that is produced.

(B) The larger the body, the less heat per pound that is lost.

(C) A small mammal produces less heat per pound than a large mammal.

(D) A large mammal produces less heat per pound that a small mammal.

14. Which of the following is NOT a homeostatic organ?

(A) Skin

(B) Small intestine

(C) Liver

(D) Kidney

15. Glycolysis and some forms of ATP creation do not involve this.

(A) Chemiosmosis

(B) Concentration gradients

(C) ATPases

(D) NAD^+

ANSWERS AND EXPLANATIONS

1. B
ATPases are specific membrane proteins that create proton channels and convert ADP to ATP when a proton passes through. These are very important for ATP production. Choice (A) is too general, and (C) and (D) are too specific.

2. B
This question asks precisely for a form of thermoregulation that involves metabolic rate. The body does regulate body temperature with the hypothalamus, but fortunately this isn't a possible response. Metabolic rate is closely tied to control by the T_3 and T_4 produced by the thyroid, and people with hypo- or hyperthyroid conditions often have trouble with body weight and body temperature fluctuations.

3. B
The second law states that heat cannot spontaneously flow from a colder location to a hotter location.

4. C
The third law of thermodynamics talks about absolute zero—the coldest possible temperature where there is no longer any heat energy in a system or substance.

5. B
Entropy, energy transferred from one form to another, is defined and used as part of the definition of the second law of thermodynamics.

6. A
The first law of thermodynamics states that energy can neither be created nor destroyed, it can only change form.

7. A
In anaerobic catabolism, there is no electron transport chain or oxygen (O_2) available to carry out a reaction.

8. C
In the absence of oxygen, only glycolysis can occur. This is anaerobic catabolism (or anaerobic respiration).

9. B
It is much more advantageous to have the ability to utilize the oxidative properties of oxygen (O_2), as reactions in aerobic catabolism do, because more energy can be generated.

10. B
Positive feedback results when the effects of feedback from a system result in an increase in the original factor that caused the disturbance. Because this cell division is just causing the other cells to divide, it is positive feedback.

11. A
The increasing levels of circulating ghrelin cause the brain to shut off ghrelin production. This is a type of negative feedback—an increase in a product causes the reaction to shut down.

12. C
Homeostasis is the process by which a stable internal environment is maintained within an organism. All the others are examples of ways in which we maintain our thermal homeostasis.

13. C
This relates to the laws about heat and body size. Choice (C) is false because a small mammal produces more heat per pound than a large mammal. A mouse releases more heat per pound than a horse.

14. B
In mammals, the primary homeostatic organs are the kidneys, the liver, the large intestine, and the skin. The small intestine is not involved in maintaining homeostasis.

15. A
Glycolysis and some other forms of ATP creation do not involve chemiosmosis.

IMMUNE RESPONSE

This Review Quiz is designed to assess your content knowledge, and is not necessarily representative of the actual AP Biology exam. So don't panic if you see a new question type or a question asking about a relatively low-yield detail from the chapter.

1. Which of the following would be LEAST able to make bacteria more virulent and infectious?

   (A) The ability to evade nonspecific and specific body defenses

   (B) The secretion of lipoteichoic acid, which contributes to septic shock

   (C) Increasing the rate of bacterial replication

   (D) Increasing the production of proteins allowing bacterial pili to extend and contact other pili

2. Which of the following is an example of passive immunity?

   (A) A nurse gets stuck with a needle containing blood from a patient infected with tuberculosis (TB), gets a brief flu-like illness, and a few years later tests positive for anti-TB antibodies.

   (B) A farmer exposed to a virus infecting some of his farm animals does not get ill when exposed to a related virus infecting his family.

   (C) An adult exposed to a certain influenza strain will not become sick again because he was exposed to that strain as a child.

   (D) A baby born to a woman who has antibodies to hepatitis may be temporarily resistant to the disease.

3. Before a specific immune response can be mounted against a foreign pathogen, the immune system launches a nonspecific defense that is active against a broad range of pathogens. Which of the following represents a nonspecific defense?

   (A) Mucous secreted by cells lining the lungs

   (B) B cell-mediated immunity

   (C) Antibody-mediated immunity

   (D) MHC-mediated hypersensitivity

4. During maturation of T cells in the thymus, large numbers of T cells that are able to bind to an individual's own MHC molecules are killed off. Which of the following is most likely to occur if such cells are allowed to survive?

   (A) T cells will launch an immune response against the individual's own cells.

   (B) T cells will be less able to recognize and bind to foreign MHC molecules.

   (C) More T cells than B cells will be produced by the individual's body.

   (D) The individual will be less susceptible to bacterial infections.

5. Which of the following would convey passive immunity?

   (A) Memory B cells formed following an infection

   (B) Maternal antibodies found in breast milk

   (C) An inactivated vaccine administered to a child

   (D) An attenuated vaccine administered to an adult

6. Which of the following is not involved in nonspecific immune defense?

(A) Skin

(B) Lysozymes

(C) Interferons

(D) B cells

7. Which of the following is responsible for the specificity of a particular antibody for a particular antigen?

(A) Constant regions

(B) Variable regions

(C) Heavy chains

(D) Light chains

8. Which of the following is not part of an inflammatory response?

(A) Increase in temperature

(B) Blood vessel constriction

(C) Blood flow increase

(D) Histamine production

Questions 9–11 refer to the following list.

(A) Specific immune response

(B) Nonspecific immune response

(C) Both specific and nonspecific immune responses

(D) Neither specific nor nonspecific immune responses

9. This response uses protein-to-protein interaction.

10. This response allows us to be immunized for a disease without actually contracting it.

11. This response involves the inflammatory response.

12. When first exposed to a microbe, antibodies are produced to slow the infection. If a second exposure occurs, the levels of antibody in the bloodstream will be

(A) lower than in the first because the body has already fought off this infection.

(B) the same as before because the body knows exactly how much antibody to produce to fight the infection.

(C) nonexistent because the body has already fought off the infection and does not need to do it

(D) again, higher than before, because the body is having a secondary response.

13. The time between exposure and B cell antibody production is called a(n)

(A) infection.

(B) primary response.

(C) secondary response.

(D) lag period.

14. B and T cells

(A) target individual microbes as part of the nonspecific immune system.

(B) work via secretion and immune system activation.

(C) do not directly kill cells, but only stimulate antibody production.

(D) function because T cells secrete antibodies while B cells kill pathogenic cells.

15. Which of the following is the classical pathway of complement activation?

(A) Complement proteins bridge the gap between adjacent antibody molecules.

(B) Complement activation occurs independently of antigen-antibody binding.

(C) Cell surface molecules of bacteria cause membrane-attack complexes.

(D) Antibodies produce several different clones of plasma B cells.

ANSWERS AND EXPLANATIONS

1. D

Of all the answer choices, the only one that does not deal with bacterial infection of other organisms' cells is (D). While the ability of bacterial cells to exchange genetic information through pili helps antibiotic resistance spread, all the other choices are more direct causes of increased virulence within a given organism. Look out for answer choices such as (D) that do not fit in with the pattern given by all the other choices.

2. D

Passive immunity results when antibodies are transferred from one individual to another. Often, this occurs when a pregnant woman passes on her own antibodies to her fetus through the placenta or to her baby via breast milk. Because antibodies are proteins, they are eventually degraded, and passive immunity may last for only a few days or weeks. All choices other than (D) are examples of active immunity, whereby one's immune system, after being exposed to a particular pathogen, mounts a response and creates memory B and T lymphocytes that can be activated very quickly in the event of re-exposure.

3. A

This question draws on your knowledge of the difference between the specific and nonspecific defenses of the body. Nonspecific defenses include physical and chemical barriers, the inflammatory response, and widely released chemicals such as cytokines. Physical barriers include the intact skin and mucous membranes. These barriers are aided by various microbe-catching fluids such as saliva and mucous.

4. A

T cells mediate rejection of foreign grafts by recognizing foreign-MHC molecules and then lysing the infected cells on which these molecules are displayed. If T cells bind to self-MHC molecules, this can result in lysis of the individual's own cells. In fact, failure to eliminate self-reactive lymphocytes is believed to be the basis for many autoimmune diseases.

5. B

Passive immunity does not involve direct infection. It is temporary and involves a transfer of antibodies. Choices (A), (C), and (D) are all active immunity caused by infections and, therefore, not correct.

6. D

This question is asking about the specific, adaptive immune system. B cells and T cells are involved in specific immune defense. None of the others listed are.

7. B

The variable regions are responsible for the specificity of a particular antibody for a particular antigen.

8. B

During an inflammatory response, blood flow is needed at the infected area. Therefore, blood vessels would not need to constrict. Instead, they would need to increase in size, or dilate. The other responses are important parts of an inflammatory response.

9. A

The specific immune system is able to attack very specific disease-causing organisms by protein-to-protein interaction.

10. A

The specific immune system "learns" and "remembers" pathogens from immunizations to fight off future exposures to the disease.

11. B

Inflammation is considered a nonspecific immune response. It is a general reaction that allows for increased blood flow to an area of injury or disease. This is not specific to a certain type of injury or pathogen.

12. D

After the first exposure to a microbe, antibodies are present in the blood to noticeably slow the infection. If you are exposed to the same antigen a second time, there is a very quick secondary response and levels of antibody in the bloodstream typically reach much higher levels than they did in the primary response.

13. D

The period after exposure to a pathogen but before helpful levels of antibodies have been made by B cells is called the lag period.

14. B

The specific immune system uses B and T cells to target individual microbes and can work by secreting antibodies into the bloodstream or by directly killing cells. While antibody secretion is the job of the B cells, direct killing of cells and overall activation of the immune system are the jobs of T cells.

15. A

The classical pathway of complement activation requires antibodies bound to antigens. Complement proteins bridge the gap between two adjacent antibody molecules and use a protein complex called the membrane-attack complex to lyse the cell membrane of the invader. Complement proteins also activate mast cells to release histamine, which brings more blood cells to the area.

CHAPTER 12: MOLECULAR GENETICS, GENE REGULATION, AND MUTATIONS

BIG IDEA 3: Living systems share, retrieve, transmit, and respond to information essential to life processes.

IF YOU ONLY LEARN FIVE THINGS IN THIS CHAPTER . . .

1. Nitrogenous base pairs make up DNA and RNA: Adenine pairs with thymine (DNA only) or uracil (RNA only) and cytosine pairs with guanine.

2. In transcription, the DNA strands separate and mRNA copies one side. The mRNA takes the information to the ribosome, where protein synthesis occurs. In translation, tRNA carries amino acids to the mRNA and assembles them into proteins based on the mRNA code.

3. Genetic expression is controlled by many factors. Operons occur in bacterial genomes. They are sets of genes that perform a function. Genes in the operon code for repressor proteins that inhibit the function of the operon. Repressor protein function is controlled by a negative feedback loop (they bind to a substrate so that the operon can do its task, which includes depleting the substrate so that repressor proteins inhibit the operon once again).

4. Mutations are the source of genetic change. Types include base-pair substitutions, which affect one amino acid, as well as insertions and deletions, which shift the genetic code and affect many amino acids.

5. Scientists can modify an organism's DNA by adding new genes.

THE DOUBLE HELIX: DNA AND THE MECHANISMS OF INHERITANCE

The mechanistic view of looking at biology at the level of molecules and cells took an abrupt historical bend with the study of the topic of heredity. The classical genetics concepts covered in this text are a more simplistic explanation for inheritance than the details of molecular genetics discussed in this chapter. Classical genetics concepts were discovered before advances in technology allowed more advanced concepts in molecular genetics to be perceived.

MOLECULAR GENETICS—THREE THEMES

There are three overall themes of molecular genetics to remember:

1. Be familiar with the structure of DNA and RNA and how these building blocks provide both a simple system for duplication and a wealth of genetic diversity that codes for variation in all existing and extinct organisms.

2. Know that ultimately DNA is a code for the translation of proteins, that these proteins impart function to all of the biological processes that make an organism alive, and that control of expression of the genes for these proteins largely controls bioactivity.

3. Before turning your study toward evolution, you must understand DNA's role as the root of genetic change in the form of mutations.

DNA and RNA Structure ★★★★

You have undoubtedly learned that **DNA** and **RNA** molecules are long strands of nucleotides linked together by their sugar-phosphate backbone between the 5′ and 3′ carbons of deoxyribose or the ribose ring. DNA has a **deoxyribose sugar backbone** while RNA has a **ribose backbone**. RNA also contains the nitrogenous base uracil instead of thymine. Both strands of complementary DNA serve as templates for duplication during mitosis and meiosis, whereas only one strand serves as a template for transcription. The simple sugar-phosphate backbone allows for exact copies of itself to be reproduced during propagation.

There are only four different nitrogenous bases that make up the nucleotides in DNA: adenine, cytosine, guanine, and thymine. These bases pair along complementary strands of DNA. Guanine pairs with cytosine, and thymine pairs with adenine. In RNA, thymine is replaced with uracil, which pairs with adenine. Therefore, four different nucleotides form a code for protein synthesis in all of the organisms that have ever existed on Earth. It is amazing that the combination of nitrogenous bases in the DNA of all organisms on Earth leads to such genetic variety.

Nucleotides are the basic building block of DNA and RNA. Millions of nucleotides make up the DNA in each cell. However, because there is so much DNA, our DNA is packaged into smaller tightly wound structures, chromosomes. Because each nucleotide has a nitrogen base (A, C, G, T) and it takes three bases to code for one amino acid, the average gene that codes for 300 amino acids is approximately 900 nucleotides long. Each chromosome may have hundreds to thousands of genes. Chromosomes are made up of millions of nucleotides wound tight and held together by histone proteins.

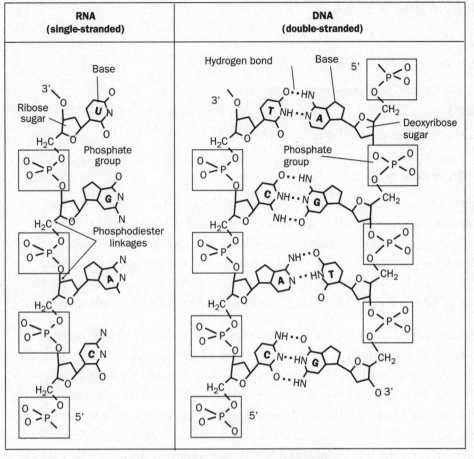

Structural Comparison of DNA and RNA

The molecular structures of the different nucleotides are outside the scope of the AP Biology exam.

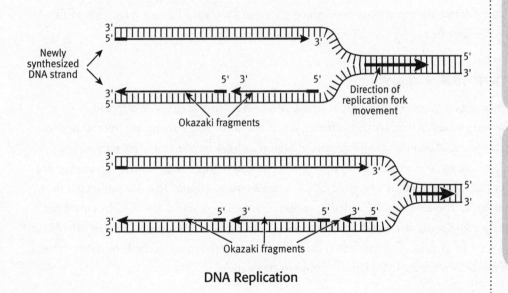

DNA Replication

KNOW THIS!

Know the basics of DNA replication, but do not get bogged down in the details.

DON'T FORGET!

Replication is always in the 5' to 3' direction—no exceptions!

FOUR TYPES OF RNA ★★★★

- mRNA: messenger RNA

- tRNA: transfer RNA; brings amino acids to ribosome

- rRNA: ribosomal RNA; component of ribosome

- hnRNA: heterogeneous nuclear RNA; synthesized from a DNA template by transcription—sometimes called pre-mRNA

REPLICATION ★★★★

It is important to understand that the strands of the DNA double helix are oriented in an antiparallel manner to each other—with one strand 5′ to 3′ and the other side 3′ to 5′. The 3′ to 5′ designations refer to the carbons that make up the sugar in each nucleotide, and the nucleotides of each strand are connected by hydrogen bonds to a nucleotide in the complementary strand. DNA polymerase II adds from the 5′ end of the incoming nucleotide to the 3′ position of the ribose in the last nucleotide on the DNA strand being synthesized.

DNA helicase unzips the DNA to ready it for replication. DNA replication is semiconservative, meaning that half of the original DNA is conserved in each daughter molecule. This means that the new DNA uses the existing DNA strands as a template. When DNA splits and copies, although the two resulting DNA strands that are created are identical to each other, one strand is made up of the "old" DNA strand while the other daughter strand is made of newly synthesized DNA. Okazaki fragments are formed because DNA polymerase can only read the older DNA in a 3′ to 5′ direction and synthesize a new strand in a 5′ to 3′ direction. It works away from the origin of replication on one side, and then jumps back to follow the unzipping DNA but can only continue to work in a direction away from the replication fork, thus leaving fragments of DNA that have to be later linked by DNA ligase.

TRANSCRIPTION ★★★★

The sequence of base pairs on the original strand of DNA is "mirrored" with a strand of **complementary bases** of mRNA (substituting uracil for thymine), starting at a special region called a promoter and ending at another special region called a terminator. This process is called transcription. The mRNA undergoes a sequence of changes before it reaches the cytoplasm. For reasons not fully understood, the original DNA contains many regions that are not coded for in protein synthesis. These regions are called introns and are excised before the mRNA enters the cytoplasm for protein translation. Exons are regions that code for protein synthesis in mRNA and are the opposite of introns. The regions on the mRNA that correspond to the introns are excised before the mRNA enters the cytoplasm for protein translation.

Transcription occurs in the nucleus. The DNA helix is unwound at the region of transcription. RNA polymerase II binds at the DNA promoter site and builds in hnRNA in the 5′ to 3′ direction. Then the hnRNA is processed by removing introns and adding the 5′ methylguanine cap and poly A tail to complete the mRNA. The mRNA then exits via the nuclear pores and goes to the ribosomes in the cytoplasm for translation.

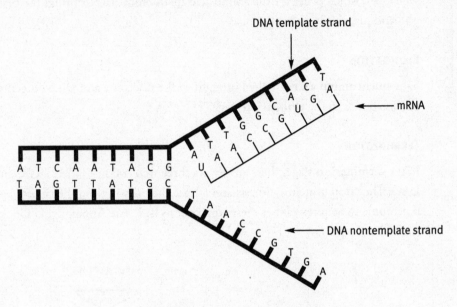

Process of Transcription

PROTEIN TRANSLATION ★★★★

In a eukaryotic organism, DNA is found in the **nucleus** of the cell. **Protein synthesis** takes place in the cytoplasm, so information from the coding regions of DNA in the nucleus must be moved to the cytoplasm. Information is moved to the cytoplasm through **mRNA (messenger RNA)**.

Once in the cytoplasm, mRNA acts as the template for protein translation. A short series of three bases, called a triplet or codon, codes for a tRNA (transfer RNA) that carries a specific amino acid. There are 64 (4^3) possible triplets, but there are only 20 amino acids. Even allowing for the start signal codon AUG as a site to begin protein translation and several codons acting as stop signals, there is considerable redundancy in the genetic code. This redundancy is sometimes described as being degenerate. Once translation is initiated at an AUG triplet, the protein continues to grow through elongation. The ribosome performs protein synthesis. Once an mRNA enters the cytoplasm, several ribosomes attach to it, creating multiple lengths of the protein simultaneously. Elongation continues until a stop codon is reached and the ribosome disengages the mRNA. The newly synthesized polypeptide usually undergoes transformation (i.e., removal of terminal amino acids) before achieving its active state.

STAGES OF TRANSLATION ★★★★

INITIATION

The first stage of translation is initiation. During this stage, the mRNA attaches itself to the ribosome. The first codon on the mRNA is always AUG (the start sequence), which codes for the amino acid methionine. tRNA brings the first amino acid and places it in its proper place.

ELONGATION

Subsequent amino acids are then brought to the ribosome and joined together by peptide bonds in the **elongation** stage.

TERMINATION

In the **termination** stage, there are always three stop codons: UAA, UGA, and UAG. They stop protein synthesis and release the protein from the ribosome. A mnemonic to help remember these stop codons is: U Are Annoying, U Go Away, U Are Going.

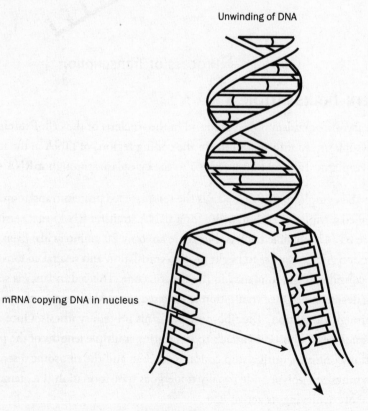

Unwinding of DNA

mRNA copying DNA in nucleus

Translation

CONTROL OF EXPRESSION ★★★★

Even a simple bacterium contains thousands of genes. If all of these genes were expressed (transcribed and translated into protein) all of the time, the cell would never be able to obtain enough resources to stay alive, not to mention the confusion that all of those proteins floating around would cause. Prokaryotes and single-celled organisms have a relatively easy time controlling the differential expression of their genomes compared with multicellular organisms that have the same genome in every cell, yet carry out very specific and different functions in different structures. How does a cell in the eye of an eagle express the proper genes to make proteins for eye development and function, while a cell in the eagle's pectoral muscle expresses proteins to build muscle tissue and use of energy? Both cells have the same genes. The answer is only partially understood, even with modern technological advances in genomics and proteomics. Furthermore, the intricacies of gene expression are too much to cover in just a few pages of this test-prep book.

The majority of what is known about the control of **gene expression** is at the **transcriptional level**. Expression of genes takes place in both prokaryotes and eukaryotes, but eukaryotes have more levels of control because of the separation of DNA within their nuclear membrane. The operon model of transcription is an important system to know. Operons appear in certain bacterial genomes.

An **operon** is a series of genes that includes a promoter and a terminator that synchronize to perform a biological function. The operon consists of the structural genes that code for proteins for other functions, but may also contain genes that code for repressor proteins that inhibit the transcription of the operon by blocking a region of the operon called the operator. The **repressor protein** is usually removed from its blocking position by the presence of a **substrate** (a sugar), which allows the structural genes to be transcribed and translated. The structural genes then perform some action on the substrate, making it unavailable. The repressor again blocks the expression of that operon. This is a classic example of a negative feedback loop. After all, there is no need for the proteins until the substrate is available. This is also an example of the control of gene expression by induction because the expression of the genes is induced by the presence of some substrate.

PROMOTERS, ENHANCERS, AND TRANSCRIPTION FACTORS ★★★★

The major difference between prokaryotic genes and eukaryotic genes is the presence of introns. Another difference is the presence of **enhancers**, noncoding regions of DNA that influence the activation of genes. These enhancers are often located a considerable distance upstream from the genes they exert control over.

When eukaryotic DNA is transcribed, RNA polymerase must bind to a promoter sequence ahead of the structural genes to be "read," just like in prokaryotes. The enhancers are regions of DNA that bring certain **transcription factors** into contact with the promoter regions of genes and act to "enhance" transcription. Experimental removal of enhancers can cause drastic decreases in gene transcription. It is thought that these enhancers, even though they may be thousands of nucleotides away from the structural genes they affect, are brought into contact with those genes or with their promoter sequence by a loop in the DNA structure.

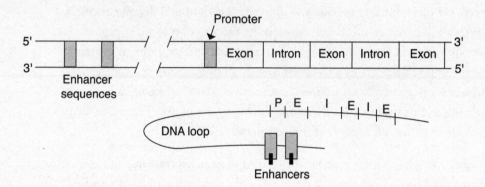

PROMOTERS ★★★

One of the most common elements in eukaryotic promoters is the **TATA box**. Although not present in "housekeeping genes" and other developmental genes, such as the homeotic genes (covered in later chapters), TATA boxes are A-T-rich regions of DNA that are involved in positioning the start of transcription. The reason for this is that regions of DNA rich in adenine and thymine tend to separate more easily than those rich in C's and G's. Adenine and thymine form only two hydrogen bonds across the double helix, while C and G form three. Separation of DNA at the TATA boxes in the promoter regions of genes allows DNA to unzip at those regions for RNA polymerase to access the DNA template. Other types of promoters known as *internal promoters* also exist that can be found within the introns of genes. These promoters occur especially in genes that encode rRNA and tRNA molecules.

TRANSCRIPTION FACTORS ★★★★

There are hundreds or thousands of proteins that exert transcriptional control over the genome, and they are known as transcription factors. Some bind to enhancer sequences, others to promoter sequences. All transcription factors help RNA polymerase find and bind to a given promoter region. Each of these proteins has a highly conserved DNA-binding domain, which allows it to attach to DNA nucleotides. Most are specific for certain enhancer or promoter sequences.

IMPORTANT

The important thing to remember is that genes are always present in the DNA, but certain environmental variables affect their expression.

Some other forms of gene regulation are those that are based on an internal clock or synchronized with other cellular events.

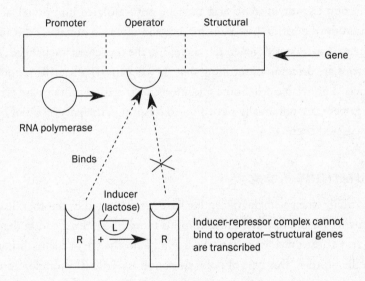

The *Lac* Operon in *E. coli*

In this illustration, a protein repressor binds to the operator in the absence of lactose in the growth medium. When lactose (the inducer) is present, it binds to the repressor and disengages from the operator. The transcription of the operon by RNA polymerase then commences. The transcribed mRNA codes for the proteins that metabolize lactose, a process which eventually leads to the removal of lactose from the medium. When lactose is gone, the repressor binds to the operator again and the operon is back to square one.

Another important concept to appreciate from this model of gene expression is that the promoter and operator of an operon are "unaware" of what happens at another part of the DNA strand, unless the DNA performs a specific activity on them. For example, if the gene to produce insulin were placed within the operon to produce the enzymes that metabolize glucose in a bacterium, the insulin gene would be translated as well as the glucose metabolizing genes if the bacterium were placed in a medium with glucose. This type of genetic manipulation has revolutionized the discipline of molecular genetics.

KINDS OF GENETIC CHANGE ★★★★

The goal of a cell is usually to maintain the same code in its DNA, base pair by base pair. Sometimes changes occur spontaneously during replication or are due to environmental factors such as irradiation. These changes in the DNA code are called **mutations**. Mutations can involve a change in only one base or several bases. Mutations are also placed in one of two categories, base-pair substitutions or insertions and deletions. **Base-pair substitutions** occur when one base pair is incorrectly reproduced and exchanged with a different base pair. An **insertion** is when any number of extra base pairs are added to the code, and a **deletion** is when any number are removed from the code.

The effect of mutations is dependent upon where they occur in the code. Mutations of introns often have no effect on an individual because they are not translated into proteins. Likewise, base-pair substitutions in third position base pairs for a codon often do not affect the amino acid it codes for (remember, the code is degenerate); therefore, the protein is unchanged. Most mutations in structural proteins are deleterious, meaning that they negatively affect the nature of the protein and usually produce a nonviable cell. Some mutations in structural proteins are viable and, if the mutation is in a gamete, are potentially passed on to offspring. The passing on of mutations to offspring will be covered later.

TYPES OF MUTATIONS ★★★★

Point mutations occur when a single nucleotide base is substituted by another. If the substitution occurs in a noncoding region, or if the substitution is transcribed into a codon that codes for the same amino acid as the previous codon, there will be no change in the resulting amino acid sequence of the protein. This type of point mutation is a "silent" mutation. However, if the mutation changes the amino acid sequence of the protein, the result can range from an insignificant change to a lethal change depending on where the alteration in amino acid sequence takes place.

Frameshift mutations involve a change in the reading frame of an mRNA. Because ribosomes and tRNAs "read" the mRNA in sections of three bases (codons), if a base is inserted or deleted due to faulty transcription or a mutation in the actual DNA, the reading of the resulting mRNA will shift, and this is called a frameshift mutation. Base insertions and deletions, particularly toward the start of the protein's amino acid sequence, can render the remaining structure nonfunctional as almost every amino acid along the sequence gets changed.

Nonsense mutations produce a premature termination of the polypeptide chain by changing one of the codons to a stop codon. Beta-thalassemia is a hereditary disease in which red blood cells are produced with little or no functional hemoglobin for oxygen carrying. The different forms of this disease can be produced by a variety of mutations, including point mutations, frameshift mutations, and nonsense mutations.

The following diagram is useful for understanding point mutations.

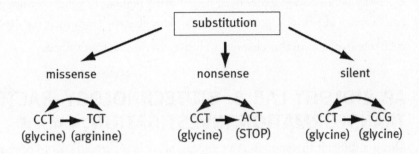

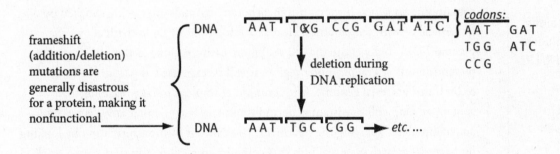

CHROMOSOME ALTERATIONS ★★★

Several types of chromosomal breakage can occur in the course of DNA replication or at other points in the cell cycle. Some of these alterations can cause drastic changes in the ability of a cell to produce certain proteins. Many birth defects can be traced to defects in chromosome structure passed down through sperm or egg cells in which one of the following alterations has taken place:

Deletion is when a chromosomal fragment, either from the end of a chromosome or from somewhere in the middle, is lost as the chromosome replicates.

If the fragment that detaches from a chromosome during a deletion event reattaches itself to the homologous chromosome, that chromosome will then have two sets of identical genes in a particular region (a **duplication**), while the original chromosome where the deletion occurred will be shorter than normal.

There is also the possibility that the deleted fragment could attach back into the chromosome from which it came, but in a reverse direction. This is known as **inversion**.

Translocation is a common alteration in which the piece of DNA that breaks off a chromosome attaches to the end of another chromosome, most commonly not the homologous chromosome. In some cases, there is a reciprocal translocation, in which the chromosome giving a piece of DNA also receives one back from the receiving chromosome that is comparable in size.

In many cases, these alterations in structure render large groups of genes useless, especially because most genes need neighboring regulatory genes to work properly. After certain alterations, genes may be far enough away from their regulatory elements (see later in this chapter) that they cannot be transcribed.

AP BIOLOGY LAB 8: BIOTECHNOLOGY–BACTERIAL TRANSFORMATION INVESTIGATION ★★★★

This is a difficult investigation to conceptualize. It is also difficult to perform because it requires a considerable amount of specialized materials, expensive equipment, and aseptic conditions. Some companies have even prepared software that replaces the lab with a virtual experiment.

This investigation deals with the transformation of bacteria using plasmids that contain known genes. Bacteria can incorporate plasmids after being shocked by a chemical or with temperature. Heat shock followed by an ice bath allows the plasmid DNA to penetrate the cell wall more easily. As long as the gene from the plasmid inserts after a promoter region, it will be expressed as part of the new, recombined bacterial genome. The common example is to take plasmids that contain an ampicillin-resistant gene and transform bacteria that are affected by ampicillin. After applying the treatment, you can test for transformation by looking for bacterial growth on a medium that contains ampicillin. You start with a stock of ampicillin-sensitive bacteria and a stock of plasmid containing the gene for ampicillin resistance. You apply the shock treatment to the bacteria and add the plasmids, then apply the bacteria to growth media with and without ampicillin. Only bacteria that have incorporated the ampicillin-resistant gene from the plasmid into their genome will survive in the medium containing ampicillin. Your goal in this part of the laboratory is to see firsthand the products of recombination and transformation that you have already learned on paper. You will see some examples of how this knowledge will be tested in the sample questions.

AP BIOLOGY LAB 9: BIOTECHNOLOGY: RESTRICTION ENZYME ANALYSIS OF DNA INVESTIGATION

This experiment involves splicing DNA using restriction enzymes and visualizing these fragments using gel electrophoresis. A gel is a matrix composed of a polymer that acts like filter paper. Think of the gel as a thick, tangled jungle. DNA fragments are negatively charged, so they will move away from the negative pole of an applied current and move toward a positive pole. As the DNA fragments move away from the negative terminal of the charge applied, they get tangled in the jungle. The smaller the fragments of DNA, the faster they can move through the tangle. The movement of the fragments can be adjusted by changing the density of the gel (the tangles in the jungle) or by changing the quantity of charge (the pushing/pulling power) applied to the gel.

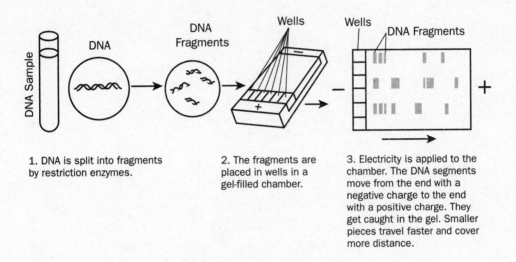

1. DNA is split into fragments by restriction enzymes.

2. The fragments are placed in wells in a gel-filled chamber.

3. Electricity is applied to the chamber. The DNA segments move from the end with a negative charge to the end with a positive charge. They get caught in the gel. Smaller pieces travel faster and cover more distance.

Diagram of Gel Electrophoresis

Restriction enzymes are manufactured by molecular genetics laboratories to have a high specificity to a series of base pairs on a strand of DNA or RNA. When the enzyme finds and binds to these base pairs, it snips the DNA. The more this series of base pairs appears on the DNA or RNA strand, the more cuts the restriction enzyme makes in the strand. After the strand is digested by the restriction enzyme, the fragments with different lengths can be separated using gel electrophoresis. This technique is commonly used to determine genetic differences between organisms, as long as the appropriate region of DNA can be isolated and the right combination of restriction enzymes can be determined. The most important thing to remember from this lab is that smaller fragments move farther on the gel. They move farther *because* they are smaller. All of the fragments move for the same reason: They are charged molecules moving away from an applied charge of the same kind (positive or negative).

CHAPTER 13: THE CELL CYCLE

BIG IDEA 3: Living systems share, retrieve, transmit, and respond to information essential to life processes.

> ## IF YOU ONLY LEARN TWO THINGS IN THIS CHAPTER . . .
>
> 1. Mitosis is the process in which a cell produces two identical daughter cells. Its stages include interphase, prophase, metaphase, anaphase, and telophase. Cytokinesis occurs immediately following mitosis and refers to the splitting of the cell into two new cells.
> 2. Meiosis refers to the process by which sexually reproducing organisms produce sex cells (gametes) with half the chromosomes (haploid) of the rest of the organism's cells (which are diploid). It has two stages, meiosis I and meiosis II, and results in the creation of four gametes. In sexual reproduction, the male and female gametes join to create a new organism with the normal number of chromosomes.

MITOSIS AND MEIOSIS OVERVIEW ★★★★

Mitosis results in the production of identical cells for growth or, in the case of single-celled organisms, **asexual reproduction**. In **sexual reproduction**, two parent organisms each contribute a cell, which combine to form an offspring that shares half of each parent's DNA. Sexual reproduction can also occur when a hermaphroditic organism fertilizes its own eggs to form offspring.

An organism that has two sets of **chromosomes** is said to be **diploid** (designated as $2N$). A diploid organism has one set of chromosomes from each parent, for a total of two sets. When a cell has one set of chromosomes, which is half the number of chromosomes that a diploid cell contains, it is said to be **haploid** ($1N$ or N). In some cases, cells have several sets of chromosomes and are called **polyploid** ($3N$, $4N$, etc.). There are a few organisms that exist in a natural state with haploid or polyploid cells, but most organisms are diploid.

When diploid organisms sexually reproduce, one of their own cells is combined with a cell from the other parent. These cells, called **gametes**, are haploid before they combine. When gametes combine, they form a diploid offspring. The formation of these special haploid cells (gametes) is called **gametogenesis**. The process that forms cells with one set of chromosomes (haploid cells) is called **meiosis**.

When considering meiosis for the AP Biology exam, you are most likely to encounter questions that compare meiosis with mitosis. Meiosis occurs in two steps, **meiosis I** and **meiosis II**. Meiosis I is different from meiosis II; meiosis II is essentially mitosis with half the number of chromosomes. The differences between meiosis I, meiosis II, and mitosis are shown in the following tables and the figure. Between the illustration and the information in the table, you should have enough repetition to nail down the differences between mitosis and meiosis before going on to the practice questions.

DIPLOID, HAPLOID, AND POLYPLOID

Ploidy	Number of Chromosomes	Designation	Examples
Diploid	2 of each, 1 from each parent	$2N$	Most organisms that reproduce sexually
Haploid	1 of each	N	Some plants
Polyploid	More than 2 of each	$3N$, $4N$, $5N$, etc.	Most bees, some fungi, some protists

COMPARISON OF MITOSIS AND MEIOSIS

	Number of Chromosomes in Parent Cell	Prophase/ Prophase I	Metaphase/ Metaphase I	Anaphase/ Anaphase I	Number of Daughter Cells	Number of Chromosomes in Daughter Cells
Mitosis	$2N$	Replicated chromosomes come into view as sister chromatids	Individual chromosomes align at metaphase plate	Centromeres separate and sister chromatids travel to opposite poles	2	$2N$
Meiosis	$2N$	Chromosomes form tetrads by synapsis; crossing over at chiasmata	Pairs of homologous chromosomes align at metaphase plate	Synapsis ends and homologous chromosomes travel to opposite poles; sister chromatids travel to same pole	4	N

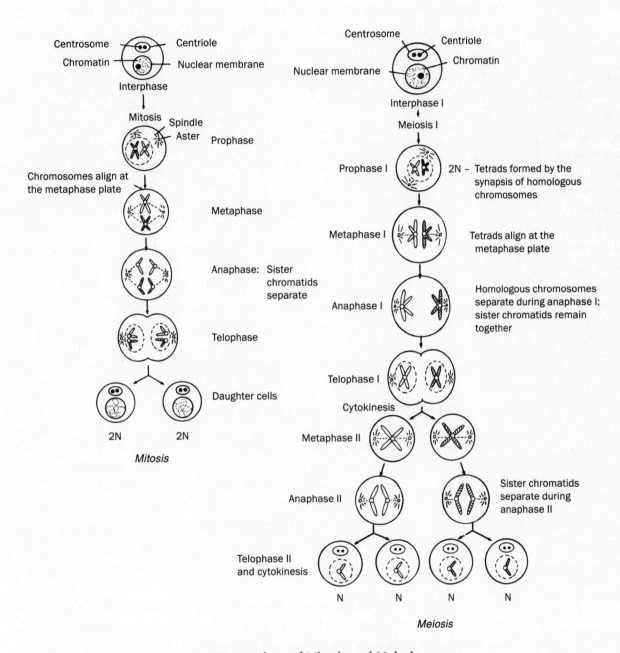

Comparison of Mitosis and Meiosis

The ultimate goal of meiosis is to produce cells with half the number of original chromosomes so that two cells with half the number of chromosomes can combine to create offspring with a complete set of chromosomes. Later, you will review how chromosomes sort to provide genetic material from each parent to the offspring.

THE CELL CYCLE AND MITOSIS ★★★

The cell cycle is the life cycle of the cell, including reproduction. **Mitosis** is the process through which a cell replicates and divides. The following two figures provide details on these two processes. The first figure shows the cell cycle, with a table following it that describes each phase of the cycle.

S = DNA replicates
G2 = Gap 2 (cell gets ready to divide)
M = Mitosis (cell division)
G1 = Gap 1 (cell grows)

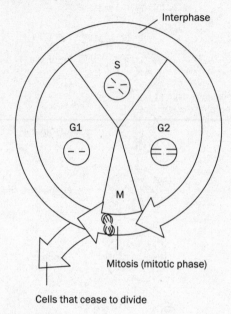

The Cell Cycle

Interphase (> 90% cell cycle)	G_1 (first gap phase)	Cell growth, protein and RNA synthesis
	S (synthesis phase)	DNA synthesis, duplication
	G_2 (second gap phase)	Cell growth, protein synthesis
M phase (mitotic phase)		Cell division

The next figure provides details on each phase of mitosis, or nuclear division. The cell starts out in **interphase**, then proceeds to **prophase**, **metaphase**, **anaphase**, and **telophase**. **Cytokinesis** is the division of the cytoplasm in animal cells following division of the nucleus, when two distinct cells are produced.

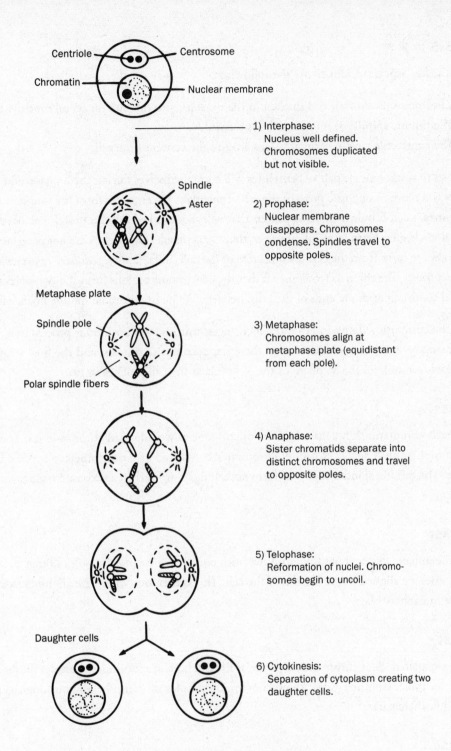

Centriole — Centrosome
Chromatin — Nuclear membrane

1) **Interphase:**
   Nucleus well defined.
   Chromosomes duplicated
   but not visible.

Spindle
Aster

2) **Prophase:**
   Nuclear membrane
   disappears. Chromosomes
   condense. Spindles travel to
   opposite poles.

Metaphase plate
Spindle pole

3) **Metaphase:**
   Chromosomes align at
   metaphase plate (equidistant
   from each pole).

Polar spindle fibers

4) **Anaphase:**
   Sister chromatids separate into
   distinct chromosomes and travel
   to opposite poles.

5) **Telophase:**
   Reformation of nuclei. Chromosomes begin to uncoil.

Daughter cells

6) **Cytokinesis:**
   Separation of cytoplasm creating two
   daughter cells.

Mitosis and Cytokinesis

Cell division is primarily controlled through genetics. The DNA in a cell controls whether a cell divides at all and at what rate. Certain hormones or chemicals need to be present for a cell to divide. Cell division can be controlled in tissue cultures by inhibiting protein synthesis or affecting nutrient availability. Cell division is reduced by crowding through contact inhibition.

MITOSIS ★★★

Among the key aspects of mitosis are the following:

- Chromosomes shorten and thicken in the nucleus, and the nuclear membrane dissolves.
- The mitotic spindle of microtubules is formed.
- The contractile ring of actin develops around the center of the cell.

For mitosis to work, a single pair of **centrioles** will copy themselves during the S phase, and the two pairs will move to opposite poles of the cell. These pairs of centrioles form the foundation for **centrosomes,** microtubule organizing centers that will shoot linked tubulin proteins across the cell as mitosis begins. Keep in mind, however, that the centrioles themselves are not necessary for microtubules to form from the centrosome areas of the cell. In fact, *plant cells have centrosomes without centrioles.* The thing to remember is that the centrosome regions form the two poles (like north and south) on opposite ends of the cells, between which microtubule spindle fibers will form.

Despite the conventional division into distinct stages, mitosis is a continuous process that does not stop between each phase. Four of these stages comprise mitosis, and the final stage—**cytokinesis**—completes the M phase of the cell cycle as the cell pinches in two.

PROPHASE

In prophase, chromatin shows up under the microscope as well-defined chromosomes. These chromosomes are an X shape, two sister chromatids connected by a centromere, a specific DNA sequence. The mitotic spindle begins to form and elongate from the centrosome regions during prophase.

METAPHASE

During metaphase, kinetochore microtubules push equally from opposite poles so that chromosomes are aligned in the middle of the cell. This center area where the alignment occurs is called the metaphase plate.

ANAPHASE

In anaphase, paired sister chromatids separate as kinetochore microtubules shorten rapidly. The polar microtubules lengthen as the kinetochore microtubules shorten, thereby pushing the poles of the cell farther apart.

TELOPHASE

During telophase, separated sister chromatids group at opposite ends of the cell, near the centrosome region, having been pulled there by the receding microtubules. New nuclear envelope re-forms around each group of separated chromosomes. At this point, mitosis has ended.

CYTOKINESIS

In cytokinesis, the contractile ring of actin protein fibers shortens at the center of the cell. A cleavage furrow, or indentation, is created as the ring contracts. As one side of the cell contacts the other, the membrane pinches off and two cells exist where one did before.

MEIOSIS ★★★

Remember that meiosis takes place only in areas where gametes are produced (the testes and ovaries in mammals, for example). The first round of cell division (meiosis I) produces two intermediate daughter cells, while the second division (meiosis II) involves the separation of sister chromatids (copied chromosomes), similar to what happens in mitosis but with haploid gametes as the result. Each meiotic division has the same stages as mitosis.

MEIOSIS I

Prophase I

The chromatin condenses into chromosomes, the spindle apparatus forms, and the nucleoli and nuclear membrane disappear. Homologous chromosomes (chromosomes that code for the same traits, one inherited from each parent) come together and intertwine in a process called synapsis. Because at this stage each chromosome consists of two sister chromatids, each synaptic pair of homologous chromosomes contains four chromatids and is therefore often called a tetrad. Sometimes chromatids of homologous chromosomes break at corresponding points and exchange equivalent pieces of DNA; this process is called crossing over. Note that crossing over occurs between homologous chromosomes, and not between sister chromatids of the same chromosome. (The latter are identical, so crossing over would not produce any change.) Those chromatids involved are left with an altered but structurally complete set of genes. The chromosomes remain joined at points, called chiasmata, where the crossing over occurred. Such genetic recombination can unlink linked genes, thereby increasing the variety of genetic combinations that can be produced via gametogenesis. Recombination among chromosomes results in increased genetic diversity within a species. Note that sister chromatids are no longer identical after recombination has occurred.

Metaphase I

Homologous pairs (tetrads) align at the equatorial plane, and each pair attaches to a separate spindle fiber by its kinetochore.

Anaphase I

The homologous pairs separate and are pulled to opposite poles of the cell. This process is called disjunction, and it accounts for a fundamental Mendelian law: independent assortment. During disjunction, each chromosome of paternal origin separates (or disjoins) from its homologue of maternal origin, and either chromosome can end up in either daughter cell. Thus, the distribution of homologous chromosomes to the two intermediate daughter cells is random with respect to parental origin. Each daughter cell will have a unique pool of alleles (genes coding for alternative forms of a given trait; e.g., yellow flowers or purple flowers) from a random mixture of maternal and paternal origin.

Telophase I

A nuclear membrane forms around each new nucleus. At this point, each chromosome still consists of sister chromatids joined at the centromere. The cell divides (by cytokinesis) into two daughter cells, each of which receives a nucleus containing the haploid number of chromosomes. Between cell divisions, there may be a short rest period, or interkinesis, during which the chromosomes partially uncoil.

MEIOSIS II

This second division is very similar to mitosis, except that meiosis II is not preceded by chromosomal replication.

Prophase II

The centrioles migrate to opposite poles, and the spindle apparatus forms.

Metaphase II

The chromosomes line up along the equatorial plane. The centromeres divide, separating the chromosomes into pairs of sister chromatids.

Anaphase II

The sister chromatids are pulled to opposite poles by the spindle fibers.

Telophase II

A nuclear membrane forms around each new (haploid) nucleus. Cytokinesis follows and two daughter cells are formed. Thus, by the completion of meiosis II, four haploid daughter cells are produced per gametocyte. (In women, only one of these becomes a functional gamete.)

AP BIOLOGY LAB 7: CELL DIVISION: MITOSIS AND MEIOSIS INVESTIGATION ★★★★

In Part 1 of this investigation, you will review and model mitosis and crossing over using simple craft materials. Though this part of the investigation is simple and straightforward, it will help you to recognize, through hands-on repetition, how genetic material is transferred to further generations. Genetic material in cells is in chromosomes that must be condensed to be easily and safely duplicated and moved. This experiment will also help you to differentiate between the sister chromatids, tetrads, and chromosomes and where the genetic material is located.

The next part of the lab provides some hands-on experience working with real organisms. You will identify different cell phases in onion bulbs. Working with squashes of those onion bulbs, you will explore what substances in the environment might increase or decrease the rate of mitosis. The data that is collected can be statistically analyzed by calculating chi-square values.

The chi-squared analysis measures the difference between the number of observations you make that meet your expectations and the number that don't. You put these values into a formula and compare the answer with a table of standards. The formula is:

$$\chi^2 = \sum \frac{(\text{Observed} - \text{Expected})^2}{\text{Expected}}$$

As an example, look at the frequencies observed if you toss a coin 100 times. You observe that the coin comes up heads 52 times and comes up tails 48 times. Because there is an equal likelihood that the coin will come up heads or tails, the expected values for heads and tails are 50 and 50. Putting these values into the formula:

$$\chi^2 = \frac{(52 - 50)^2}{50} + \frac{(48 - 50)^2}{50} = 0.16$$

Compare the value 0.16 to a chi-squared table at one **degree of freedom (d.f.)**. One degree of freedom is used because two observations were made and d.f. equals number of categories of observations minus one.

Probability (p)	Degrees of Freedom (d.f.)				
	1	2	3	4	5
0.05	3.84	5.99	7.82	9.49	11.1

At $p = 0.05$ in the chi-squared table, the value obtained from the previous equation would have to be at least 3.8 for your observations to be statistically different from what is expected. The value calculated was only 0.16, so it can be concluded that although your observations weren't exactly $\frac{50}{50}$, they were not statistically significantly different from the expected outcome. There are many considerations and assumptions to make when performing statistical analyses, which would be covered in greater depth in a class on statistical analyses. For the chi-squared analysis, there are three key points to remember:

- The formula for the chi-squared statistic: $\chi^2 = \sum \frac{(\text{Observed} - \text{Expected})^2}{\text{Expected}}$

- The formula for degrees of freedom: d.f. = # of categories of observations − 1

- Your calculated χ^2 must be greater than the corresponding p value in the table for the outcome to be significantly different than what you expected.

Part 3 of the lab asks you to explore the differences between normal cells and cancer cells. Remember that cancer typically results from a mutation in a gene or protein that controls the cell cycle. Comparing pictures of chromosomes from normal and HeLa cells, you will be able to count the number of chromosomes found in each type of cell and identify differences in appearance. The main conclusion to draw from this part of the investigation is that in normal cells, division is blocked if there is damage to or mutations in DNA. In contrast, this ability to block division is lacking in cancer cells, and they divide uncontrollably.

Finally, you will review meiosis and nondisjunction events by modeling. Based on this review, you will measure crossover frequencies and genetic outcomes in a fungus. This part of the investigation will also give you solid experience with the microscope by preparing slides and counting asci under the microscope. Though you will not need to use a microscope or describe how to use a microscope on the AP Biology exam, these experiences are helpful for answering short and long free-response questions, which often ask you to describe experimental setups and procedures.

CHAPTER 14: THE CHROMOSOMAL BASIS OF INHERITANCE

BIG IDEA 3: Living systems store, retrieve, transmit, and respond to information essential to life processes.

IF YOU ONLY LEARN THREE THINGS IN THIS CHAPTER . . .

1. Inheritance patterns include incomplete dominance, polygenic inheritance, genetic recombination, and gene transfer.

2. Chromosomes are the basic units of inheritance. How they reorganize and combine directly influences the genetic material present in an offspring.

3. Sex-linked genes occur on the X or Y chromosome. Males who inherit recessive X-linked genes from their mothers always express the trait in question, as men have only one X chromosome.

THE ROOTS OF THE FAMILY TREE: UNDERSTANDING INHERITANCE ★★★★

Even if you never took a biology class, it's hard not to notice that children look more like their parents than other adults. The passing on of characteristics or traits is controlled by the genetic information contained within the DNA in cell nuclei. The concepts in this chapter include classical genetics, the principles of inheritance through gamete formation, and sexual reproduction. Many of these concepts rely more on sequential logic and are less memory intensive, which is helpful to know when studying for the AP Biology exam. However, these concepts will also build upon those learned in the previous chapter on mitosis and meiosis.

The division and duplication of cells through the process of mitosis was previously reviewed. Mitosis results in the production of identical cells for growth or, in the case of single-celled organisms, reproduction. In sexual reproduction, two parent organisms each contribute a cell.

KEY POINT

The chromosomal basis of heredity is an important theme on the AP Biology exam, make sure you are prepared!

GENES

Instead of thinking of a gene as the DNA itself, the term *gene* can be associated with a location on the chromosome.

These cells combine to form an offspring that shares half of each parent's DNA. Sexual reproduction can also occur when a hermaphroditic organism fertilizes its own eggs to form offspring.

EUKARYOTIC CHROMOSOMES ★★★★

Chromosomes are condensed bodies of DNA molecules that store codes for the translation of several different kinds of proteins. These proteins dictate how an organism is put together and functions. In the most simplified way of looking at the **chromosomal theory of inheritance**, each section of DNA that translates a different protein is called a **gene**. A region of DNA may code for a protein that controls eye color, so this region of DNA is called the gene for eye color.

Alleles specifically code for the traits of an organism. In the case of eye color, there might be an allele for blue eye color and an allele for green eye color. The eye color that an organism ends up with (its **phenotype**) is dictated by which allele is placed in the gene location for eye color (its **genotype**). Sometimes an organism might have a genotype for one allele and express the phenotype of another.

In a typical diploid ($2N$) eukaryotic organism, each cell has two copies of each chromosome. These pairs of chromosomes are a result of the combination of two haploid gametes formed by meiosis, one from the mother and one from the father. The mother and father each provided one allele for each gene, leaving the offspring with either two of the same allele (e.g., two for blue eye color or two for green eye color) or one of each allele (e.g., one allele for blue eye color and one allele for green eye color). In most genes, there is one allele that is **dominant** over the other and hides the expression of that other allele in the phenotype of the offspring. The other allele is called **recessive**. Geneticists record the alleles for genes with uppercase, italicized letters for dominant alleles (*B*) and lowercase, italicized letters for recessive alleles (*b*). If blue eye color is the dominant allele (indicated as *B*) and green eye color is the recessive allele (indicated as *b*), an offspring could have one of three different genotypes (*BB*, *Bb*, and *bb*) when its parents' gametes combine. The phenotype of an offspring with the genotype *BB* is blue eyes. An offspring with the genotype *Bb* will also have blue eyes because the blue eye color allele is dominant. Only offspring with the genotype *bb* will have a green eye phenotype. The following figures clarify this information.

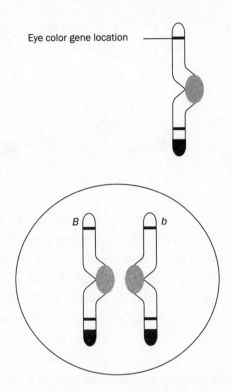

The model shown here is a cell from an animal with one allele for blue eye color (*B*) and one allele for green eye color (*b*). This animal's genotype for eye color is *Bb*, and its phenotype is blue eyes because the blue eye color allele is dominant. The chromosomes are essentially identical in appearance, but the alleles have a slightly different DNA code.

GENES

This model indicates the location of the gene for eye color on an animal chromosome. Each animal has two of each of these chromosomes, one from its mother and one from its father.

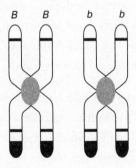

The previous figure shows chromosomes after interphase I of meiosis. The DNA strands have been doubled so that there is an exact duplicate of each original chromosome linked by a centromere to sister chromatids. There are now two copies of the *B* allele and two copies of the *b* allele.

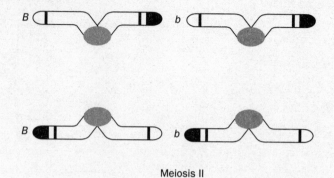

Meiosis I   Meiosis II

After the cell undergoes complete meiosis, the original cell with $2N$ chromosomes has divided into four daughter cells, each with $1N$ chromosomes. The daughter cells are four gametes, two with the B allele and two with the b allele. Because the original animal cell had both a B allele and a b allele, it can produce gametes with alleles for either blue or green eye color. This occurs in all individuals of the same species. An individual with the recessive green eye color (its genotype is bb) can only produce gametes with b alleles. When an organism has two of the same alleles (i.e., BB or bb), it is called **homozygous** for that gene. If both alleles are the dominant allele, the gene is called **homozygous dominant**. If both alleles are recessive, the gene is called **homozygous recessive**. If the organism has different alleles (i.e., Bb), it is called **heterozygous** for the gene.

Because the green eye color allele is recessive, it can be discerned that an individual in the population with green eyes has the genotype bb. An individual with blue eyes can have either a BB genotype or a Bb genotype. The genotype of a blue-eyed individual can be determined by mating the blue-eyed individual with a green-eyed individual. This is called a **test cross**. If all of the offspring have blue eyes, then the blue-eyed individual was homozygous dominant. The genotypes of all of the offspring from a mating between a BB genotype and a bb genotype can only be Bb.

If any of the offspring have green eyes, the blue-eyed adult must have been heterozygous. The genotypes of the offspring from a mating between a Bb genotype and a bb genotype are either Bb or bb. Mating between two heterozygous adults can produce offspring with three different genotypes: BB, Bb, and bb. In this case, two blue-eyed adults can produce a green-eyed offspring.

PHENOTYPE AND GENOTYPE: FACTORS TO CONSIDER

Many different factors can affect phenotype and genotype in offspring. You should be familiar with the following topics:

- **Incomplete dominance** is a form of inheritance in which *both* heterozygous alleles are expressed. This means that the offspring will display a combined phenotype that is distinct from both parent organisms. For example, a plant with red flowers and a plant with white flowers might produce offspring that have pink flowers.

- **Polygenic inheritance** is a type of inheritance in which several interacting genes control a single trait. Many traits result from the additive influences of multiple genes; skin color is one common example of a polygenic trait.

- **Genetic recombination** is a molecular process by which an organism's genes are rearranged in its offspring. Through this process, two alleles can be separated and replaced by different alleles, thereby changing the genetic makeup but preserving the structure of the gene. Chromosomal crossing over is an example of a mechanism by which this process takes place.

- **Gene transfer** can be vertical or horizontal. Vertical gene transfer occurs when an organism receives genetic material (i.e., DNA) from a parent organism or from a predecessor species. Horizontal gene transfer occurs when an organism transfers genetic material to cells that are not its offspring.

CHAPTER 15: MENDELIAN GENETICS

BIG IDEA 3: Living systems share, retrieve, transmit, and respond to information essential to life processes.

IF YOU ONLY LEARN FOUR THINGS IN THIS CHAPTER . . .

1. Different versions of a gene that code for the same trait are called alleles. In classical (Mendelian) genetics, an individual receives one allele from each parent. Individuals with matching alleles are homozygous for that trait while those with different alleles are heterozygous. Usually one version of the allele is dominant (e.g., brown eye color) and the other is recessive (e.g., blue eye color). Heterozygotes are "ruled" by the dominant allele.

2. Geneticists perform test crosses to determine the genetic makeup (genotype) of organisms displaying the dominant phenotype. A Punnett square is used to illustrate a test cross. Mendel's Law of Segregation states that an individual's alleles separate during meiosis, and either may be passed on to the offspring. Mendel's Law of Independent Assortment states that inheritance of a particular allele for one trait does not affect inheritance of other traits.

3. Chi-squared analysis refers to a means of determining if experimental observations are significantly different from the expected result.

4. Pedigrees are family trees that enable us to study the inheritance of a particular trait across many related generations. They can also help us determine the type of transmission through generations: autosomal recessive/dominant, X-linked recessive/dominant, and so on.

INHERITANCE PATTERNS ★★★★

After learning the basics of genetics, it is important to review some specific information about inheritance that might appear on the exam. You need to be familiar with mathematical principles of simple probability. Relax; it's not as hard as it sounds. Even if there are more than two alleles for a gene, an individual can only have two alleles on each pair of chromosomes. Ending up with a combination of alleles is like tossing a coin.

An organism that is homozygous for a particular gene (*AA*) produces four gametes each with an *A* allele. Each of the gametes is one out of a possible four gametes, but because all of the gametes have an *A*, four out of four of the gametes have an *A*. Four out of four is a probability of one because $\frac{4}{4} = 1$. You can look at the question more intuitively by saying that because there are only *A* alleles in the parent cell, all of the daughter cells (100 percent, or $\frac{1}{1}$) will have the *A* allele.

If an organism is heterozygous (*Aa*), it produces two gametes with an *A* allele (2 out of 4) and two gametes with an *a* allele (2 out of 4). The probability of either the *A* or *a* allele in the gametes is $\frac{2}{4} = \frac{1}{2}$. Or again, intuitively, if the parent cell has one *A* and one *a* allele, it will produce gametes that are half *A* (50 percent, or $\frac{1}{2}$) and half *a* (50 percent, or $\frac{1}{2}$).

Now consider two genes and an individual that is homozygous for both (*AABB*). Although there are two genes to consider, the organism still produces four gametes from a single parent cell. Each gamete gets an *A* allele and a *B* allele. What is the probability that a gamete will have both an *A* and *B* allele?

THE TWO LAWS OF MENDELIAN INHERITANCE

Law of Segregation—This describes the separation of alleles in the parent genotype during the process of gametogenesis. There can be a maximum of only two different alleles in a single parent; half the gametes get one allele and the other half get the other allele.

Law of Independent Assortment—This suggests that different genes sort into different gametes, independently of each other. For example, the sorting of alleles for eye color is not affected by the sorting of alleles for hair color. These laws explain the 3:1 and 9:3:3:1 ratios of phenotypes observed in monohybrid and dihybrid crosses.

Four out of four will have an A $\left(\dfrac{4}{4}, \text{or } 1\right)$ and four out of four will have a B $\left(\dfrac{4}{4}, \text{or } 1\right)$, so four out

of four will have AB $\left(\dfrac{4}{4} \times \dfrac{4}{4} = \dfrac{16}{16} = 1\right)$. If the individual is heterozygous for one of the genes

($AABb$), the probability of the A allele stays the same, but now only half of the gametes get a B

allele $\left(\dfrac{2}{4} = \dfrac{1}{2}\right)$ and the other two get a b allele $\left(\dfrac{2}{4} = \dfrac{1}{2}\right)$.

What is the probability of producing a gamete with both an A and B allele from this individual?

Four out of four will have an A $\left(\dfrac{4}{4}\right)$ and two out of four will have a B $\left(\dfrac{2}{4}\right)$, so only two out

of four will have AB $\left(\dfrac{4}{4} \times \dfrac{2}{4} = \dfrac{8}{16} = \dfrac{1}{2}\right)$. If the individual is heterozygous for both genes

($AaBb$), the probability of each allele in the gametes (A, a, B, or b) is $\dfrac{1}{2}$ so the probability of

AB is $\dfrac{1}{4} \left(\dfrac{1}{2} \times \dfrac{1}{2} = \dfrac{1}{4}\right)$, the probability of aB is $\dfrac{1}{4} \left(\dfrac{1}{2} \times \dfrac{1}{2} = \dfrac{1}{4}\right)$, and the probability of ab is

$\dfrac{1}{4} \left(\dfrac{1}{2} \times \dfrac{1}{2} = \dfrac{1}{4}\right)$.

Once you can figure out what gametes will be produced from which individuals, you can figure out the genotypes of offspring from a **Punnett square**. The following is a simple cross between two heterozygous individuals for one gene. A mating of this kind is called a **monohybrid** cross.

	A	a
A	AA	Aa
a	aA	aa

Monohybrid Cross

Typically, the gametes of the sperm are recorded along the left side of the Punnett square, and those of the egg are recorded across the top. The allele from each sperm is paired with the allele from each egg where the columns and rows meet, showing the genotypes of the offspring. The different indications of Aa and aA are used only to demonstrate the source of the alleles; they are the same genotype.

This cross results in three different genotypes (AA, Aa, and aa), but only two phenotypes because the dominant trait is expressed in both the AA and Aa individuals. The dominant trait shows up in the offspring in a ratio of 3 to 1.

You can put together a **dihybrid cross** (tracking two genes—A and B—rather than one) just as easily as a monohybrid cross. Just put the gametes of the male in the left column and the gametes of the female across the top.

	AB	*Ab*	*aB*	*ab*
AB	AABB	AABb	AaBB	AaBb
Ab	AAbB	AAbb	AabB	Aabb
aB	aABB	aABb	aaBB	aaBb
ab	aABb	aAbb	aabB	aabb

Dihybrid Cross

The dihybrid cross produces the famous phenotypic ratio of 9:3:3:1. If you consider a large number of offspring, there will be on average 9 out of every 16 that express the dominant phenotypes for both genes, 3 out of every 16 that express the dominant phenotype of one gene, 3 out of every 16 that express the dominant phenotype of the other gene, and only one that expresses the recessive phenotypes for both genes.

These are the basics of inheritance. However, genes do not operate in isolation from one another; this makes genetics more complex than Mendel's experiments and the Punnett square might suggest.

Mendel's laws aren't true for all genes. There are some genes that do not sort independently from others. These genes may be linked to other genes on the same chromosome. It can also be observed that the probability ratios expected are often not quite what is expected based on Punnett squares. This is because some chromosomes exchange genetic information with each other by **crossing over** when the homologous pairs are synapsed into **tetrads**. The farther away genes are from each other, the more likely they are to exchange DNA with another **chromatid**. The frequency of certain observed phenotypes can be used to estimate a relative rate of crossovers. The higher the frequency of occurrence of a certain phenotype, the more likely there is crossing over between genes, and the farther away the genes are from each other.

KEY STRATEGY

Notice patterns in the ratios that occur with specific types of crosses. For example, crossing a heterozygous dominant (*Rr*) with a homozygous recessive (*rr*) always gives a 1:1 ratio of dominant to recessive phenotypes (half dominant phenotype; half recessive phenotype). This can speed up the process of figuring out phenotypic ratios and probabilities.

IMPORTANT

As you study, be sure you're aware of the distinct difference between incomplete dominance and codominance.

PEDIGREES ★★★★

Pedigrees are family trees that enable us to study the inheritance of a particular trait across many related generations. Males are typically designated as squares on the pedigree, while females are designated as circles. Those who phenotypically show a trait are shaded, while those who carry a trait but do not show it are either half-shaded or given a dot inside their circle or square.

Genetic traits can be classified by whether they are transmitted on autosomal chromosomes (numbers 1–22) or on sex chromosomes (the X almost always, because the Y chromosome carries a limited number of genes). In addition, traits can be dominant or recessive. For the AP Biology exam, you should be familiar with the characteristics of different inheritance patterns so you can easily spot patterns.

AUTOSOMAL DOMINANT

- Males and females are equally likely to have the trait.

- Traits do not skip generations.

- The trait is present if the corresponding gene is present.

- There is male-to-male and female-to-female transmission.

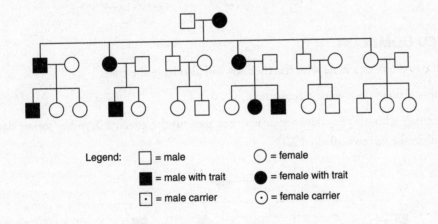

Legend: □ = male   ○ = female

⬛ = male with trait   ● = female with trait

⊡ = male carrier   ⊙ = female carrier

Autosomal Recessive

- Males and females are equally likely to have the trait.

- Traits often skip generations.

- Only homozygous individuals have the trait.

- Traits can appear in siblings without appearing in parents.

- If a parent has the trait, those offspring who do not have it are heterozygous carriers of the trait.

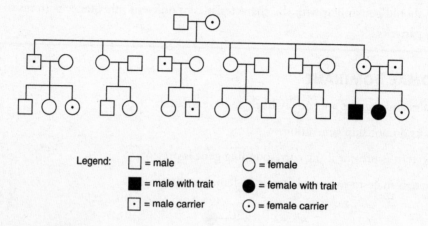

Legend: □ = male   ○ = female

■ = male with trait   ● = female with trait

⊡ = male carrier   ⊙ = female carrier

X-Linked Dominant

- All daughters of a male who has the trait will also have the trait.

- There is no *male-to-male transmission*.

- A female who has the trait may or may not pass on the affected X to her son or daughter (unless she has two affected X's).

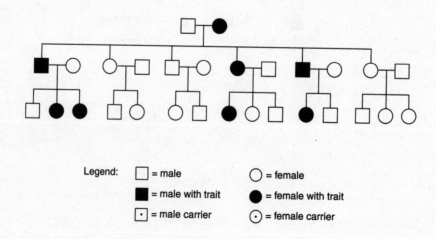

Legend: □ = male   ○ = female

■ = male with trait   ● = female with trait

⊡ = male carrier   ⊙ = female carrier

CHAPTER 16: VIRUSES

BIG IDEA 3: Living systems share, retrieve, transmit, and respond to information essential to life processes.

IF YOU ONLY LEARN FOUR THINGS IN THIS CHAPTER . . .

1. Viruses are obligate intracellular parasites that must replicate inside the cells of living organisms. Viral genomes can range from small to large and be DNA or RNA and double-stranded or single-stranded.

2. Viruses can replicate in several ways including lytic or lysogenic cycles.

3. Bacteriophages, or phages, are viruses that infect bacteria.

4. Viroids and prions are nontypical viruses. Even though they do not look like normal viruses, they can still infect humans. One of the most famous prions is responsible for causing bovine spongiform encephalopathy, or mad cow disease.

INTRODUCTION ★★★★

Viruses defy much of the logic with which we approach our study of life on Earth. Their genomes often differ strikingly from those of other living organisms, their life cycles depend upon their ability to enter and replicate within a living host cell, and they have no cell membranes or organelles as part of their structure.

Virus particles are surrounded by a **capsid**, or protein shell, which can come in a variety of shapes (cones, rods, and polyhedrons). Capsids are built out of proteins, many of which have various sugars attached to them, poking upward from the viral surface. These glycoproteins are used to gain entry into a living cell by binding with surface proteins on the living cell's membrane. Most viral capsids are made up of only one or two different types of protein. In addition, some viruses are able to surround themselves with an envelope of cell membrane as they burst out of a cell they have just infected. This **viral envelope** can help them avoid detection by the host's immune system, because the viral particles resemble (at least on the outside) the host's own cells.

VIRAL GENOMES AND VIRAL REPLICATION ★★★★

Viral genomes can range from quite small (five to ten genes in all) to fairly large (several hundred genes). The genetic material found within a virus may be held on DNA or RNA, both of which can be found in either a double-stranded or single-stranded state. The nucleic acid can be linear or circular, and although the DNA is always found together in a single chromosome, the RNA can be in several pieces. Viruses are considered to be **obligate intracellular parasites**, meaning that they must live and reproduce within another cell, where they act as a parasite, using the host cell's machinery to copy themselves or to make proteins encoded for by their DNA or RNA. Independent of host cells, viruses conduct no metabolic activity of their own.

Lytic Cycle ★★★

Some viruses contain the enzymes they need for replication of their genome, stuffing these proteins into each "baby" virus as it is produced by the host cell. Others do not travel with their proteins but rather make them only when they are inside a host. The typical viral growth cycle includes the ordered events of attachment and penetration, uncoating, viral mRNA and protein synthesis, replication of the genome, and assembly and release.

Attachment and penetration by parent virion (viral particle): The specificity of the proteins coating the capsid of the virus determines the host range of the particular virus, or how many kinds of cells the virus can infect. Those viruses with a wide range of surface proteins (unusual) or with proteins that can bind to many kinds of cell surface receptors (more common) are said to have a wide host range.

Uncoating of the viral genome: As the viral particle penetrates, the cell traps it in a vesicle, at which point the virus will break open its capsid. When the vesicle breaks due to viral uncoating, the inner core of the virus with its genetic material can dump itself into the cytoplasm.

Viral mRNA and protein synthesis: DNA viruses replicate their DNA in the nucleus of the cell and use the host cell's RNA polymerases to make their proteins. There are a few exceptions to this (e.g., the pox family of viruses that causes smallpox, chicken pox, etc.) that cannot enter the nucleus and actually carry the necessary RNA polymerase with them. RNA viruses replicate in the cytoplasm, sometimes using their RNA as mRNA directly, and sometimes using their RNA as a template for mRNA synthesis. One group of RNA viruses, the retroviruses (e.g., HIV), converts its RNA into DNA first, using the enzyme *reverse transcriptase*. The DNA copy is then transcribed into mRNA by the host's RNA polymerases. Many of the viral proteins that are first synthesized on the host's ribosomes are the ones needed for replication of the genome.

Replication of the genome: All virus particles will first replicate the DNA or RNA they came into the host cell with; the complementary copy that is created is then used as a template for all new copies of the genome. This is analogous to making a negative from a photograph and using the negative for all subsequent copies rather than using the photograph itself. Replication proceeds using host cell DNA and RNA polymerases as cells normally would.

Assembly and release: Viral nucleic acid is packaged within a protein capsid, and "baby" virus particles are released from the cell either en masse or slowly. *En masse* release ruptures and kills the host cell; *slowly* means the particles bud out of the host cell surface and envelop themselves in host cell membrane as they push their way out. Budding in a slow and deliberate fashion will usually allow the host cell to live.

LYSOGENIC CYCLE ★★★

The cycle described in the previous section is typical of most viruses: It is called a **lytic cycle** (because the host cells usually lyse, or burst, in the end). Some viruses, however, use a **lysogenic** cycle, whereby the viral DNA gets integrated into the host cell's DNA and can remain there indefinitely, allowing viral DNA to be replicated for generations alongside the host cell DNA. The infection continues so that all offspring of the host cell carry the viral DNA. This DNA can simply remain a part of the host cell's genome forever. For example, certain bacteria (e.g., *Corynebacterium diphtheriae*, which causes diphtheria, and *Clostridium botulinum*, which causes botulism) secrete toxins that are coded for by viral genes that were acquired from a **bacteriophage** (a virus that infects bacteria). In some cases, however, certain environmental events may cause this integrated viral DNA to begin a full replicative cycle (lytic cycle), accompanied by the production of baby viruses and the eventual death of the cell.

BACTERIOPHAGES ★★★

The viruses that infect bacteria are known as bacteriophages, or phages for short. They are a diverse group of organisms that are best characterized by their head and tail structure, which is unique to phages:

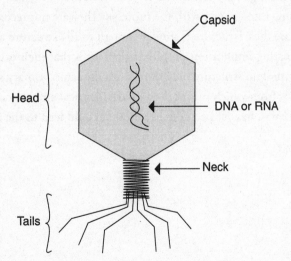

The tails are used to latch on to the host cell surface, after which enzymes digest a portion of the cell membrane to insert viral genetic material (either DNA or RNA) into the host cell. The capsid and the tails are left outside of the host cell membrane, differentiating phage infection from that of other viruses. Often, the viral DNA or RNA is rapidly destroyed by powerful bacterial enzymes called **restriction enzymes**. These enzymes are a primitive type of immune system in bacteria and chew up foreign genetic material.

For the viral genomes that do survive, they will either cause the host cell to produce new viral particles as described previously or will integrate into the bacterial chromosome. Phages that infect bacteria in a lytic way and cause active viral replication are called **lytic phages**, while those whose DNA gets integrated into the host cell's DNA (in a lysogenic fashion) are called **temperate phages**. While integrated, the bacteriophage is known as a **prophage**.

ATYPICAL VIRUSLIKE FORMS ★

There are a few exceptions to the typical virus as described here, and these particles (which certainly cannot be considered living organisms according to the standard definition of life) are infectious to many living cells:

Viroids are viruslike particles that are composed of a single molecule of circular RNA without any surrounding capsid or envelope. The RNA can replicate using a host's machinery, but it does not seem to code for any specific proteins. Despite this, viroids have been *implicated in some plant diseases*, though not in diseases in any other organisms.

Prions are simply pieces of protein that are infectious. They have gained fame through the recent spread of mad cow disease, and they are connected with diseases that are typically slow to form, taking years before symptoms develop. Despite once being called slow viruses, prions are not viruses at all and do not contain any DNA or RNA. Recent evidence suggests that prions are infectious because they change the structure of one's own normal proteins, a switch not encoded in genes at all, by coming into contact with the proteins. They are not recognized by the immune system, probably because their structure is similar enough to the structure of normal proteins. Perhaps the most fascinating implication of studying prions is that their existence shows that more than one tertiary protein structure may form from the same primary structure. Both prions and the normal proteins from which prions derive have identical amino acid sequences, and no traceable DNA mutation has yet been discovered that could lead to the formation of prion particles.

CHAPTER 17: CELL COMMUNICATION

BIG IDEA 3: Living systems share, retrieve, transmit, and respond to information essential to life processes.

IF YOU ONLY LEARN THREE THINGS IN THIS CHAPTER . . .

1. There are two major ways animal cells communicate: through signaling molecules secreted by cells and molecules that rest on the cell surface. Signaling molecules can be further divided into paracrine, synaptic, and endocrine signaling molecules.

2. Cell communication occurs through cell junctions and is important for complex tissues to be able to function properly.

3. There are three major types of cell junctions: tight junctions, desmosomes, and gap junctions.

CELL JUNCTIONS ★★★★

For cells to form complex tissues, secure cell-to-cell bonds must hold them together. These cell junctions come in a variety of forms and serve many different purposes. Occluding junctions, otherwise known as **tight junctions**, seal spaces between cells; anchoring junctions, which include desmosomes, connect one cell's cytoplasm to another through anchoring proteins; and communicating junctions, including gap junctions and **plasmodesmata** in plants, allow cells to directly exchange cytoplasmic material via channels that cross both cells' membranes.

GENERAL PRINCIPLES OF SIGNALING ★★★★

Two main ways that animal cells communicate with one another are via signaling molecules that are secreted by the cells and through molecules that rest on the cells' surfaces and remain attached to the cell even as they signal other cells. The target cell receives information through receptors on its surface, which are generally membrane-spanning proteins with binding domains on the outside of the plasma membrane.

KEY STRATEGY

Hormones and hormone-specific signaling will be discussed in the endocrinology chapter. Use the information in this chapter as a general overview of cell signaling.

Although some signaling molecules may act far away from the cell that secreted them, many bind only to receptors on cells in the immediate vicinity. **Paracrine** signaling refers to the process of signaling only nearby cells, with signaling molecules quickly pulled out of the extracellular matrix. **Synaptic** signaling occurs only in nerve cells, over extremely short distances, as electric signals reach axon terminals and cause the release of chemicals called neurotransmitters. These chemical messengers bind to nearby nerve cells and cause an electrical signal to be propagated or continued. Synaptic signaling within the nervous system will be discussed in detail later. By sending information from one neuron to another, nervous systems can produce rapid communication over long distances without a diminished effect, which can happen when sending individual molecules over long distances, such as through the bloodstream. **Endocrine** signaling refers to the secretion of chemical messengers into the bloodstream (or other liquid medium, such as is the case with plants) for widespread distribution throughout the entire organism.

Cells communicate with each other using a wide variety of molecules. Some can pass through the cell membranes of the target cells and act directly within the cell's cytoplasm. Others that bind to plasma membrane receptors and act through second messengers. All of these molecules can be considered **hormones**, circulating signals that are released by specialized cells and travel throughout the bloodstream.

OCCLUDING (TIGHT) JUNCTIONS ★★

These connections are thought to be formed by proteins that tightly wind between the adjacent plasma membranes of neighboring cells, binding the cells together at those points so tightly that *nothing can diffuse between cells or past the junction*. In such a way, tight junctions form a total barrier to transport and diffusion where they exist.

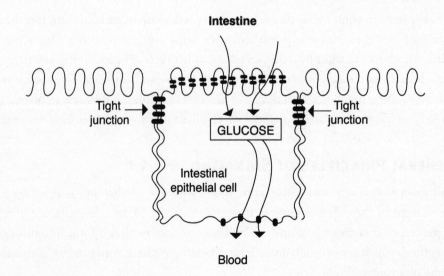

Where they are most useful, as shown in the previous diagram, is in places such as the *intestines*, where specialized cells absorb nutrients from one side of the cell (the intestinal side) and transport them through the cell and out the other end (into the bloodstream). Certain transporters (e.g., sodium ion-driven glucose transporters) exist on the intestinal side of the cell but not on the bloodstream side, and glucose that enters the bloodstream is prevented from diffusing back into the intestinal tract by tight junctions.

ANCHORING JUNCTIONS ★★

Found between cells subjected to fair amounts of stress, either from shearing forces or contacting forces, these junctions connect one cell's cytoplasm to another's via a series of proteins. Regardless of classification, anchoring junctions not only allow cells to adhere to neighbors, but also may allow them to contract themselves into large tubelike tissues. This occurs when the fibers holding the junctions together contract across the cells' cytoplasms.

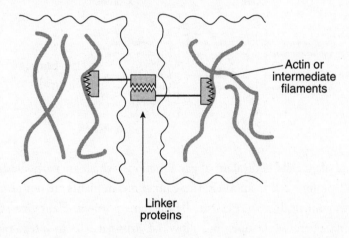

Actin or intermediate filaments

Linker proteins

Notice in the previous diagram that anchoring junctions involve proteins that attach to actin filaments within the cell cytoplasm and also attach to linker proteins across the intercellular space. Thus, the junction is not an actual linking of cytoplasm, where material can be freely exchanged, but rather a *physical joining so that the cells do not shear away from each other*.

The most often mentioned of these attachments are the desmosomes, found in heart cells and between epithelial cells in the skin. Although they function in the same way as other anchoring junctions, they attach two cells using **intermediate filaments** within the cytoplasm rather than actin filaments.

COMMUNICATING JUNCTIONS ★★

The best known of these cell-to-cell connections are the gap junctions, formed by proteins called **connexins**, which build tubes or pores between two adjacent cells' cytoplasms. It is through these pores that ions and other material can pass from one cell to the other. In cells that rapidly transmit chemical or electrical signals across tissues, gap junctions are everywhere. Because chemical and electrical transmission is mediated through the movement of ions and other messengers, gap junctions *allow for undisrupted and very fast signal transmission* across wide areas of tissue. In heart cells, gap junctions allow for rhythmic contractions of large sections of the heart all at once. In the digestive tract, gap junctions allow for waves of muscle contraction such as those found in the esophagus. These junctions also allow for coordination of *rapid and complex movements,* such as a fish's tail-flip to escape an oncoming predator. The following diagram shows a generic gap junction with many connexin proteins that form a **connexon**.

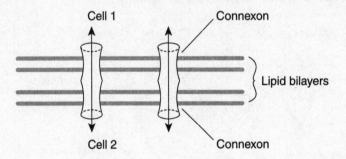

Plasmodesmata are plant cells' equivalent of gap junctions, which are particularly useful for the free flow of nuclei from one cell to another. These junctions in plants are not nearly as complex in structure as those in animals, but serve essentially the same purpose. Plant viruses, however, often exploit plasmodesmata because the openings allow the virus particles to spread rapidly from one section of the plant to others.

CHAPTER 18: HORMONES AND THE ENDOCRINE SYSTEM

BIG IDEA 3: Living systems share, retrieve, transmit, and respond to information essential to life processes.

IF YOU ONLY LEARN THREE THINGS IN THIS CHAPTER . . .

1. The endocrine system works with the nervous system to produce certain reactions as needed in response to changes in the body's organ systems.

2. Hormones are one of the body's methods of internal communication. Hormones are secreted by glands and travel throughout the body. They activate receptor cells that produce receptor proteins, thereby creating a reaction.

3. There are two major types of hormones: steroid hormones and nonsteroid hormones.

THE ENDOCRINE SYSTEM AND ITS MESSENGERS ★★★★

The endocrine system acts along with the nervous system as a means of internal communication, coordinating the activities of organ systems around the body. Endocrine glands synthesize and secrete chemical hormones that are dumped directly into the bloodstream and affect specific target organs or tissues. Compared with the nervous system, hormones take much longer to communicate with cells of the body. After all, they can travel only as fast as the blood flows. Yet hormonal signaling can cause behavioral effects that last for days, far longer than any nervous signal can last.

If an animal were injured, it would release the hormone epinephrine (adrenaline) from the adrenal glands. The epinephrine would travel in the bloodstream throughout the entire body. Any cell in the body that is accessible by blood flow and has **receptor proteins** for epinephrine responds to the "message" of the injury. Blood vessels respond by constricting blood flow, the heart and lungs increase their rate of activity, and the liver releases sugar to the bloodstream. The key issues of any endocrine response are the widespread release of hormones, specificity controlled by the receptor's ability to recognize the signal, and the location of the cells that contain the receptor. Some specific hormones, their target tissues, and their activity are discussed later in the chapter.

HORMONES

Hormones are defined as chemical signals that:

- Are synthesized by specialized cells
- Travel throughout a multicellular organism by some kind of bodily fluid
- Coordinate systemic (total body) responses by activating specific **receptor cells**

Hormonal responses take time and are indirect. This is a reasonable form of communication for a plant that doesn't move, but for a bird in flight, a fish swimming from a predator, or a gibbon brachiating through the trees, a quicker, more intricately coordinated form of communication inside the body is necessary. Coordinating environmental signals with movement responses and complex processing is the job of the **central nervous system**. The **endocrine (hormone) system** interacts with the **nervous system**, but the mechanism by which nerves communicate messages is entirely different.

It's important to understand that built into the endocrine system are several controls based on **negative feedback** so that the oversecretion of hormones can be avoided. In general, high levels of a particular hormone in the bloodstream inhibit further production of that hormone. In some cases, the hormone itself acts back on the cells that first produced it and blocks the hormone biosynthesis pathways in those cells. In other cases, an antagonistic hormone will act to counter another's effects. We see this in the case of the pancreatic hormones **insulin** and **glucagon**, which together regulate the concentration of glucose in the bloodstream.

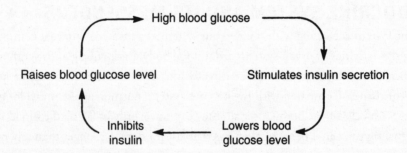

While insulin stimulates the uptake of glucose by muscle and adipose cells around the body, glucagon stimulates the breakdown of glycogen in the muscles and liver to increase blood glucose levels. In addition, glucagon stimulates **gluconeogenesis**, the process by which nonhexose substrates such as amino acids and fatty acids are converted into glucose. The combination of these two hormones working together in an antagonistic function serves to regulate blood glucose levels in a precise manner.

> **REMEMBER**
>
> You will not need to memorize the names, molecular structures, and specific effects of any hormones. These are not tested on the AP Biology exam.

IMPORTANT HORMONES IN VERTEBRATES

Compound	Type	Origin	Action
Androgens	Steroid	Testis	Secondary sex characteristics
Epinephrine (adrenaline)	Catecholamine	Adrenal gland	Fight-or-flight response
Estrogens	Steroid	Ovary	Secondary sex characteristics
Follicle-stimulating hormone (FSH)	Glycoprotein	Anterior pituitary	Regulate gonads
Glucagon	Polypeptide	Pancreas	Increase blood glucose
Insulin	Polypeptide	Pancreas	Decrease blood glucose
Melatonin	Catecholamine	Pineal gland	Circadian rhythm
T_3 and T_4	Amino acids	Thyroid	Stabilize metabolic rate
Thyroid-stimulating hormone (TSH)	Glycoprotein	Anterior pituitary	Regulate thyroid

STEROID HORMONES ★★★

Steroid hormones, such as the sex hormones testosterone and estrogen, are lipids with cholesterol-based structures and can pass through the cell membrane to act directly on the DNA in the cell nucleus. Often, steroid hormones must bind to intracellular receptor proteins to cross the nuclear membrane or regulate transcription. The steroid-receptor complex is generally considered to be a **transcription factor**, as it is able to bind to "enhancer" regions on the DNA, turning on the transcription of certain genes.

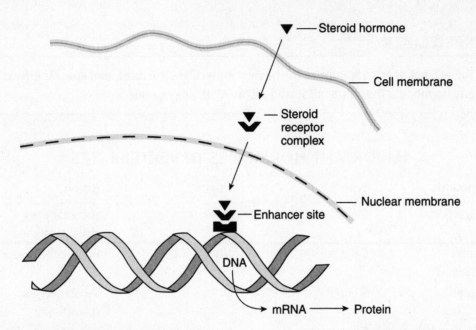

The basis for the production of steroid hormones, as well as compounds such as vitamin D and the thyroid hormones, is cholesterol. Steroid hormones travel through the bloodstream bound to carrier molecules, from which they dissociate when entering a cell. *Because these hormones are lipids and do not dissolve in water, their effects can last for hours or days* in the bloodstream after being released, a much longer period of time than water-soluble nonsteroid hormones can last.

Another molecule worth mentioning here is not a steroid, though it can easily pass through a cell's lipid bilayer because it is so small. **Nitric oxide** (NO), recently recognized as an important signaling molecule, passes into the cell's cytoplasm, acting on the enzyme guanylyl cyclase, which in turn produces **cyclic-GMP**. Cyclic-GMP is an important molecule that, in the case of NO, causes smooth muscle cells in blood vessels to relax when stimulated by acetylcholine. Keep in mind, too, that different cells can respond to the same hormone or neurotransmitter in different ways. For example, although acetylcholine can cause smooth muscle cells to relax, it causes skeletal muscle cells to contract. It is postulated that these differences are caused by the variety of receptor complexes that can respond to the same hormone.

Nonsteroid Hormones ★★★

Many compounds that are unable to cross the plasma membrane and enter the cytoplasm still act as powerful signaling molecules. Examples include peptides, such as **atrial natriuretic peptide**, released by the heart to influence water absorption in the kidneys, **calcitonin**, involved in calcium regulation, and **glucagon**, involved in blood sugar regulation.

These nonsteroid hormones act via signal-transduction pathways, whereby binding of the hormone to a cell-surface receptor induces a conformational change in the receptor protein, setting off an intracellular cascade to alter the cell's behavior in some way. **Cell-surface receptors** come in three different forms: ion-channel-linked, G-protein-linked, and enzyme-linked. All of these receptors can bind to the hormone or chemical signal (also referred to as the **ligand**) very accurately and very tightly and can turn on intracellular signals.

Ion-channel-linked receptors are also called ligand-gated channels. These membrane-spanning proteins undergo a conformational change when a ligand binds to them so that a "tunnel" is opened through the membrane to allow the passage of a specific molecule. These ligands can be neurotransmitters or peptide hormones, and the molecules that pass through are often ions, such as sodium (Na^+) or potassium (K^+), which can alter the charge across the membrane. The ion channels, or pores, are opened only for a short time, after which the ligand dissociates from the receptor and the receptor is available once again for a new ligand to bind.

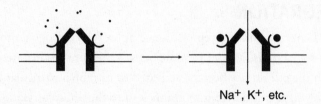

Na^+, K^+, etc.

G-protein-linked channels cause G-proteins to dissociate from the cytoplasmic side of the receptor protein and bind to a nearby enzyme. This enzyme continues the signaling cascade by inducing changes in other intracellular molecules; in addition, it can also cause other membrane channels to open in areas some distance from the originating receptor.

Most G-proteins activate what are known as second messengers, small intracellular molecules like **cyclic AMP** (cAMP), calcium, and phosphates, which in turn activate key enzymes or transcription factors involved in essential reactions. Signaling cascades involving G-proteins can be very complex and involve many different enzymes and conversions, which prevents the reactions from running out of control.

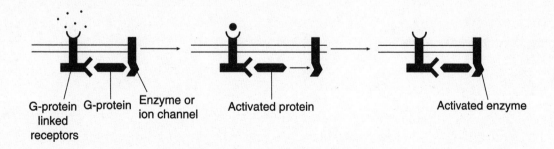

G-protein   G-protein   Enzyme or   Activated protein   Activated enzyme
linked   ion channel
receptors

Enzyme-linked receptors can act directly as enzymes, catalyzing a reaction inside the cell, or they can be associated with enzymes that they activate within the cell. Most enzyme-linked receptors turn on a special class of enzymes called protein **kinases**, which add free-floating phosphate groups to proteins, regulating their activity. Protein **phosphorylation** is an essential means of intracellular signaling and control, as proteins become activated or deactivated simply by the addition or removal of phosphates. Protein kinases often target other protein kinases, initiating a cascade of kinase activity. Protein phosphatases reverse the action of protein kinases.

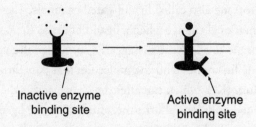

Inactive enzyme binding site

Active enzyme binding site

SIGNAL INTEGRATION ★

G-protein-linked and enzyme-linked receptors use complex relays of signal proteins to amplify and/or regulate their signal transduction. In some cases, a measure of safety requires that two different receptors on the cell surface become activated to turn on a particular intracellular protein. In other cases, signals at different receptors lead to the phosphorylation of different proteins that activate together only when *both* proteins have been phosphorylated. This signal integration leads to a measure of control over reactions and the ability to use multiple inputs to cause a certain effect that can vary in degree.

CHAPTER 19: NEURONS AND THE NERVOUS SYSTEM

BIG IDEA 3: Living systems share, retrieve, transmit, and respond to information essential to life processes.

IF YOU ONLY LEARN FOUR THINGS IN THIS CHAPTER . . .

1. The nervous system is broken down into two main parts: the central nervous system and the peripheral nervous system. The central nervous system contains the brain and spinal column and is used to control animal behavior based on input from the environment. The peripheral nervous system connects the central nervous system to all other parts of the body.

2. A neuron (or nerve cell) processes and transmits information via electrical and chemical signaling. Chemical signals occur across a synapse. Electrical signals along the axon depend upon the flow of ions across the axon membrane and cell body membrane. The opening and closing of sodium and potassium volted-gated channels is responsible for the transmission of electrical signals.

3. The autonomic nervous system (ANS) regulates the body's internal environment without the aid of conscious control. It is composed of two subdivisions, the sympathetic and the parasympathetic nervous systems. The sympathetic division is responsible for the "fight-or-flight" responses that ready the body for action. The parasympathetic division acts to conserve energy and restore the body to resting activity levels following exertion.

4. The body has several types of special sensors and sensory receptors to monitor its internal and external environment: interoceptors, proprioceptors, and exteroceptors. The chemical senses are taste and smell. The ear transduces sound energy (pressure waves) into impulses and is responsible for maintaining equilibrium while the eye uses light input for vision.

INTRODUCTION ★★★★

Nerves are like electrical wiring throughout the body, connecting stimuli from both the external and internal environment with the central processing units, the brain and spinal cord (if the animal has them). A **neuron** (nerve cell) is made up of a **soma** (cell body), **dendrites**, and an **axon**. Neurons are closely associated with each other at gaps called **synapses** or **synaptic junctions**. Electrochemical signals travel down the length of the nerve cell in a process called an action potential, and it is the action potential that is the basis for all nervous activity.

ORGANIZATION OF THE VERTEBRATE NERVOUS SYSTEM ★★

There are many different kinds of neurons in the vertebrate nervous system. Neurons that carry information about the external or internal environment to the brain or spinal cord are called **afferent neurons**. Neurons that carry commands from the brain or spinal cord to various parts of the body (e.g., muscles or glands) are called **efferent neurons**. Some neurons (**interneurons**) participate only in local circuits; their cell bodies and their nerve terminals are in the same location.

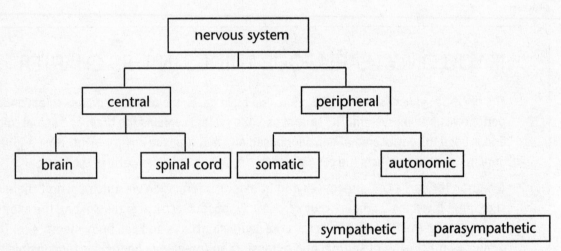

Nerves are essentially bundles of axons covered with connective tissue. A nerve may carry only sensory fibers (**a sensory nerve**), only motor fibers (**a motor nerve**), or a mixture of the two (**a mixed nerve**). Neuronal cell bodies often cluster together; such clusters are called **ganglia** in the periphery; in the central nervous system, they are called **nuclei**. The nervous system itself is divided into two major systems, the **central nervous system** and the **peripheral nervous system**. These two divisions will be discussed later in this chapter. First, we will look at the structure and function of neurons.

NOTE

The types of nervous systems, details of the various structures and features of the brain parts, and details of specific neurologic processes are beyond the scope of the AP exam. Focus on the detection, transmission and integration of stimuli and the production of a response.

NEURONS ★★★★

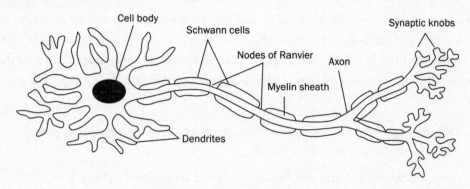

The nerve cell, or neuron, is the basic functional unit of the mammalian nervous system. Each neuron contains a cell body, dendrites, Schwann cells, nodes of Ranvier, an axon, and synaptic knobs. The **cell body** contains the nucleus and organelles and is the site of protein synthesis. **Dendrites** are extensions of cytoplasm from the cell body that receive chemical signals from nearby neurons and can initiate a new electrical signal. **Schwann cells** are separate cells from the neurons. They secrete myelin, an insulation for nerve cells that helps make signals move faster down the axon. **Nodes of Ranvier** are spaces along myelinated axons where Schwann cells have not laid down myelin covering. The **axon** is the main elongated extension of the cell body through which the electrical signal travels; the signal is in one direction only, from the cell body toward the synaptic knobs. **Synaptic knobs** release neurotransmitters, chemicals that communicate with surrounding nerve cells.

Dendrites are multi-branched and can receive signals from many different neurons at the same time. In fact, some specialized neurons such as the Purkinje cells of the brain can receive signals from over 1,000 different neurons at once into a single cell body. The amount of integration that takes place in a cell such as this, filtering out and summing together all of this input, is enormous. Most neurons have a single long axon enclosed in layers of myelin. **Myelin** insulates the axon, much like electrical tape insulates bare wire, and neurons that lose their myelin sheaths cannot transmit signals fast enough to give appropriate stimuli to muscles or organs. Each Schwann cell can contribute a single internodal segment of myelin. Many Schwann cells are needed to coat an entire axon with myelin. In the central nervous system (brain and spinal cord), the equivalent cell to a Schwann cell is known as an oligodendrocyte. Whereas a single Schwann cell can myelinate only a single axon, oligodendrocytes can send off myelin sheaths in several directions at once.

Schwann cells and oligodendrocytes are not the only "support cells" present around neurons. Astrocytes in the central nervous system (CNS) far outnumber the neurons of the CNS. The extensions of astrocytes stick to various parts of neurons and help to break down and remove certain neurotransmitter chemicals as well as engulf debris. Ependymal cells line the fluid-filled cavities of the brain and spinal cord and secrete cerebrospinal fluid (CSF) that helps cushion the CNS.

RESTING AND ACTION POTENTIALS ★★★★

Signals along the axon are electrical in nature and depend upon the flow of ions across the axon membrane and cell body membrane. How does electrical voltage arise in a cell and how does a signal arise from ion flow? All living cells have an electrical charge difference across their membranes, the inside of the cells being more negatively charged than the outside. The reasons for this are:

- DNA is a negatively charged molecule due to copious negative phosphate groups (remember the basic units of nucleotides).

- Many proteins (amino acid side chains) in the cell are negatively charged.

- The Na^+/K^+ (sodium/potassium) pumps in the cell membrane kick out three positive sodium (Na^+) ions for every two positive potassium (K^+) ions they move into the cell. Overall, that means that one positive charge is leaving the cells every time one Na^+/K^+ pump "turns."

- While sodium is kicked out of the cell and potassium is brought in due to the action of the Na^+/K^+ ATPase pump (ATPase means that active transport is involved here—the pump uses ATP to work), some potassium leaks out. This passive diffusion of potassium through leakage channels works along with the sodium-potassium pump to create an overall charge difference across the axon or cell body membrane.

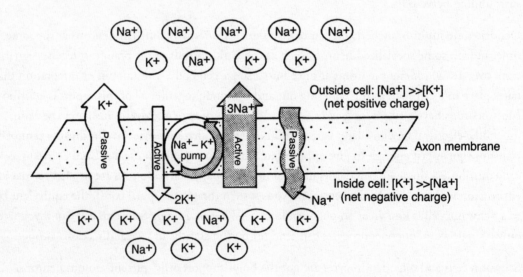

Resting Potential of a Neuron

An action potential is a temporary discharging of the battery power stored in neurons. It is also the up-and-down signal transmitted on a heart monitor that is commonly seen in movies and on television shows. Almost all cells in the body are constantly pumping Na^+ ions out and K^+ ions in, creating an electrical potential between the more positively charged outside matrix and the more negatively charged cytoplasm. The **resting membrane potential** of most neurons is about –70 mV (the minus sign indicates a negative charge). When some kind of stimulus depolarizes the membrane potential to a specific, predetermined threshold, a gate opens and lets all of the Na^+ into the cell, down the **electrochemical gradient** (a combination of the attraction of unlike charges and the diffusion gradient). After the neuron completes its depolarization, the Na^+ gates close and the K^+ gates open, reestablishing the resting membrane potential. The **action potential** takes place locally across the cell membrane, but propagates down the axon as the original depolarization causes Na^+ gates in an adjacent part of the membrane to depolarize as well. The depolarization travels down the length of the axon like the current in an electrical wire.

The following graph sums up this action potential, or nerve impulse:

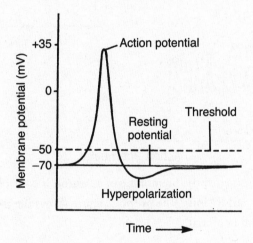

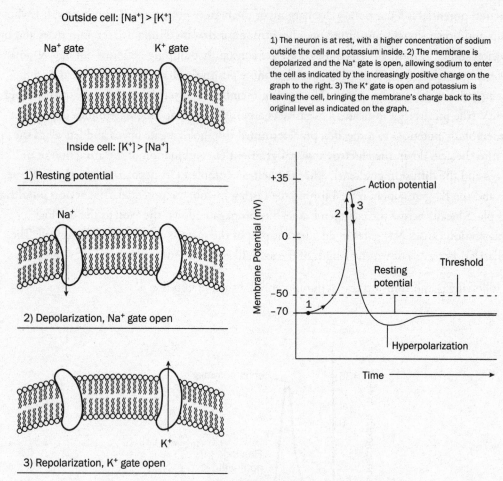

Outside cell: [Na⁺] > [K⁺]

Na⁺ gate   K⁺ gate

Inside cell: [K⁺] > [Na⁺]

1) Resting potential

Na⁺

2) Depolarization, Na⁺ gate open

K⁺

3) Repolarization, K⁺ gate open

1) The neuron is at rest, with a higher concentration of sodium outside the cell and potassium inside. 2) The membrane is depolarized and the Na⁺ gate is open, allowing sodium to enter the cell as indicated by the increasingly positive charge on the graph to the right. 3) The K⁺ gate is open and potassium is leaving the cell, bringing the membrane's charge back to its original level as indicated on the graph.

Action Potential

The depolarization that occurs across the membrane of a neuron is controlled by special gates that control the diffusion of Na⁺ and K⁺ ions. The previous figure shows the timing of the ion gates during the course of an action potential.

When the action potential reaches the end of the cell, it initiates the release of special chemical messengers called **neurotransmitters** that travel across the synaptic gap between two neurons. When the neurotransmitter binds to receptor proteins on the other neuron, it causes an action potential in that neuron. The process repeats along several neurons until the original signal reaches a processing area where a signal is often sent back to the point of origin, causing some kind of reaction. To enhance conductivity, some neurons are surrounded by an insulating sheath called myelin, which speeds up the movement of the action potential. Action potentials can travel very quickly through a neuron and are directed down the length of an axon to a specific location, like the soma (cell body) of another neuron.

Ions do *not* readily travel across cell membranes because of their charges, and they must be carried across by either membrane-spanning proteins or specific protein channels that let only certain types of ions through. Some channels let only Na⁺ through (sodium channels), some only K⁺, some only Cl⁻, and so on.

Sodium and potassium channels are open for a very short time (milliseconds) only, yet the rush of positive charge into one region of a neuron can set off a cascade of channel opening along the entire length of the axon. Voltage-gated sodium channels adjacent to where the axon first depolarizes (usually at the axon hillock) will open, causing sodium channels farther down the axon membrane to do the same. In such a way, the action potential is propagated down the entire length of the axon. Once opened, sodium channels inactivate for a brief period of time, which means that the action potential (flow of positive charge down the axon) occurs in one direction only, down toward the axon terminal. The inability of channels to open again creates a refractory period in which another impulse cannot travel along the axon. This period limits the number of signals that can travel through the axon in a given period of time.

Different kinds of axons will propagate the action potential at different speeds. The *larger diameter axons have faster transmission than small diameter ones* (recall your physics knowledge: Resistance to the flow of electrical current is inversely proportional to the cross-sectional area of the wire carrying the current). The membrane of the neuron is not a perfect conductor of electricity, and the current of the action potential will diminish over time as the signal travels down the axon if it is not replenished. That is why successive depolarizations down the axon membrane succeed in bringing the charge all the way from the cell body of the neuron to the axon terminal. Myelin increases the speed at which the signal travels, as well as holds the temporary influx of positive charge inside the axon. Because only the spaces in between the myelin, called nodes of Ranvier, are permeable to ions, action potentials do not propagate smoothly through myelinated axons but rather jump from node to node in saltatory fashion. This allows for much quicker signal transmission down the axon.

The **synapse** is the gap between the axon terminal of one neuron (called the presynaptic neuron because it is before the synapse) and the dendrites of another neuron (called the postsynaptic neuron). Neurons can also communicate directly with other postsynaptic cells, such as those on glands or muscles. The vast majority of synapses are chemical, whereby the electrical signal in the presynaptic neuron is converted into a chemical signal in the synapse that then incites a new electrical signal in the postsynaptic neuron. Electrical synapses can occur where gap junctions join the cytoplasms of neighboring cells, directly transmitting ionic signals.

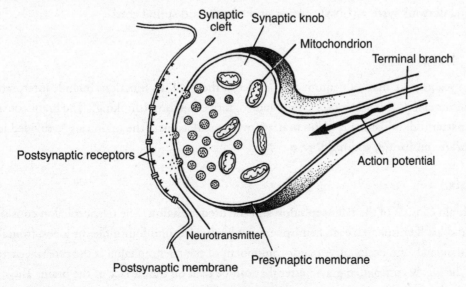

Synaptic Knob

As the action potential reaches the synaptic knob (terminal), voltage-gated calcium (Ca^{2+}) channels at the terminal end open up, and the rapid influx of calcium ions causes membrane-bound vesicles filled with neurotransmitters to merge with the presynaptic membrane, releasing their contents into the synapse. The synapse is a very small space, comprising at most a few micrometers from the presynaptic neuron to the postsynaptic one. Excitatory neurotransmitters, such as **acetylcholine** (ACh), will bind to membrane receptors on the postsynaptic dendrites or cell body and cause the opening of sodium channels (ligand-gated channels) on the postsynaptic cell, starting the process of an action potential all over again.

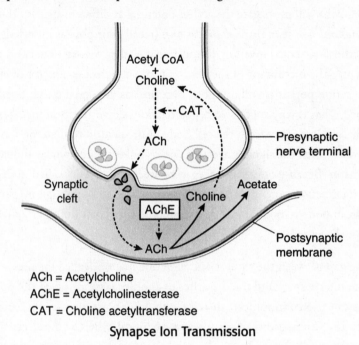

ACh = Acetylcholine
AChE = Acetylcholinesterase
CAT = Choline acetyltransferase

Synapse Ion Transmission

CENTRAL NERVOUS SYSTEM

The central **nervous system (CNS)** consists of the **brain** and **spinal cord**.

BRAIN ★★★

The brain is a jellylike mass of neurons that resides in the skull. Its functions include interpreting sensory information, forming motor plans, and cognitive function (thinking). The brain consists of **gray matter** (cell bodies) and **white matter** (myelinated axons). The brain can be divided into the **forebrain**, **midbrain**, and **hindbrain**.

FOREBRAIN

The forebrain consists of the **telencephalon** and the **diencephalon**. The telencephalon consists of right and left hemispheres; each hemisphere can be divided into four different lobes: frontal, parietal, temporal, and occipital. A major component of the telencephalon is the cerebral cortex, which is the highly convoluted gray matter that can be seen on the surface of the brain. The cortex processes and integrates sensory input and motor responses and is important for memory and creative thought. Right and left cerebral cortices communicate with each other through the corpus callosum. The diencephalon contains the thalamus and hypothalamus. The thalamus is a relay and integration center for the spinal cord and cerebral cortex. The hypothalamus controls visceral functions such as hunger, thirst, sex drive, water balance, blood pressure, and temperature regulation. It also plays an important role in the control of the endocrine system.

MIDBRAIN

The midbrain is a relay center for visual and auditory impulses. It also plays an important role in motor control.

HINDBRAIN

The hindbrain is the posterior part of the brain and consists of the **cerebellum**, the **pons**, and the **medulla**. The cerebellum helps to modulate motor impulses initiated by the motor cortex and is important in the maintenance of balance, hand-eye coordination, and the timing of rapid movements. One function of the pons is to act as a relay center to allow the cortex to communicate with the cerebellum. The medulla (also called the medulla oblongata) controls many vital functions such as breathing, heart rate, and gastrointestinal activity. Together, the midbrain, pons, and medulla constitute the **brain stem**.

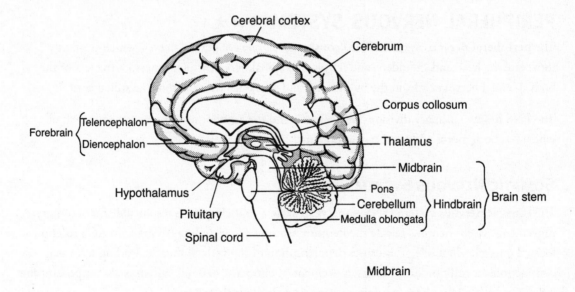

Diagram of the Brain

SPINAL CORD ★★★

The spinal cord is an elongated structure continuous with the brain stem that extends down the dorsal side of vertebrates. Nearly all nerves that innervate the viscera or muscles below the head pass through the spinal cord, and nearly all sensory information from below the head passes through the spinal cord on the way to the brain. The spinal cord can also integrate simple motor responses (e.g., reflexes) by itself. A cross-section of the spinal cord reveals an outer white matter area containing motor and sensory axons and an inner gray matter area containing nerve cell bodies. Sensory information enters the spinal cord dorsally; the cell bodies of these sensory neurons are located in the dorsal root ganglia. All motor information exits the spinal cord ventrally. Nerve branches entering and leaving the cord are called roots. The spinal cord is divided into four regions (in order from the brain stem to the tail): cervical, thoracic, lumbar, and sacral.

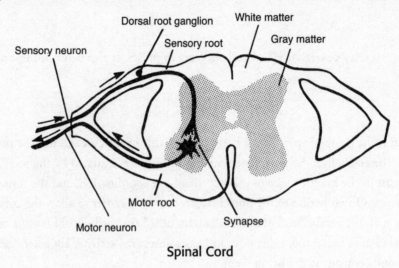

Spinal Cord

PERIPHERAL NERVOUS SYSTEM ★★★

The **peripheral nervous system** (**PNS**) consists of 12 pairs of cranial nerves, which primarily innervate the head and shoulders, and 31 pairs of spinal nerves, which innervate the rest of the body. Cranial nerves exit from the brain stem, and spinal nerves exit from the spinal cord.

The PNS has two primary divisions: the somatic and the autonomic nervous systems, each of which has both motor and sensory components.

SOMATIC NERVOUS SYSTEM

The **somatic nervous system** (**SNS**) innervates skeletal muscles and is responsible for voluntary movement. Motor neurons release the neurotransmitter acetylcholine (ACh) onto ACh receptors located on skeletal muscle. This causes depolarization of the skeletal muscle, leading to muscle contraction. In addition to voluntary movement, the somatic nervous system is also important for reflex action. There are both **monosynaptic** and **polysynaptic reflexes**.

Monosynaptic reflex pathways have only one synapse between the sensory neuron and the motor neuron. The classic example is the knee-jerk reflex. When the tendon covering the patella (kneecap) is hit, stretch receptors sense this and action potentials are sent up the sensory neuron and into the spinal cord. The sensory neuron synapses with a motor neuron in the spinal cord, which in turn stimulates the quadriceps muscle to contract, causing the lower leg to kick forward.

In polysynaptic reflexes, sensory neurons synapse with more than one neuron. A classic example of this is the **withdrawal reflex**. When a person steps on a nail, the injured leg withdraws in pain, while the other leg extends to retain balance.

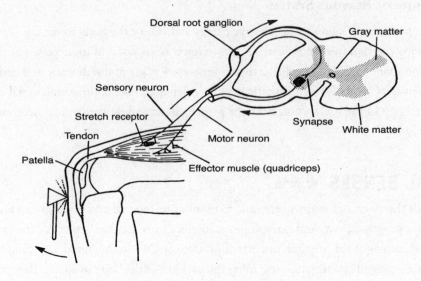

Reflex Arc for Knee Jerk

AUTONOMIC NERVOUS SYSTEM ★★★

The **autonomic nervous system (ANS)** is sometimes also called the involuntary nervous system because it regulates the body's internal environment without the aid of conscious control. Whereas the somatic nervous system innervates skeletal muscle, the ANS innervates cardiac and smooth muscle. Smooth muscle is located in areas such as blood vessels, the digestive tract, the bladder, and bronchi, so it isn't surprising that the ANS is important in blood pressure control, gastrointestinal motility, excretory processes, respiration, and reproductive processes. ANS pathways are characterized by a two-neuron system. The first neuron (preganglionic neuron) has a cell body located within the CNS and its axon synapses in peripheral ganglia. The second neuron (postganglionic neuron) has its cell body in the ganglia and then synapses on cardiac or smooth muscle. The ANS is composed of two subdivisions, the **sympathetic** and the **parasympathetic nervous systems**, which generally act in opposition to one another.

SYMPATHETIC NERVOUS SYSTEM

The sympathetic division is responsible for the fight-or-flight responses that ready the body for action. It basically does everything you would want it to do in an emergency situation. It increases blood pressure and heart rate, increases blood flow to skeletal muscles, and decreases gut motility. The preganglionic neurons emerge from the thoracic and lumbar regions of the spinal cord and use acetylcholine as their neurotransmitter; the postganglionic neurons typically release norepinephrine. The action of preganglionic sympathetic neurons also causes the adrenal medulla to release adrenaline (epinephrine) into the bloodstream.

PARASYMPATHETIC NERVOUS SYSTEM

The parasympathetic division acts to conserve energy and restore the body to resting activity levels following exertion (rest and digest). It acts to lower heart rate and to increase gut motility. One very important parasympathetic nerve that innervates many of the thoracic and abdominal viscera is the **vagus** nerve. Parasympathetic neurons originate in the brain stem (cranial nerves) and the sacral part of the spinal cord. Both the preganglionic and postganglionic neurons release acetylcholine.

SPECIAL SENSES ★★★

The body has three types of sensory receptors to monitor its internal and external environment: interoceptors, proprioceptors, and exteroceptors. Interoceptors monitor aspects of the internal environment such as blood pressure, the partial pressure of CO_2 in the blood, and blood pH. Proprioceptors transmit information regarding the position of the body in space. These receptors are located in muscles and tendons to tell the brain where the limbs are in space, and are also located in the inner ear to tell the brain where the head is in space. Exteroceptors sense things in the external environment such as light, sound, taste, pain, touch, and temperature.

THE EYE

The eye detects light energy (as photons) and transmits information about intensity, color, and shape to the brain. The eyeball is covered by a thick, opaque layer known as the **sclera**, which is also known as the white of the eye. Beneath the sclera is the **choroid** layer, which helps to supply the retina with blood. The innermost layer of the eye is the **retina**, which contains the photoreceptors that sense the light.

The transparent **cornea** at the front of the eye bends and focuses light rays. The rays then travel through an opening called the **pupil**, whose diameter is controlled by the pigmented, muscular **iris**. The iris responds to the intensity of light in the surroundings (light makes the pupil constrict). The light continues through the lens, which is suspended behind the pupil. The lens, the shape of which is controlled by the **ciliary muscles**, focuses the image onto the retina. In the retina are photoreceptors that **transduce** light into action potentials. There are two main types of photoreceptors: **cones** and **rods**. Cones respond to high-intensity illumination and are sensitive to color, while rods detect low-intensity illumination and are important in night vision. The cones and rods contain various pigments that absorb specific wavelengths of light.

The cones contain three different pigments that absorb red, green, and blue wavelengths; the rod pigment, **rhodopsin**, absorbs one wavelength. The photoreceptor cells synapse onto **bipolar cells**, which in turn synapse onto **ganglion cells**. Axons of the ganglion cells bundle to form the right and left **optic nerves**, which conduct visual information to the brain. The point at which the optic nerve exits the eye is called the **blind spot** because photoreceptors are not present there.

There is also a small area of the retina called the fovea, which is densely packed with cones, and is important for high acuity vision.

The eye also has its own circulation system. Near the base of the iris, the eye secretes aqueous humor, which travels to the anterior chamber of the eye from which it exits and eventually joins venous blood.

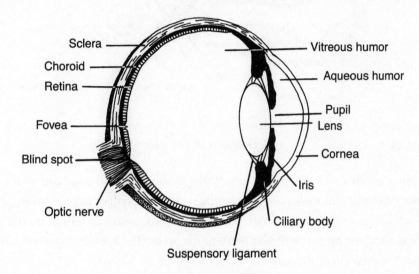

Sclera
Choroid
Retina
Fovea
Blind spot
Optic nerve
Suspensory ligament
Vitreous humor
Aqueous humor
Pupil
Lens
Cornea
Iris
Ciliary body

Diagram of the Eye

THE EAR

The ear transduces sound energy (pressure waves) into impulses perceived by the brain as sound. The ear is also responsible for maintaining equilibrium (balance) in the body.

Sound waves pass through three regions as they enter the ear. First, they enter the **outer ear**, which consists of the **auricle** (pinna) and the **auditory canal**. At the end of the auditory canal is the **tympanic membrane (eardrum)** of the **middle ear**, which vibrates at the same frequency as the incoming sound. Next, the three bones, or ossicles (**malleus, incus,** and **stapes**), amplify the stimulus and transmit it through the **oval window**, which leads to the fluid-filled **inner ear**. The inner ear consists of the **cochlea** and the **semicircular canals**. The cochlea contains the **organ of Corti**, which has specialized sensory cells called hair cells. Vibration of the ossicles exerts pressure on the fluid in the cochlea, stimulating the hair cells to transduce the pressure into action potentials, which travel via the **auditory (cochlear) nerve** to the brain for processing.

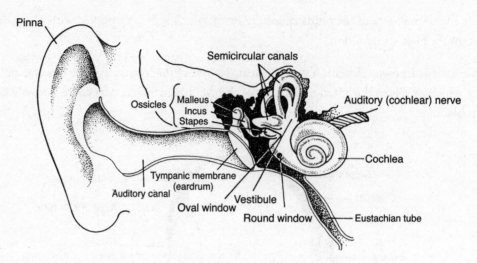

Diagram of the Ear

The three semicircular canals are each perpendicular to the other two and filled with a fluid called **endolymph**. At the base of each canal is a chamber with sensory hair cells; rotation of the head displaces endolymph in one of the canals, putting pressure on the hair cells in it. This changes the nature of impulses sent by the vestibular nerve to the brain. The brain interprets this information to determine the position of the head.

THE CHEMICAL SENSES

The chemical senses are taste and smell. These senses transduce chemical changes in the environment, specifically in the mouth and nose, into **gustatory** and **olfactory** sensory impulses, which are interpreted by the nervous system.

TASTE

Taste receptors, or **taste buds**, are located on the tongue, the soft palate, and the epiglottis. Taste buds are composed of approximately 40 epithelial cells. The outer surface of a taste bud contains a **taste pore**, from which microvilli, or **taste hairs**, protrude. The receptor surfaces for taste are on the taste hairs. Interwoven around the taste buds is a network of nerve fibers that are stimulated by the taste buds. These neurons transmit gustatory information to the brainstem via three cranial nerves. There are four kinds of taste sensations: sour, salty, sweet, and bitter. Although most taste buds will respond to all four stimuli, they respond preferentially; i.e., at a lower threshold, to one or two of them.

SMELL

Olfactory receptors are found in the olfactory membrane, which lies in the upper part of the nostrils over a total area of about 5 cm². The receptors are specialized neurons from which **olfactory hairs**, or cilia, project. These cilia form a dense mat in the nasal mucosa. When odorous substances enter the nasal cavity, they bind to receptors in the cilia, depolarizing the olfactory receptors. Axons from the olfactory receptors join to form the **olfactory nerves**. The olfactory nerves project directly to the **olfactory bulbs** in the base of the brain.

APPLYING THE CONCEPTS: BIG IDEA THREE REVIEW QUESTIONS

MOLECULAR GENETICS, GENE REGULATION, AND MUTATIONS

This Review Quiz is designed to assess your content knowledge, and is not necessarily representative of the actual AP Biology exam. So don't panic if you see a new question type or a question asking about a relatively low-yield detail from the chapter.

1. All of the following are considered posttranscriptional modifications that occur in the nucleus EXCEPT

   (A) 5′ capping with methylated guanines.

   (B) the addition of a 3′ poly-adenine tail.

   (C) the excision of introns from mRNA via spliceosome formation.

   (D) mRNA attachment to polyribosomes.

2. A scientist places free strands of DNA, which contain a gene that codes for the protein allowing the metabolism of glucose, in a medium containing bacteria that can only survive on the sugar lactose. The scientist heat shocks the bacteria in $CaCl_2$ and lets them recover before plating them in several petri dishes with only glucose as a nutrient source. After several days, there are no signs of bacterial growth in the glucose medium. All of the following are possible explanations for the results EXCEPT:

   (A) All of the bacteria died from the heat shock treatment.

   (B) The gene for glucose metabolism was not incorporated into any of the bacteria.

   (C) The genes inserted into the bacteria didn't have the code for the metabolism of glucose.

   (D) The free strands of DNA were from a sheep, so they will not function in bacteria.

3. The complementary strand of DNA to the DNA fragment 5′-GGC ATA CAT-3′ is

   (A) 3′-CCG UAU GUA-5′.

   (B) 3′-GTA TAT CCG-5′.

   (C) 3′-ATG TAT GCC-5′.

   (D) 3′-CCG TAT GTA-5′.

4. An operon in a bacterial genome is composed of

   (A) several structural genes that express proteins with similar function.

   (B) one or more structural genes, a promoter, an operator, and a terminator.

   (C) incorporated DNA from plasmids.

   (D) viral DNA from infection and reverse transcription.

5. The following diagram shows the results of gel electrophoresis of several fragments of DNA obtained from a restriction enzyme digestion. What can be concluded from the diagram?

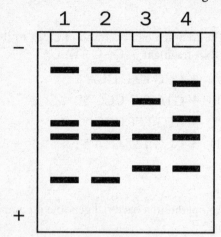

(A) The DNA in wells 1, 2, 3, and 4 are likely from the same organism.

(B) The DNA in wells 1 and 2 are likely from the same organism.

(C) The DNA in wells 1 and 3 are likely from the same organism.

(D) The DNA in wells 1 and 4 are likely from the same organism.

6. Which of the following types of mutation involves only one base pair?

(A) An inversion

(B) A deletion

(C) An addition

(D) A substitution

7. What type of bonds join one strand of DNA to another strand of DNA, forming a double helix?

(A) Ionic bonds

(B) Hydrogen bonds

(C) Polar covalent bonds

(D) Hydrophobic interactions

8. Which of the following statements is FALSE?

(A) Mutations are important in evolution.

(B) Errors in DNA replication can cause mutations.

(C) Mutagens are substances that cause cancer.

(D) Products of cellular metabolism can cause mutations.

9. In gel electrophoresis, DNA fragments migrate toward the electrode that is

(A) smaller.

(B) negative.

(C) positive.

(D) larger.

Questions 10–13 refer to the following list.

(A) Uracil nucleotide
(B) Guanine nucleotide
(C) Transcription
(D) Splicing

Questions 14–16 refer to the following list.

(A) Conjugation
(B) Replication
(C) Transcription
(D) Transformation

10. This takes place only in eukaryotes.

11. This forms three hydrogen bonds when linked with a cytosine nucleotide.

12. DNA → RNA

13. This is found only in RNA.

14. This is the process in which free DNA is incorporated into the genome of a bacterium.

15. This is the process that creates duplicate strands of DNA.

16. This is the process in which bacteria engage in the sexual exchange of DNA.

ANSWERS AND EXPLANATIONS

1. D

All of the choices listed describe posttranscriptional modifications, those that take place *in the nucleus* after transcription, except (D), the attachment of mRNA to polyribosomes. Polyribosomes are groups of multiple ribosomes that can attach to mRNA to read the codons and produce multiple polypeptide chains at the same time. This takes place outside the nucleus during translation.

2. D

Choices (A), (B), and (C) describe possible problems associated with the procedure or materials given. The correct answer is (D). All DNA is composed of base pairs; it doesn't matter what organism it comes from. DNA can even be manufactured from scratch and not ever exist in an organism.

3. D

All you have to do here is remember that bases pair up between G and C and A and T in DNA. Remember this is different with RNA, where there are no T's. The sample fragment is given from the 5′ to the 3′ end, so your complementary strand will be from 3′ to 5′ and be in the same order but with the appropriate base pair. Choices (B) and (C) have the correct code in reverse, individually, (C), or by triplets, (B). Choice (A) contains uracil, so it can't be a DNA complement. The correct answer is (D). If the question had asked for the complementary strand of RNA, the answer would have been (A), but it asked for the complementary strand of DNA.

4. B

An operon has several different types of DNA regions. You may consider answer choice (A) because an operon has the structural genes that code for proteins for similar function, but there is more to an operon. Choices (C) and (D) can be part of an operon but do not define it completely. The most correct answer is (B) because it lists all of the parts of the operon.

5. B

You can approach this question two ways. First, you can guess that because the possible responses are about similarity and dissimilarity, (B) is most likely correct because the banding pattern is the same in wells 1 and 2. This is the correct answer. The other approach is to remember that restriction enzymes are very specific and, therefore, will cut DNA only at sites with the same series of base pairs. The only way the base pairing can be identical is if the DNA code that is being digested is identical.

6. D

First, eliminate the answer choices that are obviously wrong. An inversion by nature must involve more than one base pair, otherwise it would not mean anything to reverse it. Both (B) and (C) can involve only one base pair and are the tricky choices to eliminate, but they can also involve any number of base pairs. Choice (D) is a direct exchange of only one base pair, so it is the correct answer.

7. B

In forming the double helix, one strand of DNA is joined to another strand via hydrogen bonds.

8. C

Mutagens are substances that cause mutations; carcinogens are substances that cause cancer. Not all mutations cause cancer. Also, if a mutation occurs in a gene that is not expressed, the mutation may not result in cancer. The other statements are true: Mutations are important in evolution because without mutations, there would be no new alleles that could be selected for; errors in DNA replication can cause mutations; and products of cellular metabolism, such as superoxide, can cause mutations.

9. C

DNA fragments are negatively charged, so in gel electrophoresis, DNA fragments migrate toward the positive electrode.

10. D

Splicing is the removal of introns, or sections of DNA that are not expressed. Splicing takes place only in eukaryotes.

11. B

A guanine nucleotide forms three hydrogen bonds when linked with a cytosine nucleotide.

12. C

Transcription is the process of transferring genetic information in DNA to an RNA message that is decoded to produce protein.

13. A

Uracil nucleotides are found only in RNA.

14. D

Scientists have used this phenomenon in genetic engineering research.

15. B

DNA replication involves the unwinding of DNA and production of a complementary strand for each strand.

16. A

You may be able to remember this because the word *conjugal* is used to refer to marriage.

THE CELL CYCLE

This Review Quiz is designed to assess your content knowledge, and is not necessarily representative of the actual AP Biology exam. So don't panic if you see a new question type or a question asking about a relatively low-yield detail from the chapter.

1. When does independent assortment take place?

   (A) Telophase in meiosis I

   (A) Telophase in meiosis II

   (B) Metaphase in meiosis I

   (C) Metaphase in meiosis II

2. Which of the following is FALSE with regard to mitosis and meiosis?

   (A) Crossing over takes place in prophase I of meiosis.

   (B) Chromosomes line up on the metaphase plate in prophase I of meiosis.

   (C) DNA is replicated during the S phase of the cell cycle.

   (D) Mitosis is the production of identical daughter cells.

3. A mutation has occurred on one chromosome during prophase II. This mutation will show up in how many of the gametes produced at the end of meiosis?

   (A) None

   (B) Two

   (C) All

   (D) At least one

4. At what stage(s) are cells haploid?

   (A) Interphase only

   (B) Telophase I

   (C) Telophase II

   (D) Prophase II and telophase II

5. During what phase does the cell grow and replicate its organelles and chromosomes?

   (A) G_1

   (B) G_2

   (C) S

   (D) None of the above

6. When do spindle fibers attach to kinetochores?

   (A) Interphase

   (B) Telophase

   (C) Metaphase

   (D) Anaphase

7. When are nucleoli present?

   (A) Interphase of mitosis

   (B) Metaphase of mitosis

   (C) Anaphase of meiosis

   (D) Prophase of mitosis and meiosis

8. One of two identical *sister* parts of a duplicated chromosome is called what?

   (A) Gamete

   (B) Chromatid

   (C) Chromatin

   (D) Somatic cell

Questions 9–13 refer to the following list.

(A) Prophase

(B) Metaphase

(C) Anaphase

(D) Telophase

9. This is identical in meiosis I and II.

10. This may contain haploid cells.

11. Spindle movement occurs during this.

12. Chromosomes coil during this.

13. Centromeres attach during this.

14. All of the following occur during both mitosis and meiosis I EXCEPT:

(A) Chromosomes have sister chromatids.

(B) Nuclear DNA is duplicated during interphase.

(C) Chromosomes condense and the nuclear membrane disappears.

(D) Chromosomes synapse into tetrads.

15. The cancer drug Taxol interferes with the separation of chromosomes during anaphase of mitosis. The likely mechanism of Taxol's action is

(A) interference with the synthesis of cyclin-Cdk complexes.

(B) prevention of kinetochore microtubule breakdown.

(C) excitation of inhibitory transcription factors.

(D) methylation of select areas of DNA on metaphase chromosomes.

ANSWERS AND EXPLANATIONS

1. A

Independent assortment is the formation of random combinations of chromosomes. It happens in telophase of meiosis I because that is when homologous chromosomes separate from each other into different cells.

2. B

 Choice (A) is correct: Crossing over occurs during prophase I of meiosis. Choice (C) is correct: DNA is replicated during the S phase of interphase. Choice (D) is correct: Mitotic division retains the $2N$ chromosome count in both daughter cells. Choice (B) is false: The chromosomes line up on the metaphase plate during metaphase I of meiosis.

3. D

If a mutation occurred during prophase II, then that cell will be affected as it undergoes telophase II. At least one cell will contain the mutation. However, it is uncertain if both daughter cells will have the mutation because the mutation may or may not be present in both sister chromatids.

4. C

Cells are haploid after they have undergone telophase II.

5. C

S phase is the synthesis phase. This is where DNA duplicates itself. Both G phases are times of cell growth and protein synthesis, but the DNA is already replicated.

6. C

Spindle fibers attach to kinetochores during the early stages of metaphase—sometimes called prometaphase.

7. A

Interphase is the only time where nucleoli are present. After interphase, the nucleoli are broken down for cellular replication.

8. B

Sister chromatids are created when a single chromosome is replicated into two copies of itself. Sister chromatids then separate during meiosis II to form two haploid cells.

9. D

In telophase, cytokinesis is identical in meiosis I and II.

10. D

Telophase II produces haploid cells.

11. A

Spindle fibers travel to opposite poles during prophase.

12. A

Chromosomes coil and condense during prophase.

13. B

Centromeres attach to spindle fibers in metaphase.

14. D

Remember that you are comparing mitosis with the first part of meiosis, which is where the real differences between mitosis and meiosis occur. The two main differences are that in meiosis, the chromosomes undergo synapsis as well as crossover (sometimes), and homologous pairs of chromosomes align at metaphase I instead of individual chromosomes.

15. B

Chromosomes separate to opposite poles of the cell during anaphase of mitosis due to the rapid shortening of kinetochore microtubules, those parts of the spindle that are attached to the centromere regions of the chromosomes. Taxol interferes with the breakdown of these microtubules, thereby halting the separation of the chromosomes and stopping mitosis in its tracks, perfect for halting the spread of a tumor. No new transcription is required at this point; therefore, (C) and (D) are incorrect.

CHROMOSOMAL BASIS OF INHERITANCE

This Review Quiz is designed to assess your content knowledge, and is not necessarily representative of the actual AP Biology exam. So don't panic if you see a new question type or a question asking about a relatively low-yield detail from the chapter.

1. Nondisjunction can occur in _____ and may result in _____.

    (A) anaphase; extra chromosomes

    (B) prophase; extra chromosomes

    (C) anaphase; extra chromosomes or missing chromosomes

    (D) prophase; extra chromosomes or missing chromosomes

Questions 2–4 refer to the following list.

    (A) Genotype

    (B) Phenotype

    (C) Allele

    (D) Heterozygous

2. This is the observable appearance reflecting the expression of genes.

3. What is the genetic makeup of an individual?

4. What is an organism that has two different forms of a gene?

5. Which of the following statements is true?

    (A) There are three hydrogen bonds linking adenine to thymine.

    (B) Transcription takes place at the ribosome.

    (C) DNA is composed of a deoxyribose sugar and a hydrogen replacing the –OH group at the 3′ carbon.

    (D) Translation takes place at the ribosome.

Questions 6–9 refer to the following list.

    (A) Incomplete dominance

    (B) Polygenic inheritance

    (C) Genetic recombination

    (D) Horizontal gene transfer

6. Crossing over is an example of this.

7. With this, offspring display a combined phenotype that is different from both parents.

8. What occurs when an organism gives genetic material to cells that are not its offspring?

9. Several interacting genes control a single trait in this.

10. Which of the following statements is true?

    (A) In asexual reproduction, a female organism fertilizes her own eggs.

    (B) In sexual reproduction, mitosis results in the production of gametes.

    (C) In asexual reproduction, a male organism fertilizes his own eggs.

    (D) A full cycle of meiosis produces four cells.

11. Which of the following statements is true?

    (A) Chromosomes can be haploid, diploid, and polyploid.

    (B) All chromosomes are diploid.

    (C) All chromosomes are polyploid.

    (D) All chromosomes are haploid.

12. In most genes, there is one allele that is _____ over the other and hides the expression of that other allele in the offspring phenotype.

    (A) recessive

    (B) incompletely dominant

    (C) dominant

    (D) incompletely recessive

Questions 13–15 refer to the following list.

    (A) Polyploid

    (B) Haploid

    (C) Diploid

13. This has the designation $2N$.

14. This is only found in certain plants.

15. This is found in almost all organisms that use sexual reproduction.

ANSWERS AND EXPLANATIONS

1. C
Nondisjunction can occur in anaphase and may result in extra copies of chromosomes or missing chromosomes.

2. B
The observable appearance of an organism that reflects the expression of genes is called the organism's phenotype.

3. A
The genetic makeup of an individual is the individual's genotype.

4. D
An organism that has two different forms of a gene is heterozygous.

5. D
Choice (A) is incorrect because there are only two hydrogen bonds formed between adenine and thymine. Choice (B) is incorrect because transcription takes place in the nucleus while the ribosome is the site of translation. Choice (C) is incorrect because in DNA the hydroxyl group replacement takes place at the 2′ carbon of the ribose.

6. C
Crossing over is an example of genetic recombination.

7. A
Incomplete dominance is a form of inheritance in which both heterozygous alleles are expressed. For example, a black mouse and a white mouse create gray offspring.

8. D
Horizontal gene transfer occurs when an organism transfers genetic material to cells that are not its offspring.

9. B
Polygenic inheritance is a type of inheritance in which several interacting genes control a single trait.

10. D
Choices (A) and (C) are incorrect because a hermaphrodite fertilizes its own eggs. Choice (B) is incorrect because mitosis results in the production of two identical cells. Choice (D) is correct because meiosis I and II will produce four gametes from an original cell.

11. A
There are some organisms that exist in a natural state with haploid or polyploid cells, but most organisms are diploid. Therefore, without more specific information, chromosomes can be any of the three.

12. C
In most genes, there is one allele that is dominant over the other and hides the expression of that other allele in the phenotype of the offspring. This is basic terminology that should be quick and easy to work through.

13. C
In general, when we say a cell is diploid, that means it has one chromosome from each parent, or $2N$ for short.

14. A
Polyploidy is very rare and is only found in some plants. If an animal, especially a human, has polyploid genetics, then that is the result of major problems during cell division, and those cells will not function properly.

15. C
Diploid chromosomes are the most common type. They occur in organisms that reproduce sexually and get half of their genetics from each parent.

MENDELIAN GENETICS

This Review Quiz is designed to assess your content knowledge, and is not necessarily representative of the actual AP Biology exam. So don't panic if you see a new question type or a question asking about a relatively low-yield detail from the chapter.

1. In tabby cats, black coloration is caused by the presence of a particular allele on the X chromosome. A different allele at the same locus can result in orange color. The heterozygote calico cat, however, with splotches of both black fur and orange fur, is due to

   (A) polar body formation and inactivation.

   (B) crossing over during gamete formation.

   (C) the formation of thymine dimers.

   (D) Barr body formation.

2. Assume that genes *A* and *B* are not linked. If an organism is heterozygous for both genes, what is the probability that the organism will produce a gamete with both the *A* and *B* alleles?

   (A) $\frac{1}{4}$

   (B) $\frac{1}{2}$

   (C) 1

   (D) 0

3. A man and woman have two sons and would like to have a little girl. As the couple's genetic counselor, you explain to the parents that the chance their third child will be a girl is

   (A) 1 in 1.

   (B) 1 in 2.

   (C) 1 in 3.

   (D) 1 in 4.

4. Which of the following defines a test cross?

   (A) A cross between two heterozygous individuals

   (B) A cross between two homozygous recessive individuals

   (C) A cross between an individual with the dominant phenotype and one with the recessive phenotype

   (D) A cross between two individuals with the dominant phenotype

5. What is the percentage of homozygous yellow individuals in the F2 progeny resulting from the original cross of a homozygous yellow pea plant and a homozygous green pea plant?

   (A) 0%

   (B) 25%

   (C) 50%

   (D) 75%

6. In *Drosophila*, the traits for red eye color (*R*) and straight wings (*W*) are dominant, and the traits for white eye color (*r*) and curly wings (*w*) are recessive. A cross between two flies produces progeny comprised of 607 red-eyed flies with straight wings and 202 red-eyed flies with curly wings. Assuming neither of these two genes is sex-linked, which of the following are most likely to be the genotypes of the parents?

   (A) $RRWW \times RRWW$

   (B) $RRWW \times RRWw$

   (C) $RrWw \times RrWw$

   (D) $RrWw \times RRWw$

7. Red-green color blindness is a sex-linked recessive trait in humans. Which of the following is true if a colorblind woman and a man with normal vision have children?

   (A) None of their children will be colorblind.

   (B) Half of their male children will be colorblind.

   (C) Half of their female children will be colorblind.

   (D) All of their male children will be colorblind.

8. Multiple crosses involving genes known to occur on the same chromosome produce frequencies of phenotypes that suggest there is a high rate of crossover between these two genes. Which of the following is the most likely explanation for the phenotypic frequencies observed due to crossing over?

   (A) The two genes are far apart from one another.

   (B) The two genes are both recessive.

   (C) The two genes have incomplete dominance.

   (D) The two genes are both located far from the centromere.

9. Which of the following is true of meiosis II?

   (A) It produces two identical daughter cells.

   (B) It produces haploid daughter cells.

   (C) Chromosomes create chiasmata during crossing over.

   (D) Homologous chromosomes travel to opposite poles.

10. Which of the following phenotypic ratios describes the results of a dihybrid cross between two genes that follow Mendel's Laws of Segregation and Independent Assortment?

    (A) 3:1

    (B) 1:1

    (C) 7:2:2:1

    (D) 9:3:3:1

11. The following pedigree traces the incidence of the hereditary disease β-thalassemia in a particular family. β-thalassemia is a hemolytic anemia that arises from the abnormal synthesis of β-chains in hemoglobin.

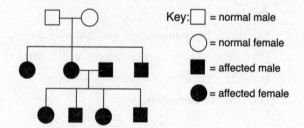

According to the pedigree, the gene for β-thalassemia is inherited as an

    (A) autosomal recessive trait.

    (B) autosomal dominant trait.

    (C) X-linked recessive trait.

    (D) X-linked dominant trait.

12. What is the probability that a mother who is a carrier for cystic fibrosis (an autosomal recessive disorder) will pass the diseased allele on to her child whose father is genotypically normal?

(A) 50%

(B) 25%

(C) 100%

(D) 0%

13. Given the following pedigree, what is the likelihood the offspring is a carrier?

(A) 25%

(B) 50%

(C) 66%

(D) 75%

14. What is the inheritance pattern of the observed trait indicated by the pedigree shown?

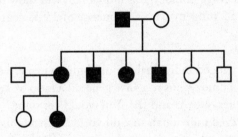

(A) Autosomal recessive

(B) Autosomal dominant

(C) X-linked recessive

(D) Cannot be determined

15. What is the inheritance pattern of the observed trait indicated by the pedigree shown?

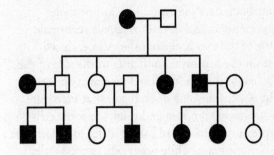

(A) Autosomal recessive

(B) Autosomal dominant

(C) X-linked recessive

(D) Cannot be determined

16. What rule describes the separation of alleles in the parent genotype during the process of gametogenesis?

(A) Law of Independent Assortment

(B) Law of Segregation

(C) Nondisjunction

(D) Crossing over

ANSWERS AND EXPLANATIONS

1. D

Barr bodies are inactivated X chromosomes commonly found in cells of female mammals. One of the two X chromosomes in each cell within the organism condenses to the side of the nucleus and cannot be used for transcription. The X chromosome of each pair that inactivates is apparently random and results in some cells expressing traits coded for by alleles found on one X chromosome, while other cells express traits coded for by alleles on the other X. This leads certain patches of calico tabby fur to express orange coloration while other patches remain black.

2. A

Because the organism is heterozygous for both alleles, it has an *AaBb* genotype. This means that the probability is $\frac{1}{2}$ that any gamete will get an *A* allele and $\frac{1}{2}$ that any gamete will get a *B* allele. The probability that a gamete will get both alleles is $\frac{1}{2} \times \frac{1}{2}$, or $\frac{1}{4}$. Thus, the correct answer is (A). The correct answer can't be (B), because $\frac{1}{2}$ is the probability that any gamete will get one of the alleles, not both. If the organism is heterozygous for both genes, it can't have *AB* in all of the gametes, so the answer can't be (C). Similarly, the heterozygous organism will be able to produce some *AB* gametes, so the answer is not (D).

3. B

Don't get trapped here. The chance of having either a boy or girl is always 50 percent regardless of the gender of the previous children. The mother is always going to produce eggs with an X chromosome. Half of the father's sperm will carry an X chromosome and half will carry a Y chromosome. This makes it even odds for either a boy or girl.

4. C

A test cross is a way of discovering the genotype of an individual expressing the dominant phenotype. The individual can either be homozygous dominant or heterozygous. Because at least one individual has to express the dominant phenotype, (B) can be ruled out. The way to discover the unknown genotype is to cross the individual with an individual that is homozygous for the recessive allele. The individual that is homozygous for the recessive allele will express the recessive phenotype.

5. B

This question sounds complicated but it isn't—you don't even need to know which color is dominant. The gametes produced by the homozygous yellow plant are all going to have the yellow allele, and the homozygous green plant will only produce gametes with the green allele. This means that all of the F1 progeny will be hybrids for seed color. The phenotypic ratio of the progeny of a monohybrid cross (in this case, the F2 generation) is 3:1 with one homozygous dominant individual, two heterozygous individuals, and one homozygous recessive individual (on average for every four individuals). It doesn't matter whether yellow is dominant or recessive because both the dominant homozygous and recessive homozygous individuals will show up in the F2 progeny with a frequency of 25 percent.

6. D

Curly winged individuals would only be expressed with a homozygous recessive gene, in a ratio of 1:3. This rules out (A) and (B). You would get some white-eyed individuals in a dihybrid cross, so this rules out (C). The correct answer must be (D) by process of elimination. You can test your answer with a Punnett square.

7. D

You are given two crucial pieces of information. One is that the gene is sex-linked recessive and the other is that the mother expresses the trait. This means that the mother must be homozygous for the recessive allele, which is expressed in the male children. All of the children will get a colorblind allele from the mother. The female children will get their second X chromosome from the father, who has a normal allele for seeing colors, and because the colorblind allele is recessive it won't be expressed.

Choice (C) can't be correct. The male children will get a Y chromosome from their father, so they must get a colorblind allele from their mother. All of the male children will be colorblind. This rules out (A) and (B).

8.   A
The question specifically asks for the effect of crossing over, which relates to the location of genes on the chromosome. This rules out (B) and (C) because while dominance can affect phenotypic frequencies, it has nothing to do with crossing over or gene locus. Choice (D) sounds appealing because it makes sense that there will be more crossover events farther from the centromere. However, (D) does not address the relative location of the two genes to one another. If both genes are far from the centromere but very close to each other, there is little chance that the genes will cross in relation to each other. High rates of crossover occur between genes when they are located far from each other on the chromosome, which leads to (A).

9.   B
This should be an easy question after you review meiosis. The only tricky choices are (C) and (D), but both of these things occur in meiosis I. Choice (A) is the result of mitosis, not of meiosis. There are four daughter cells produced at the end of meiosis, but they are all haploid, which makes (B) the correct answer.

10.   D
The question tells you that the genes are autosomal, not linked, and not subject to varying probabilities in their phenotypes due to crossover. One of the possible answers, (A), is a familiar phenotypic ratio from a standard cross. The other ratios in the list shouldn't look familiar at all. Don't let them confuse you. If you simply can't remember that the results of a dihybrid cross give you the ratio 9:3:3:1, you might try to figure out the correct answer through logic. The question deals with two genes so there are four different gamete genotypes possible. A Punnett square with four different gametes from the parents has 16 boxes for progeny phenotypes. Adding up the numbers in the aforementioned ratio gives you 9 + 3 + 3 + 1 = 16. No other ratio listed gives a cumulative total of 16.

11.   A
Because males can pass the trait to males, this allows us to rule out that the trait is sex-linked. Fathers will only pass Y chromosomes to their sons. Because parents who do not show the disease can have kids who do show the disease, this allows us to rule out that the trait is dominant. Choice (A) is correct here.

12.   A
The question asks you to find the probability that a mother who is a carrier for the recessive cystic fibrosis trait (*Cc*) will pass the recessive allele (*c*) on to her offspring if she mates with a genotypically normal father (*CC*). Fifty percent will be a carrier for the allele for cystic fibrosis. Students often guess 0 because none of the offspring actually have the disease. Be careful—the question asks for the chance that the mother will pass her allele to her offspring, not the chance that the offspring will actually have the disease.

13.   C
Carriers have the *Rr* genotype. If you do not have the genotype ratio of 1:2:1 memorized (which is a big time-saver), just make a Punnett square. Of the four possible offspring in the Punnett square, notice that *rr* would express the recessive trait, but the generation 2 offspring are not affected, so this is not an option. Two-thirds of the remaining offspring are carriers. Therefore, (C) is correct. Hint: (B) is a miscalculation obtained if you do not eliminate *rr* as an incorrect option.

14. D

Although X-linked recessive, (C), can be eliminated, either autosomal dominant, (B), or autosomal recessive, (A), could be true. You would not want to spend too much time on Test Day labeling every individual's genotype. Instead, look for the patterns of generation skipping or two parents with the same phenotype giving birth to an offspring with the opposite phenotype. Nevertheless, if you did label the original pedigrees, you would find both recessive and dominant patterns would work.

15. A

Generational skipping occurs in the center of the pedigree when two unaffected parents produce an affected offspring. Autosomal dominant, (B), cannot skip generations, and X-linked recessive traits, (C), can only affect a woman with an affected father, unlike what is seen from generation 1 to generation 2.

16. B

The Law of Segregation describes the separation of alleles in the parent genotype during the process of gametogenesis. There can only be a maximum of two different alleles in a single parent; half the gametes get one allele and the other half get the other allele. The Law of Independent Assortment suggests that different genes sort into different gametes, independently of each other. Nondisjunction and crossing over are meant to distract you because they are discussed with Mendelian genetics, but they are not correct answers to this question.

VIRUSES

This Review Quiz is designed to assess your content knowledge, and is not necessarily representative of the actual AP Biology exam. So don't panic if you see a new question type or a question asking about a relatively low-yield detail from the chapter.

1. Rotavirus is an encapsulated virus possessing a double-stranded RNA genome, and it is the major cause of diarrhea in infants. Rotavirus probably causes the diarrhea by

   (A) decreasing the absorption of sodium and water from the intestinal lumen due to damaged intestinal villi.

   (B) secreting RNA-encoded proteins that pull water from intestinal cells into the intestinal lumen.

   (C) increasing the absorption of water into damaged intestinal villi.

   (D) causing damage to renal glomeruli so that excess water and salts are allowed to filter out of the bloodstream into the excretory system.

2. HIV (human immunodeficiency virus) is classified as a retrovirus. This means that

   (A) pieces of its RNA genome are spliced into a single strand of RNA before translation.

   (B) it reverses its morphology from type C to type D as it enters cells.

   (C) it transcribes its RNA genome into a DNA genome.

   (D) the promoter regions for RNA polymerase binding are located downstream of genes rather than upstream of them.

3. What shape of virus is represented in the image shown?

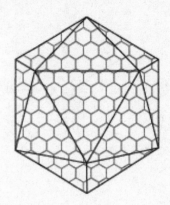

   (A) Helical

   (B) Polyhedral

   (C) Envelope

   (D) Complex

4. The HSV-1 virus, responsible for cold sores, commonly causes latent infection of cranial nerve cells. When reactivated, the HSV may travel either way along the nerve to infect either the lip or the eye and brain. This reactivation most clearly indicates

   (A) the virus has entered the lysogenic cycle.

   (B) the virus has entered the lytic cycle.

   (C) an additional infection with a different microbe has occurred.

   (D) the virus has undergone a mutation.

Questions 5–8 refer to the following diagram.

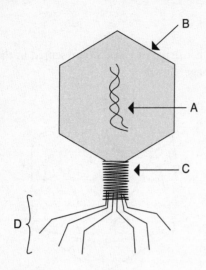

5. This is used to latch onto the host cell surface.

6. Which letter represents the protein shell?

7. This is unique to phages.

8. Which letter represents genetic material?

9. Which of the following is true?

   (A) Viruses have metabolic activity similar to that of bacteria.
   (B) Viral DNA is always found as circular DNA.
   (C) Viral genomes can be 200 genes long.
   (D) Viral RNA is always found as a single strand.

10. Which of the following is the correct progression of the lytic cycle?

    (A) Attachment, assembly, replication, uncoating
    (B) Attachment, uncoating, replication, assembly
    (C) Replication, penetration, uncoating, assembly, release
    (D) Uncoating, attachment, replication, release

11. Which of the following is true?

    (A) Viral DNA can remain indefinitely in the host.
    (B) Viral DNA cannot be directly passed from the host to offspring.
    (C) Viruses cannot choose whether to use the lytic or the lysogenic cycle.
    (D) In the lysogenic cycle, host cells lyse in the final step.

Questions 12–15 refer to the following list.

   (A) Prophage
   (B) Prion
   (C) Viroid
   (D) Lytic phages

12. This is an infectious protein.

13. This may be responsible for some plant diseases but causes no known diseases in other organisms.

14. This causes viral replication after lytic infection.

15. This is bacteriophage integrated into the host DNA.

ANSWERS AND EXPLANATIONS

1. A

Given that diarrhea is an excess of water in the stool, usually caused by improper water and salt absorption in the intestines, the most probable mechanism of rotavirus action is that it damages intestinal epithelial cells so that they can no longer absorb proper amounts of water and salt. Choice (C) can be eliminated because an increase in absorption of water and salt could lead to constipation but not to diarrhea; and (D) can be eliminated because the kidneys are not part of the digestive system, and that is where diarrhea arises. Finally, eliminate (B) because viruses have no protein-making machinery.

2. C

Retroviruses transcribe their RNA genome into DNA after they infect a host cell. They do this by carrying the enzyme reverse transcriptase into the host cell with them, so that they are able to turn ssRNA into dsDNA. This DNA can then integrate into the host cell genome or can be used for immediate transcription of viral genes.

3. B

The 2D shape is hexagonal. The 3D shape is polyhedral with many faces. This image represents the hepatitis A virus and is polyhedral.

4. B

A virus has started actively infecting new tissue. The lytic cycle involves penetration of a virus into new cells. Therefore, this virus is in the lytic cycle.

5. D

The tails are used to latch onto the host cell surface, after which enzymes digest a portion of the cell membrane to insert viral genetic material (either DNA or RNA) into the host cell.

6. B

Virus particles are surrounded by a capsid. Capsids are built out of proteins, many of which have various sugars attached to them.

7. D

Only bacteriophages have the head and tail structures.

8. A

DNA or RNA (depending on the virus) is stored inside the capsid. It can be linear or circular.

9. C

Choice (A) is incorrect because viruses conduct no metabolic activity of their own. Choices (B) and (D) are incorrect because nucleic acid can be linear or circular, and although the DNA is always found together in a single chromosome, the RNA can be in several pieces. Viral genomes can range from five to ten genes to several hundred genes long.

10. B

First is attachment and penetration by the virus. Then uncoating of the viral genome, followed by viral mRNA and protein synthesis. Next is replication of the genome followed by assembly and release of the new virus particles.

11. A

In the lysogenic cycle, viral DNA is integrated into the host cell's DNA and can remain there indefinitely, allowing viral DNA to be replicated for generations alongside the host cell DNA. In the lysogenic cycle, DNA can remain a part of the host's genome and can be replicated and passed down to offspring. Certain environmental events may cause this integrated viral DNA to enter the lytic cycle. Choice (D) is incorrect because that is the final step in the lytic cycle.

12. B

Prions are pieces of protein that are infectious.

13. C

Viroids have been implicated in some plant diseases, though not in diseases in any other organisms.

14. D

Phages that infect bacteria in a lytic way and cause active viral replication are called lytic phages.

15. A

While integrated into the host cell's DNA in a lysogenic fashion, the bacteriophage is known as a prophage.

CELL COMMUNICATION

This Review Quiz is designed to assess your content knowledge, and is not necessarily representative of the actual AP Biology exam. So don't panic if you see a new question type or a question asking about a relatively low-yield detail from the chapter.

1. Junctions between animal cells that resist shearing forces and attach cells through linker proteins are known as

    (A) tight junctions.

    (B) plasmodesmata.

    (C) anchoring junctions.

    (D) occluding junctions.

2. Which of the following is true of paracrine signaling?

    (A) It only exists in plants.

    (B) It requires a neurotransmitter to be released into a synapse.

    (C) Cells secrete a chemical into extracellular fluid.

    (D) The cells must be touching to be paracrine.

Questions 3–5 refer to the following list.

(A) Desmosomes

(B) Tight junctions

(C) Gap junctions

(D) Plasmodesmata

3. These are also called occluding junctions.

4. These use anchoring proteins.

5. These are only found in plants.

6. Which of the following is true?

    (A) Nothing can diffuse between cells held together with occluding junctions.

    (B) Occluding junctions are formed via proteins and sugars.

    (C) Tight junctions are only found in the digestive system.

    (D) Diffusion can occur through occluding junctions.

7. Which of the following is not a characteristic of a hormone?

    (A) Released by specialized cells

    (B) Dispersed via the bloodstream

    (C) Act as second messengers

    (D) Can pass through all cell membranes

8. Which of the following is true of endocrine signaling?

    (A) It acts the same as paracrine.

    (B) It can act without the bloodstream.

    (C) It does not exist in plants.

    (D) It includes nerve cells.

9. Which of the following is NOT true of anchoring junctions?

    (A) They allow cells to adhere to neighboring cells.

    (B) They allow for cell contraction.

    (C) They connect the membrane of a cell to the cytoplasm of another.

    (D) They use actin filaments for anchoring.

10. Connexins

    (A) build connections between adjacent cell membranes.

    (B) allow for fast signal transmission.

    (C) are connections only found in animal cells.

    (D) do not allow for the transmission of viruses.

ANSWERS AND EXPLANATIONS

1. C

Anchoring junctions are those that are found between cells subjected to fair amounts of stress, either from shearing forces or contacting forces. They connect two cells via a series of linker proteins often involving actin filaments.

2. C

Paracrine signaling occcurs when cells communicate by secreting chemicals into the extracellular matrix. None of the other responses reflect the paracrine system.

3. B

Occluding junctions are also called tight junctions because they seal the spaces between the cells.

4. A

Anchoring junctions, also called desmosomes, connect one cell's cytoplasm to another via anchoring proteins.

5. D

Plasmodesmata are the plant versions of gap junctions and allow cells to directly exchange cytoplasmic material.

6. A

Only (A) is a correct statement. Occluding junctions, also called tight junctions, act as a complete barrier and do not allow anything to pass through. These are found in many systems; the digestive system is just one of the examples used in the text.

7. D

By definition, hormones are released by specialized cells, travel via the bloodstream, and bind to plasma membrane receptors to act as second messengers. They cannot pass through all cell membranes. For example, most hormones cannot pass the blood-brain barrier.

8. B

Endocrine signaling in animals refers to the secretion of chemical messengers into the bloodstream, but in plants it can be another form of liquid medium. Plants have hormones similar to animals and, thus, (B) is correct. Synaptic signaling is used for the nervous system. Paracrine signaling involves adjacent cells and, thus, is different from the endocrine signaling process.

9. C

Anchoring junctions connect the cytoplasm of a cell to another's via proteins. Therefore, (C) is false and thus the correct answer. They do allow cells to adhere to neighboring cells and contract into tubelike tissues. The linker proteins adhere to actin or other intermediate filaments for support.

10. B

Connexins are tubes or pores between two cytoplasms of two adjacent cells—not connections between cell membranes; thus, (A) is incorrect. Plasmodesmata are the plant cell equivalent of gap junctions of this type, so (C) is not correct. Similarly, plant viruses can use this connection to pass virus particles from one cell to another.

HORMONES AND THE ENDOCRINE SYSTEM

This Review Quiz is designed to assess your content knowledge, and is not necessarily representative of the actual AP Biology exam. So don't panic if you see a new question type or a question asking about a relatively low-yield detail from the chapter.

1. Prostaglandins are modified fatty acids derived from a molecule called arachidonic acid. They are released by cells into the extracellular fluid and act as local mediators to regulate blood vessel contraction and pain. In such a way, prostaglandins would be considered what kind of regulators?

   (A) Endocrine

   (B) Merocrine

   (C) Paracrine

   (D) Synaptic

2. Hormones

   (A) travel via messenger cells.

   (B) are synthesized by most cell types.

   (C) are the same in the endocrine and nervous systems.

   (D) coordinate responses via receptors.

3. Insulin and glucagon

   (A) are examples of negative feedback systems.

   (B) regulate blood sugar levels.

   (C) are controlled by the nervous system.

   (D) coordinate pancreatic functions.

Questions 4–7 refer to the following list.

   (A) Melatonin

   (B) Epinephrine

   (C) Glucagon

   (D) Follicle-stimulating hormone

4. This regulates gonads.

5. This is secreted by the adrenal gland and is responsible for the fight-or-flight response.

6. This is also known as catecholamine that is responsible for normal sleep cycles.

7. This is secreted by the pancreas to help regulate blood glucose.

8. Which of the following is NOT true?

   (A) Sex hormones are made from lipids.

   (B) Steroid hormones can easily cross the nuclear membrane.

   (C) Steroid hormones can act on DNA in the cell's nucleus.

   (D) Steroid hormones can regulate transcription.

9. Why can steroid hormones last for a long time in the bloodstream?

   (A) They have strong bonds with the carrier proteins.

   (B) Steroid hormones degrade more slowly than nonsteroid hormones.

   (C) The body is slower to metabolize steroid hormones.

   (D) Steroid hormones are lipids.

10. Which of the following is true?

    (A) Nitric oxide causes smooth muscle cells to relax.

    (B) Acetylcholine stimulates smooth muscles to contract.

    (C) A single hormone can bind to several different receptor types.

    (D) Acetylcholine, via nitric oxide, stimulates skeletal muscles to relax.

11. Which of the following is able to cross the plasma membrane?

    (A) Peptides

    (B) Calcitonin

    (C) Glucagon

    (D) Androgen

12. Hormones and transcription factors

    (A) do not interact.

    (B) only turn on transcription of genes.

    (C) use the steroid-receptor complex.

    (D) are not regulated.

Questions 13–16 refer to the following list.

    (A) Ion-channel-linked receptors

    (B) G-protein-linked channels

    (C) Enzyme-linked receptors

    (D) Voltage-gated channels

13. These turn on protein kinases.

14. These cause specific proteins to dissociate from the receptor protein and bind to an enzyme.

15. These are also called ligand-gated channels.

16. These activate proteins that activate second messengers.

ANSWERS AND EXPLANATIONS

1. C

Paracrine signaling results when a secreting cell affects nearby target cells by the production and secretion of various compounds that act only locally. Here, the question stem strongly suggests that prostaglandin secretion is done in a paracrine, or local, manner. Recall that endocrine is secretion into the bloodstream, and exocrine is secretion into ducts that lead out of the body (e.g., sweat glands, digestive system).

2. D

Hormones travel by body fluid, usually blood. They are synthesized and released by specialized cells. The nervous system does not have hormones in the same way that the endocrine system does. Therefore, (A), (B), and (C) are all incorrect. Hormones travel from specialized cells and bind to receptors. The binding of hormones to receptors causes the specific reaction that the hormones were intended to elicit.

3. B

High blood glucose stimulates insulin secretion to lower circulating blood sugar levels. In a negative feedback system, the presence of blood glucose would stop the production of insulin; thus, (A) is not correct. Choice (C) is incorrect because insulin and glucagon are controlled by the endocrine system. Choice (D) is incorrect because the pancreas is one of the controlling organs. It is not controlled by the presence of insulin but instead secretes insulin to combat glucose levels.

4. D

While you do not have to memorize all the hormones, you should be able to reason through them and eliminate answers based on logic. For example, this question asks about regulating gonads. Choices (A), (B), and (C) do not have words or word roots that have to deal with gonads or follicles (another word for gametes). Melatonin (A) should make you think of melanin, both of which are responsible for skin or hair color and sleep cycles.

Epinephrine, (B), should be familiar as a stimulant, especially when having a severe allergic reaction. Glucagon, (C), has been discussed in many ways and you should see the root here (*gluc-*) that is related to sugars. Follicle-stimulating hormone, (D), is the only answer remaining.

5. B

You have probably heard of epinephrine before. If not, you can use the same eliminating techniques as discussed in question 4.

6. A

Both sunlight and the pineal gland can help your body regulate melatonin and, thus, your sleep cycle. People who live in climates that have extreme days and nights during different times of year can struggle to regulate their melatonin levels.

7. C

While it may not be directly tested on the exam, you should understand the regulation of blood glucose levels is based on insulin, glucagon, and the pancreas. It is a great example to use in your short- and long-answer free-response questions.

8. B

Steroid hormones must bind to intracellular receptor proteins to cross the nuclear membrane, so (B) is not true. All the other statements about steroid hormones are true.

9. D

Steroid hormones are lipid-based and, therefore, they do not dissolve in water—water and oil do not mix! Therefore, their effects can last for hours, or even days, in the bloodstream once they are released. Water-soluble, nonsteroid hormones cannot last very long, as they are quickly dissolved.

10. C

Nitric oxide passes into the cell's cytoplasm, acting on the enzyme guanylyl cyclase, which produces cyclic-GMP. Cyclic-GMP is an important molecule that causes smooth muscle cells in blood vessels to relax when stimulated by acetylcholine. Therefore, (A) is not true. It is not the nitric oxide, but rather the things that nitric oxide acts on that causes smooth muscle to relax. Choices (B) and (D) are not true because acetylcholine causes smooth muscle cells to relax and skeletal muscle cells to contract. Choice (C) is true because a variety of receptor complexes can respond to the same hormone. This is also one of the reasons it is difficult to isolate a single hormone as responsible for a function.

11. D

Androgen is the only steroid hormone on this list and, therefore, it is the only one able to cross the plasma membrane. Peptides such as calcitonin and glucagon are powerful signaling molecules, but they do not cross the plasma membrane.

12. C

Be careful with the wording in these answers. Steroid hormones bind to intracellular receptor proteins to cross the nuclear membrane or to regulate transcription. The steroid-receptor complex is often called a transcription factor because it is able to bind to enhancer regions to turn on gene transcription. Choices (A) and (D) are incorrect. Choice (B) is incorrect because transcription factors can turn transcription on or off, or cause it to speed up or slow down.

13. C

The word *kinase* should clue you in that we are talking about enzyme-linked receptors. Voltage-gated channels are not valid here.

14. B

G-protein-linked channels cause G-proteins to dissociate from the cytoplasmic side of the receptor protein and bind to a nearby enzyme.

15. A

Ion-channel-linked receptors are also called ligand-gated channels. They undergo a conformation change when a ligand binds to them so that an opening is created through the membrane to allow the passage of a specific molecule.

16. B

Many G-proteins activate second messengers—small intracellular molecules that activate key enzymes, or transcription factors, which are involved in important reactions.

NEURONS AND THE NERVOUS SYSTEM

This Review Quiz is designed to assess your content knowledge, and is not necessarily representative of the actual AP Biology exam. So don't panic if you see a new question type or a question asking about a relatively low-yield detail from the chapter.

1. Messages delivered by the nervous system are conveyed by a series of neurons. A message is passed from one neuron to the next by

   (A) the release of Ca^{2+} ions by the presynaptic neuron that travel across the synaptic gap and initiate the uptake of Na^+ ions in the postsynaptic neuron.

   (B) the release of neurotransmitters by the presynaptic neuron that travel across the synaptic gap and initiate an action potential in the postsynaptic neuron.

   (C) the release of Na^+ ions by the presynaptic neuron that travel across the synaptic gap and initiate an action potential in the postsynaptic neuron.

   (D) the mechanical deformation of the postsynaptic neuron by the presynaptic neuron.

2. All of the following statements about transmission along neurons are correct EXCEPT:

   (A) The rate of transmission of a nerve impulse is directly related to the diameter of the axon.

   (B) The intensity of a nerve impulse is directly related to the size of the voltage change.

   (C) A stimulus that affects the nerve cell membrane's permeability to ions can either depolarize or hyperpolarize the membrane.

   (D) Once initiated, local threshold depolarization stimulates the propagation of an action potential down the axon.

3. Neurotransmitters characterized as inhibitory would not be expected to

   (A) open K^+ channels.

   (B) open Na^+ channels.

   (C) bind to receptor sites on the postsynaptic membrane.

   (D) open Cl^- channels.

4. Which statement about the nervous system is FALSE?

   (A) A neuron fires an action potential only when the threshold potential is exceeded.

   (B) A ganglion is a group of nerve cell bodies.

   (C) Action potentials travel in two directions due to hyperpolarization.

   (D) Myelin is laid down by Schwann cells and causes accelerated transmission of action potentials.

5. After the learning process of habituation, neurons receiving a stimulus respond with less response than they would have prior to habituation. Which of the following effects of repeated stimulation is a possible explanation for habituation?

   (A) Permanent closure of calcium channels in the terminal membrane

   (B) Increase in the number of neurotransmitter receptors in the postsynaptic membrane

   (C) Decrease in the concentrations of neurotransmitter-degrading enzymes in the synapse

   (D) Neurotransmitter vesicles fusing with the terminal membrane in response to lower excitatory potentials

Questions 6–8 refer to the following diagram showing changes in the membrane potential of a neuron as it carries an impulse.

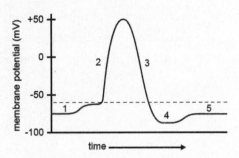

(A) 1

(B) 2

(C) 3

(D) 4

6. This letter represents the phase of the action potential in which only Na^+ channels are open.

7. This letter represents the phase of the action potential representing the end of the absolute refractory period.

8. This is the part of the diagram representing the initial resting membrane potential.

9. A model of the simple nervous system of the sea snail, *Aplysia*, is shown here.

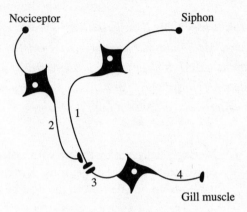

Which of the following points indicate neural axons?

(A) 1 and 2 only

(B) 1 and 3 only

(C) 2, 3, and 4 only

(D) 1, 2, and 4 only

10. Sensitization allows neurons to respond more to a stimulus than they would have prior to sensitization, for a short period of time. What change in the axon terminal could explain this phenomenon?

   (A) The axon terminal is permanently hyperpolarized.

   (B) The axon terminal is permanently depolarized.

   (C) The axon terminal remains polarized longer following an action potential.

   (D) The axon terminal remains depolarized longer following an action potential.

11. The disease multiple sclerosis is characterized by the lack of or damage to myelin on neurons, resulting in

   (A) total paralysis.

   (B) uncoordinated movement of the limbs.

   (C) seizures.

   (D) increased risk of heart attack.

Questions 12–16 refer to the following list.

   (A) Forebrain

   (B) Midbrain

   (C) Hindbrain

   (D) Spinal cord

12. Along with the pons and medulla, it constitutes the brain stem.

13. This is important for visual, auditory, and motor control.

14. This is vital for the function and control of hormones.

15. This contains the integration center for the spinal cord and the cerebral cortex.

16. This is the integration center for simple motor responses.

ANSWERS AND EXPLANATIONS

1. B

This question is asking about the transmission of a message in the nervous system between two neurons, so the message is passing across a synapse. This means that the signal is not electrical, but chemical and mediated by a neurotransmitter, not ions or electrical current. This rules out (A) and (C). Be careful to note that Ca^{2+} ions play a part in the uptake of the neurotransmitters in the postsynaptic neuron, but it is the neurotransmitters that continue to relay the message. Some sensory neurons initiate action potentials by being mechanically deformed, but this is not a mechanism for transmission across neurons, so (D) is incorrect.

2. B

Neurons send their signals in an all-or-none fashion, meaning that once they initiate an action potential, the internal voltage of the neuron climbs rapidly to a certain level no matter the scale of the initial stimulus. In other words, most neurons reach an internal voltage of +50 mV for each action potential no matter how intense the stimulus is. The reason that you can feel one stimulus as being more intense than another has to do with the number of different neurons involved or with the frequency of firing, not the voltage change of any individual neuron. All other statements regarding nerve cell potentials are correct.

3. B

Neurotransmitters that are inhibitory would not be expected to open sodium channels, because the opening of sodium channels rapidly increases the flow of positive ions into the axon, resulting in depolarization. Again, if this depolarization is great enough, the axon's membrane will pass a threshold point and an action potential will be initiated. Inhibitory neurotransmitters will, however, open chloride ion or potassium ion channels, both of which will hyperpolarize the inside of the axon relative to the outside. Chloride channels will cause the inflow of negative chlorine ions, making the inside of the axon more negative and, thus, harder to depolarize. The opening of potassium channels will cause the outflow of positive potassium ions, again causing the inside of the axon to become more negative and harder to depolarize.

4. C

A neuron will only fire an action potential when the threshold potential is exceeded, resulting in a unidirectional signal down the axon, which can be accelerated by insulating with myelin. Groups of neurons work together in ganglia. Choice (C) is correct because it is the only false statement. Hyperpolarization occurs after a neuron has just fired an action potential and occurs when the cell's membrane potential is lower than resting potential, inhibiting further action potentials because the stimulus required to surpass the threshold potential is drastically increased. This forces the action potential to go in one direction and allows the neuron to regenerate the neurotransmitters just released.

5. A

After habituation, neurons do not respond as strongly, indicating that action potentials are less frequent or that less neurotransmitter is released per action potential. Choice (A) is correct. If calcium channels in the terminal membrane are closed, an action potential would not lead to release of neurotransmitter. Choices (B), (C), and (D) describe mechanisms for increasing neuronal response rather than decreasing it. Increasing number of receptors, (B), makes it easier for neurotransmitters to bind and cause an action potential. Decreasing the concentration of neurotransmitter-degrading enzymes, (C), would allow the neurotransmitter to be present longer, causing the same effect. If anything happens at a lower excitatory potential, (D), that means it happens more readily and with less stimulation, rather than more.

6. A

Sodium channels are open during depolarization, the period of time at the start of an action potential where the inside of the neuron becomes rapidly more positive. The action potential can cause a change in membrane voltage from approximately –70 mV (millivolts) at rest to approximately +50 mV or more as sodium channels open up due to a particular stimulus. These sodium channels rapidly close as potassium channels open and bring the membrane potential back down toward the resting potential.

7. C

The absolute refractory period is the time during which sodium channels in a particular location on the axon membrane will not open, no matter how great the stimulus. It is a brief period lasting only milliseconds, but it forces the action potential to travel in one direction only—down the axon toward the axon terminal. The absolute refractory period ends during the falling period of the action potential (labeled as 3 on the graph), and the relative refractory period begins. During the relative refractory period, the membrane potential in areas of closed sodium channels is often much less than that at rest.

8. A

The resting membrane potential of most neurons is approximately –70 mV, depicted here on the graph by regions 1 and 4. Be careful of the word *initial* in the question stem. Number 4 also represents a time period during which the neuron is at the resting potential, yet region number 1 is the initial resting potential.

9. D

You don't have to be an expert in sea snails to reason through this question. The axon is the cell projection that transmits signals away from the cell body toward the dendrites, which receive messages and transmit them toward the cell body. The axons are 1, 2, and 4 because the siphon and nociceptor neurons receive sensory information and transmit it to the muscle neuron. Although 3 is not visually distinct from an axon in this image, it is receiving information and, therefore, must be a dendrite.

10. D

This question is asking what would have to occur at the axon terminal for a sensitized neuron to respond more to a stimulus than an unsensitized neuron. Choice (D) is correct: If the axon terminal remains depolarized for a longer period of time following an action potential, the sensitized neuron will continue releasing neurotransmitter, causing more activation of the postsynaptic neuron. Longer polarization, (C), is opposite to what occurs because this would make it more difficult to generate an action potential again rather than easier. Choices (A) and (B) are extreme answers because these effects would not be for a short period of time but, in fact, could disable the neuron permanently.

11. B

Myelin is an insulating material produced by the body and present in the CNS and PNS. The main function of myelin is to insulate neurons to speed up their action potentials. When myelin is damaged, relays between neurons become slow or are completely disrupted. This is especially evident in neurons with long axons. Total paralysis does not occur because not all neurons are demyelinated, and some signals can occur even if myelination is absent where normally present. Seizures result from excessive neuronal stimulation and, therefore, are generally not associated with MS. Heart attacks are the result of a blockage of the arteries and are unrelated.

12. B

The midbrain, pons, and medulla together make the brain stem.

13. B

The midbrain is a relay center for visual and auditory impulses. It also plays an important role in motor control.

14. A

The forebrain contains the hypothalamus, which is important for the control of the endocrine system.

15. A

The forebrain includes the thalamus, which is a relay and integration center for the spinal cord and the cerebral cortex.

16. D

The spinal cord integrates simple motor responses such as reflexes. Reflexes are neural responses that do not require the use of the brain. Therefore, a reflex can occur faster than a normal nerve impulse because there is less distance to travel.

CHAPTER 20: ENZYMES

BIG IDEA 4: Biological systems interact, and these systems and their interactions possess complex properties.

IF YOU ONLY LEARN FOUR THINGS IN THIS CHAPTER . . .

1. Enzymes lower the activation energy of a reaction, do not get used up in the reaction, catalyze millions of reactions per second, do not affect the overall free-energy change of the reaction, increase the reaction rate, and do not change the equilibrium of reactions.

2. Enzymes can be influenced by reaction conditions such as high temperatures, detergents, or acidic/basic conditions.

3. Enzymes can be regulated by inhibitors, molecules that bind to the enzyme either at the active site or the allosteric (regulatory) site. Feedback inhibition is when the end product of a biochemical reaction works to block the original enzyme.

4. All enzymes possess an active site, a 3-D pocket within their structure in which substrate molecules can be held in a certain orientation to facilitate a reaction. The two models of enzyme-substrate interaction are lock and key and induced fit.

INTRODUCTION ★★★★

Most enzymes are proteins with specific 3-D structures that allow them to bind to very particular molecules (called **substrate** molecules) and increase the rate of reactions between these molecules. In many cases, enzymes bind to larger molecules and break them into smaller ones. Enzymes can synthesize or break down molecules at the rate of thousands or millions per second—they are extremely fast-acting! They allow reactions to occur that either would not take place or would take place far too slowly under normal conditions to be useful.

Enzymes are very specific for the molecules they bind to and the reactions they catalyze. Each enzyme has a name that usually indicates exactly what it does, and the name often ends in *-ase*. For example, lactase enzyme breaks the complex sugar lactose into the simple sugars glucose and galactose. The enzyme pyruvate decarboxylase removes a carbon from the three-carbon molecule pyruvate. Thinking about enzyme names in this way may be helpful on the AP Biology exam.

For the AP Biology exam, you should be familiar with the term *catalyst*, which refers to any chemical agent that accelerates a reaction without being permanently changed in the reaction. Enzymes are biological catalysts, which can be used over and over again.

Enzyme specificity means that an enzyme will only catalyze one specific reaction. Molecules upon which an enzyme acts are called substrates, and the substrate binds to the active site on the enzyme, speeding up the conversion from substrate to product.

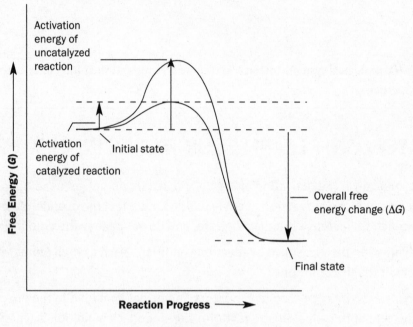

Enzymatic Energy

Physiological reactions can take place without enzymes, but they would take much longer to proceed. Though enzymes are neither changed nor consumed during the reaction, reaction conditions such as high temperatures, detergents, or acidic/basic conditions can cause enzymes to denature (lose their 3-D structure) and thereby lose their activity. Enzymes can be regulated by inhibitors, molecules that bind to the enzyme either at the active site or the allosteric (regulatory) site.

Competitive inhibition occurs when the inhibitor and the substrate compete for the active site. This type of inhibition can be overcome by increasing the concentration of substrate.

Noncompetitive inhibition occurs when the inhibitor binds to the allosteric site, inducing a conformation change in the enzyme, rendering the active site inactive. This type of inhibition cannot be overcome by adding more substrate because the enzyme's shape has been altered. Therefore, noncompetitive inhibition may or may not be irreversible. Keep in mind that enzymes do not alter reaction equilibrium or affect the free energy of a reaction. They accelerate the forward and reverse reactions by the same factor.

As a general rule, the rate of an enzyme will increase with increasing **temperature** but only up to a point. If the temperature increases too much, then the enzyme will become denatured.

Enzymes are active only within a specific **pH** range. In the human body, most enzymes work best around neutral (pH = 7).

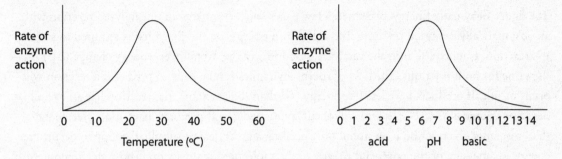

Temperature- and pH-Dependent Enzyme Reaction Rates

CONCENTRATION OF SUBSTRATE AND ENZYME ★★★★

Reaction rates increase as more and more enzyme is added to a particular environment. If the enzyme concentration is kept constant, the reaction rate will plateau at a maximum speed as substrate concentration increases, because the enzymes can only work so fast. This maximum reaction rate is termed V_{max} and is illustrated in the following graph as the flat part of the rate versus substrate concentration curve for an enzymatic reaction. This point occurs when the enzymes become saturated with substrate. Enzymes become saturated at high substrate concentrations because substrate must bind to enzymes at a particular place: the active site. While the entire process is somewhat complex, the main idea is that increasing the substrate can produce the same V_{max} if competitive inhibitors are present, but noncompetitive inhibitors lower V_{max} regardless of the amount of substrate.

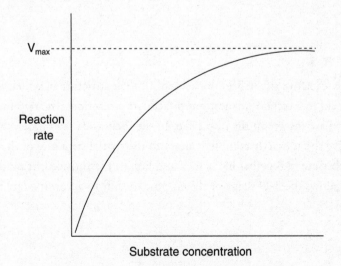

Concentration And Enzyme Reaction Rates

Most enzymes require cofactors to become active. **Cofactors** are nonprotein (inorganic) species that either play a role in binding to the substrate or stabilize the enzyme's active conformation. Two examples of cofactors are zinc and the iron in hemoglobin. **Coenzymes** are other organic molecules that play a similar role. Most coenzymes cannot be synthesized by the body but are obtained from the diet as vitamins.

The figure Enzymatic Energy, presented a few pages back, compares an uncatalyzed reaction with an enzymatically catalyzed reaction. The **activation energy,** or energy which is required to start up the reaction, is much lower in the catalyzed reaction, yet the overall **free-energy** change (ΔG) is the same for both reactions. The laws of thermodynamics can be used to predict if a reaction will occur or not. If products have less free energy (G) than the reactants, the reaction has an overall negative change in G ($\Delta G < 0$) and will occur spontaneously. If products have more free energy than reactants, the reaction is an uphill one, needing a great deal of supplied energy to occur. Free energy is a measure of the **potential energy** of the molecules in a reaction. Those starting out with high potential energy, or higher G, are more likely to react and lower their G through the reaction than vice-versa. What that means is that reactions having a $\Delta G < 0$ are deemed favorable. Keep in mind that most biosynthetic reactions have $\Delta G > 0$ and will not occur spontaneously without the help of both enzymes and ATP.

However, although thermodynamics and ΔG alone may predict that a reaction is favorable or can occur spontaneously, the **kinetics**, or rate, of the reaction may be so slow that these reactions are not feasible for living systems. Sure, a hamburger will break down eventually if exposed to enough acid in your stomach, but without digestive enzymes to help speed up this breakdown, the hamburger might take months to break down sufficiently to be useful to you. Thus, although hamburger breakdown may be spontaneous, the limiting factor in the reaction is the reaction rate, which is dependent on energy being provided to start this breakdown. This energy is the activation energy, and enzymes provide a foundation on which molecules can react so that the energy needed to start a reaction is not as great as it would have to be without the enzyme's presence.

BINDING ★★★

All enzymes possess an active site, a 3-D pocket within their structure in which substrate molecules can be held in a certain orientation to facilitate a reaction. The two models of enzyme-substrate interaction are shown on the next page. In the **lock-and-key** model, the spatial structure of an enzyme's active site is exactly complementary to the spatial structure of the substrate so that the enzyme and substrate fit together like a lock and key. In the **induced fit** model, the active site has flexibility that allows the 3-D shape of the enzyme to shift to accommodate the incoming substrate molecule.

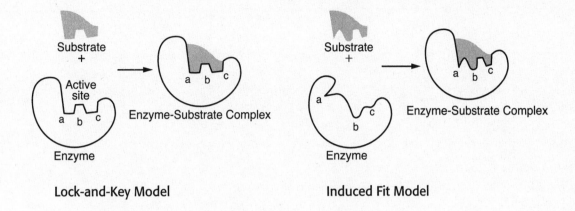

Lock-and-Key Model   Induced Fit Model

REGULATION OF ENZYMES ★★★

Cells must regulate enzyme action to keep these rapid reactions under control. Most enzymes are inactive most of the time, and enzyme pathways are regulated in complex fashions to ensure efficiency and safety. There are several kinds of inhibition that cells use to control enzymes.

Feedback inhibition occurs when the end product of an enzyme-catalyzed reaction works to block the original enzymes that started the reaction in the first place. In the following diagram, this would occur if product C could bind to the active site of enzyme 1, thereby preventing any more of compound A from becoming compound B.

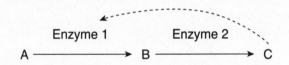

CHAPTER 21: PLANT VASCULAR AND LEAF SYSTEMS

BIG IDEA 4: Biological systems interact, and these systems and their interactions possess complex properties.

IF YOU ONLY LEARN FOUR THINGS IN THIS CHAPTER . . .

1. Plants produce energy through photosynthesis and lose water via transpiration. As water evaporates from the leaves, it pulls water up through channels in the xylem. The phloem carries nutrients throughout the plant.

2. Plants can reproduce asexually via vegetative propagation. Sexual reproduction in plants takes place in the flower.

3. Plants with vascular tissues usually have three types of structures: leaves, roots, and branches.

4. Plants have specialized structures to deal with water and nutrients. These include stomata controlled by guard cells, a loosely packed spongy layer, the palisade layer, xylem, and phloem.

INTRODUCTION ★★★

Members of the plant kingdom are photoautotrophs that utilize the energy of the sun, carbon dioxide, water, and minerals to manufacture chemical energy used in respiration and stored in carbohydrates. They are multicellular, eukaryotic, and autotrophic organisms. The direct ancestor of modern-day plants was likely a green algae. Many of the green algae species possess the same types of chlorophyll and pigments (chlorophyll b and beta-carotene) that plants possess. The first land plants were the **bryophytes**, in the division Bryophyta. Small and living close to a source of water, this group includes the mosses, liverworts, and hornworts. These nonvascular plants have two key adaptations to help them live on land: They are covered by a protective waxy cuticle that prevents desiccation (drying out), and they produce flagellated sperm cells within their male gametangium (reproductive structure) that can swim some distance through droplets of water to fertilize eggs produced by the female gametangium. As plants evolved, vascular tissues such as xylem and phloem allowed them to grow taller and live farther away from abundant water sources. Other adaptations besides a waxy cuticle to protect stems and leaves from water loss include stomata, tiny holes in the undersides of leaves for gas exchange. To continue the move onto land, it was necessary to protect embryos from desiccation and to be able to spread gametes in a nonaqueous environment.

PLANT STRUCTURE AND SYSTEMS ★★

Terrestrial plants left an aquatic environment during their evolution, and in the process lost some of the benefits that a liquid medium had previously provided. **Cell walls**, which provided simple rigidity for structures in small plants, became aligned to allow for the growth of trees over 100 meters in height and to create channels for the delivery of important soil nutrients. Water loss became a constant challenge in environments where rainfall could be unpredictable or nearly nonexistent. The sun, formerly relied upon solely for light-giving energy, became a source of dehydration and overheating. The terrestrial environment also provided a great degree of variation, in contrast to the constant temperatures and cycles of aquatic systems. Seasonal changes and the variability of weather required plants to synthesize new chemical messengers that would allow plant tissues to respond to both acute and chronic environmental cues. **Novel structures**, primarily for reproduction, required new forms of control mechanisms to function in time with a changing environment.

One of a plant's many challenges is providing all of its structures with the water needed to complete photosynthesis and to maintain an aqueous solution for all biological reactions. Water is heavy and a plant cannot rely on muscular contractions to move materials, as animals do. Instead, a plant has narrow, lifeless channels in its **xylem** tissue that take advantage of the cohesive properties of water. A plant loses the majority of its water during **transpiration** while its stomata are open for the exchange of CO_2 and O_2 in photosynthesis. As water is lost through the leaves, it creates a negative pressure in the xylem channels, just like sucking on a straw. The cohesive force of water keeps a steady flow of water moving through the plant, pulled by the negative pressure developed by transpiration occurring in the leaves. The force exerted in the water column due to cohesion is as strong as steel wire of the same diameter.

Plants need a mechanism to deliver the stored energy created during photosynthesis to all of their structures. While water travels upward from the roots through the xylem via the pull of transpiration and the cohesion of water, nutrients flow both up and down through the **phloem** via the pull of **osmotic pressure.** Osmotic pressure is built up between areas that have high concentrations of nutrients and those that have low concentrations. **Translocation** of materials in the phloem is like long-range diffusion. Similar to the cohesion of water, lower solute pressure on one end of a phloem vessel is translated along the narrow vessel to an area where the solute pressure is greater. The adaptive significance of these transport systems is the colonization of a terrestrial environment. Plants also harnessed a physical disadvantage, water loss, and turned it into a benefit, with the force of water loss powering water transport through the xylem.

All plants exhibit alternation of generations in their life cycle between diploid and haploid forms. Within the plant kingdom are several phyla, with one of the key distinguishing characteristics between phyla being the presence or absence of vascular tissue. Within the tracheophytes, which have vascular tissue, two of the important modern phyla are the gymnosperms and the angiosperms. The gymnosperms, such as the conifers, have "naked seeds" that do not have endosperm and are not located in true fruits. The angiosperms are the flowering plants. They have flowers, true fruits, and a double fertilization system to create endosperm to nourish the plant embryo.

Monocots vs. Dicots:
Two Angiosperm Classes

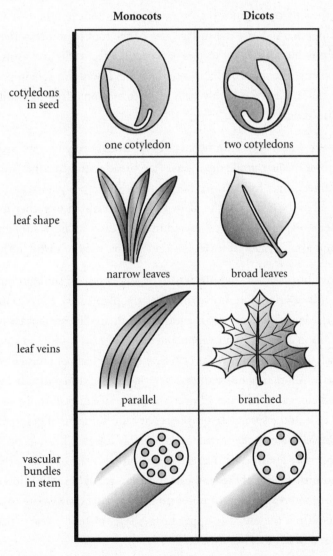

Angiosperms are the most successful form of plant life on the planet, and the previous diagram distinguishes between two main classes of angiosperm: **monocots** and **dicots**.

ROOTS, SHOOTS, AND TISSUES ★★

STRUCTURE OF PLANTS

Plants with vascular tissues usually have three types of structures, or organs: **leaves, roots,** and **branches**. The leaves provide most of the photosynthesis of the plant; the roots provide support in the soil, water, and minerals; and the branches hold the leaves up to light and convey nutrients and water between the leaves and the roots. Each of these structures can be specialized in many ways.

Plants with taproots have long roots with a single extension deep into the soil while other plants have highly branched roots. Cells on the surface of roots often have long extensions called **root hairs** that increase the surface area of roots. Some plants without root hairs have a symbiotic relationship with fungi that increase the surface area of the root to absorb water and minerals. In legumes, nitrogen-fixing *Rhizobium* species of bacteria infect roots and form root nodules in symbiosis with the plant. The roots of plants often play an important role in preventing erosion. Tropical rain forest that is cleared is highly vulnerable to erosion of the thin soil if the plants and their root systems are absent.

Leaves can have a variety of shapes. Monocot leaves are usually very narrow with veins that run parallel to the length of the leaf, while dicot leaves are broad with veins that are arranged in a net in the leaf. Modified leaves form thorns in cacti, tendrils in pea plants, and petals in flowers. The leaves produce all of the energy of the plant and are specialized to gather sunlight. The broad shape helps to gather sunlight for themselves and in some cases to block sunlight for competing plants. The shape and arrangement of leaves are key features used to distinguish plant species.

Terrestrial plants have broad leaves that maximize the absorption of sunlight but also tend to increase loss of water by evaporation. Leaves of terrestrial plants have a waxy cuticle on top to conserve water. The lower epidermis of the leaf is punctuated by **stomata**: openings that allow diffusion of carbon dioxide, water vapor, and oxygen between the leaf interior and the atmosphere. A loosely packed **spongy layer** of cells inside the leaf contains chloroplasts with air spaces around cells. Another photosynthetic layer in the leaf, the **palisade layer**, consists of more densely packed elongated cells spread over a large surface area. A moist surface that lines the photosynthetic cells in the spongy layer is necessary for diffusion of gases into and out of cells. Air spaces in leaves increase the surface area available for gas diffusion by the cells. The size of stomata is controlled by **guard cells** that can open and close the opening. These cells open during the day to admit CO_2 for photosynthesis and close at night to limit loss of water vapor through transpiration, the evaporation of water from leaves that draws water up through the plant's vascular tissues from the soil. The upper surface layer of cells in leaves has no openings, an adaptation that reduces water loss from the leaf.

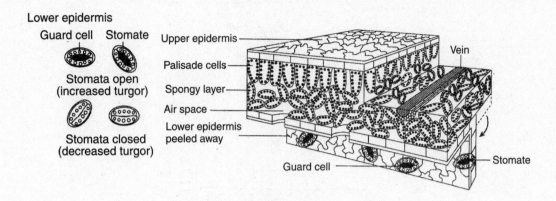

Leaf Structure

Rapid removal of potassium ions (K^+) from cells in the Venus flytrap allows the closing of a leaf trap on unsuspecting insects. The shuttling of potassium ions also seems to be the mechanism for the opening and closing of the guard cells surrounding the stomata. The guard cells are kidney-shaped cells in dicots and dumbbell-shaped cells in monocots that change their shape according to the amount of water that exists within them. This water exerts a pressure called turgor pressure. When water is abundant, the guard cells swell, and when water is sparse, they clamp down and shut the stomata. This makes perfect sense, as the guard cells allow transpiration when there is plenty of water in the leaves and water conservation when there is not. Guard cells absorb water in response to potassium ions that are driven into the cells. Water follows these ions due to the osmotic gradient that is created by the presence of extra particles inside the cell as compared with outside the cell.

The presence of blue-light receptors on the guard cell membranes is what drives the movement of K^+ into the guard cell. Sunlight, particularly blue wavelengths, causes K^+ ion channels to open and potassium to flood in. This explains why guard cells are usually open during the day and closed at night, helping photosynthesis to take place as carbon dioxide gas can then enter through the open stomata.

Growth in higher plants is restricted to areas of perpetually embryonic and undifferentiated tissue known as **meristems**. Meristems are self-renewing populations of cells that divide and cause plant growth either in height or width. **Apical meristems** exist at the tips of roots and stems, whereas **lateral meristems** (also known as **cambium**) are found within the stem between layers of xylem and phloem on the sides of the plant. The trunk of the plant thickens each year because the embryonic cells of the cambium produce more and more xylem and phloem, supporting the growth of a larger tree with more leaves. Growth upward that occurs as a result of cell division within apical meristems is called **primary growth**, while growth outward is called **secondary growth**. It is secondary growth that causes the well-known concentric tree rings that one sees when a large woody tree is cut down.

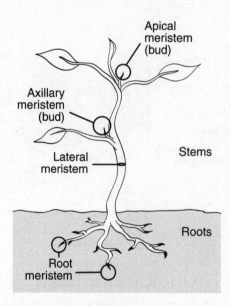

General Plant Anatomy

TRACHEIDS AND VESSEL ELEMENTS

These are specialized cells that make up the xylem, which is discussed in detail in the next part of this chapter. Xylem tissues transport water throughout the plant, and these cells form long, often spiral tubes that can stretch as the plant elongates. At maturity, tracheids and vessel elements are dead, but water continues to flow through them and they retain their capability to stretch.

SIEVE-TUBE MEMBERS

Phloem, also discussed in the next part of the chapter, is made from these specialized cells. Usually losing most of their organelles (including their nuclei) as they mature, these cells carry glucose-rich fluid from cell to cell and from leaves down to the roots.

Dermal tissue (think epidermal) forms the outer coverings of the plant, such as the waxy cuticle of leaves that protects them from water loss. Root hairs, discussed later in more detail, are extensions of dermal cells in the roots of plants. Xylem and phloem, however, are **vascular tissues** (think cardiovascular system—veins and arteries) because these tissues are used for transport of material around the plant.

XYLEM, PHLOEM, AND TRANSPORT ★★

XYLEM

The vascular tissue of the xylem contains a continuous column of water from the roots to the leaves, extending into the veins of the leaves. The leaves regulate the amount of water that is lost through transpiration. Water is transported from the roots to the shoots. In other words, from roots to stems, water is transported through two different mechanisms: **root pressure** and **cohesion-tension.**

Osmotic pressure in the roots tends to build up due to water absorption (see the next section on cation exchange for more details). This pressure pushes water up from the roots into the stem of the plant. Root pressure functions best in extremely humid conditions when there is abundant water in the ground or at night. The drops of water, or dew, that appear on blades of grass or other small plants in the morning result from this root pressure. During the day, transpiration occurs at a high enough rate that dew is generally not seen, because the water that leaves the tips of the grass is evaporating before it can build up.

Transpiration

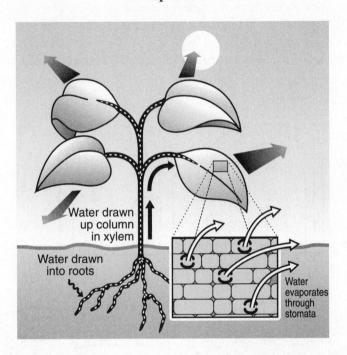

Water Gain and Loss in Plants

Root pressure is not sufficient to move water up tall plants. For taller plants, rather than being pushed from the bottom as occurs from root pressure, water in the xylem must be pulled from above by water evaporating from the leaves. The pull of water leaving the leaves by transpiration provides a negative pressure (tension), while the tendency for water molecules to stick together (cohesion) due to their polarity transmits water up the plant toward the leaves. Transpiration occurs because there is less water in the air surrounding the leaves than there is within the leaves, yet on humid days, transpiration is limited or nonexistent. Water molecules are also attracted to the walls of the xylem tubes, and the thin tubes coupled with water's cohesion help to fight gravity and lift water to the tops of trees.

PHLOEM

Phloem tubes are much thinner-walled than xylem and are found toward the outside edges of stems. They begin up in the leaves, where sugars are made by photosynthesis and stored as starch within mesophyll cells. Much of this sugar and starch, though, is transported down phloem tubes from shoots to roots (from leaves and stems to the roots). Active transport between the mesophyll cells, where sugar is made, and the nearby phloem tubes is what allows sugar to move into the phloem's sieve tubes. The active transport pumps are symport pumps that move one H^+ ion along with every sucrose molecule. The H^+ that is used to push the sucrose across into the phloem is rapidly transported back out to the mesophyll cells by an H^+ ion ATPase pump, so that more sucrose can be moved. As sugar is loaded into the phloem, the **water potential**, or the amount of water pressure, in the phloem is effectively reduced. The influx of sugar into the phloem creates an osmotic potential that pulls more water into the phloem, generating a water pressure that forces sap (essentially sugar-rich water) down the phloem toward the roots. The xylem will recycle this water back from the roots.

As you can see in the following figure, vascular cambium (C) divides laterally to create new phloem (P) on the outer edges of the tree and new xylem (X) near the inner core. This results in the formation of annual tree rings.

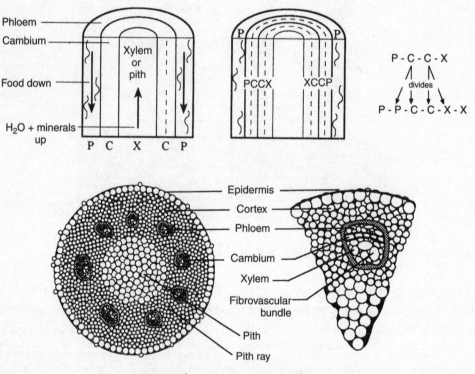

Internal Anatomy of a Plant

PLANT CHEMICAL COMPOUNDS

Plants require certain nutrients and compounds that are important for normal physiological functions. For example, the photorecptor, phytochrome, is an important pigment that plants use to detect light. Plants can use phytochrome to regulate flowering based on day length or to set up other daily rhythms. Large amounts of macronutrients such as potassium, calcium, magnesium and phosphorous are required for normal plant functions.

PLANT CHEMICAL COMPOUNDS

Compound	Type	Origin	Action
Phytochrome	Photopigment	Systemic	Detection of light to control photoperiodism
Potassium	Nutrient	Root uptake in soil	Protein synthesis, operation of stomata
Calcium	Nutrient	Root uptake in soil	Cell wall stability, enzyme activation
Magnesium	Nutrient	Root uptake in soil	Chlorophyll synthesis, enzyme activation
Phosphorous	Nutrient	Root uptake in soil	Nucleic acid and ATP synthesis

REPRODUCTION AND DEVELOPMENT ★★★

Plants possess a reproductive advantage in that they can reproduce vegetatively while very few groups of animals can reproduce without some form of sexual reproduction. Plants also live dual lives in that they have specific structures for reproduction (flowers, pollen, and fruit), while the rest of the plant continues life as usual. Female animals' entire bodies are affected by the reproductive process, so reproduction has to be an entire stage of existence for them. Some animals' adult life stage is solely for the purpose of reproduction; they don't even have mouthparts for the ingestion of nutrients. Some of these intricacies of plant reproduction and development will be discussed next.

Plants can **reproduce vegetatively** as a function of their **indeterminate growth**. Plant cells can differentiate when isolated from the rest of the plant, so one could place the stem of a rose in the soil and it would grow **adventitious roots** and become a whole plant. By default, this produces a series of plants that are all genetically identical and reduces variation in the population but is advantageous for success when **sexual reproduction** is not suitable for any given environmental situation.

Most terrestrial plants live their lives as **diploid sporophytes** and produce haploid tissue in the form of **ovules** and **pollen** (**gametophytes**), sometimes within the same flower. Pollen is delivered to the female gametophyte either by an animal, such as a bird, or by the wind, and fertilization takes place. The flower functions as the entire reproductive organ, completely dependent on the sporophyte plant. This is

DID YOU KNOW?

All plants exhibit a definite alternation of generations in which the plant life cycle alternates between haploid and diploid stages.

NOTE

You do not need to know the details of the sexual reproduction cycles in different plants and animals. Focus instead on the similarities among the processes and how this is important for genetic variation.

a departure from the aquatic condition in which spores travel independently from the plant that produced them. Flowering plants seem to have evolved a specialized region where a plant can be "pregnant." These "seed babies" are housed within a seed coat and often a fruiting body. The seed coat and fruiting body aid in plant propagation by either acting as bait to an animal or providing a source of nutrition to the developing plant embryo.

AP BIOLOGY LAB 11: TRANSPIRATION INVESTIGATION ★★★★

This investigation is another example of a controlled experiment to evaluate the effects of environmental variables on the rate of transpiration for a plant. It is easy to complete and requires simple equipment. The objectives are threefold: another opportunity to explore the scientific method through controlled experimentation, a chance to make direct observations on the phenomena immediately related to the physical properties of water, and an occasion to observe the environment's effects on an actual organism.

A suitable plant species is obtained and set up with a potometer, a fancy name for a tube that measures the amount of water taken up by a plant during transpiration. The water-filled tube is connected to the bottom of a plant so that the water leaving the tube and entering the plant due to transpiration can be measured. The potometer tube takes the place of soil as a source for water. The plant is then subjected to a number of treatment effects that might include increased light, darkness, heat, or air movement. Changes in transpiration, if any, are compared to the rate of transpiration of a control plant that is under stable conditions, usually room temperature and room lighting. The control plant serves as the benchmark with which all other treatment effects are compared.

Transpiration is affected by environmental conditions in a way very similar to human skin. Increasing the amount of heat around skin causes increased sweat and air movement, which increases evaporation on the skin surface; the result is increased water loss. The same is true for the leaf surface of a plant—increased heat causes an increase in water loss. Light or darkness in the absence of changes in temperature only affects a leaf by modifying the rate of photosynthesis. Increasing the light intensity should increase the rate of photosynthesis and the opening of stomata, thereby increasing transpiration and water loss. As light levels decrease in intensity, water loss decreases. Water loss at the leaf surface causes greater uptake of water through the potometer, which can be measured as movement of the meniscus through the tube.

CHAPTER 22: RESPIRATION AND THE CIRCULATORY SYSTEM

BIG IDEA 4: Biological systems interact, and these systems and their interactions possess complex properties.

IF YOU ONLY LEARN THREE THINGS IN THIS CHAPTER . . .

1. Lungs are the primary structure for respiration. The circulatory system connects the lungs and the cells around the organism's body. All mammalian lungs end in blind sacs called alveoli that are surrounded by a dense net of capillaries into which oxygen diffuses and from which carbon dioxide is excreted.

2. Blood enters the heart from the body through the right atrium, then travels through the tricuspid valve into the right ventricle. From the right ventricle, it flows through the pulmonary arteries into the lungs and back to the heart through the pulmonary veins. Blood enters through the left atrium and goes through the bicuspid valve into the left ventricle. Blood leaves the heart via the aorta.

3. Blood pressure is the force per area that blood exerts on the walls of blood vessels and is expressed as a systolic number over a diastolic number. Blood pressure drops as blood moves farther away from the heart because the distance from the pump has increased and because of the muscular construction of arteries versus veins.

INTRODUCTION ★★

Gas exchange and transport depend upon the movement of gases or fluids around an organism's body and the passage of these gases or material contained within the fluids into and out of cells. All gas exchange is based upon simple diffusion across cell membranes, and the key characteristic that has evolved across almost all species for maximal gas exchange is moist membranes that cover a *large amount of surface area*.

WHAT IS THE DIFFERENCE BETWEEN POSITIVE AND NEGATIVE PRESSURE BREATHING?

Frogs and other amphibians can ventilate themselves by literally pushing air down their windpipes. They open and lower the floor of their mouths and gulp in air. This is positive pressure breathing. Mammals have a muscular band of tissue called the **diaphragm** that separates the chest cavity from the abdominal cavity. As this muscle contracts and pulls down, pressure in the chest cavity decreases, sucking air in from outside. This drop in air pressure from within, called negative pressure breathing, allows breathing to take place.

While organisms such as amphibians, reptiles, and even some fishes (e.g., lungfish) may possess **lungs** to aid in breathing and gas exchange, these organisms all use other structures as their primary means of oxygen and carbon dioxide exchange. In mammals, the lungs are the primary structure. Because the lungs are restricted to one location in the body, a circulatory system is needed to bridge the gap between the lungs and the cells around the organism's body. All mammalian lungs end in blind sacs called **alveoli** that are surrounded by a dense net of capillaries into which oxygen diffuses and from which carbon dioxide is excreted.

In mammals, the path that air takes once inhaled is as follows: once brought into the nose or mouth, air enters the **pharynx** (throat) and passes across the **larynx** (voicebox) that lies within the **trachea** (windpipe). The trachea branches into two **bronchi** (singular: bronchus), which further branch into **bronchioles** and then alveoli. Cells that line the bronchioles and alveoli keep the lungs moist, secreting mucus and other watery fluids into the lung openings. There are also a great many cells covered with short hairlike cilia to keep mucus and other particles flowing across the inside epithelial surface. The millions of alveoli where the gas exchange actually takes place are separated from capillaries by only a thin layer or two of cells.

Deoxygenated blood enters the pulmonary (lung) capillaries from the systemic circulation having a **low partial pressure of oxygen** (it has lost its store of O_2 at the tissues). Inhaled air in the alveoli has a much higher partial pressure of oxygen (there is much more oxygen relative to other gases in the alveoli than within the capillaries). Therefore, oxygen diffuses down its concentration gradient into the capillaries, where it binds to hemoglobin molecules in red blood cells and returns to the heart to be pumped out to the body. In contrast, the partial pressure of CO_2 in the capillaries is greater than that of the inhaled alveolar air; thus, CO_2 diffuses from the capillaries into the alveoli, where it is exhaled.

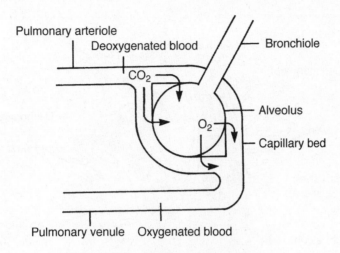

Gas Exchange in the Alveoli

At high altitudes, the partial pressure of O_2 in the atmosphere declines, making it more difficult to get sufficient oxygen to diffuse into the capillaries. The body often compensates for this by increasing the rate of breathing (hyperventilation) and by increasing the number of red blood cells available to carry oxygen. **Erythropoietin**, a hormone released by the kidneys, is involved in the increased production of red blood cells: It travels to the bone marrow to stimulate red blood cell production there.

Within each new erythrocyte (red blood cell), there are approximately one million **hemoglobin (Hb)** molecules. Each hemoglobin molecule is made of four subunits capable of binding to four different oxygen molecules. The binding of the first oxygen molecule *induces a conformational change in the hemoglobin* molecule as a whole so that the binding of the other three molecules becomes progressively easier. Similarly, the unloading of one oxygen facilitates the unloading of the other oxygens. This allosteric effect is reflected in the S-shaped oxygen dissociation curve for hemoglobin. The hemoglobin is able to hold on to its oxygen molecules with great affinity until partial pressures reach about 40 mm Hg (hemoglobin's oxygen saturation remains over 70 percent until that point). Then it quickly dumps off its oxygen. This is quite useful, for it makes sure that the oxygen in the red blood cells is not completely grabbed up by other cells until the red blood cells reach the farthest extremities of the body. Active tissues use up oxygen much faster and, therefore, have a lower partial pressure of O_2. Yet the Hb molecules still hold on to enough oxygen to supply these needy tissues with some oxygen.

In mammals and birds, blood follows a double circuit. As the diagram shows, the right and left sides of the heart are each their own separate pumps: The right side pumps deoxygenated blood to the lungs, while the left side pumps oxygenated blood to the aorta and out to the body. The two upper (receiving) chambers are atria, and the two lower chambers are ventricles. While the atria are fairly thin-walled and not very muscular, the ventricles, particularly the left ventricle, are thick with muscle tissue for sustained, forceful pumping.

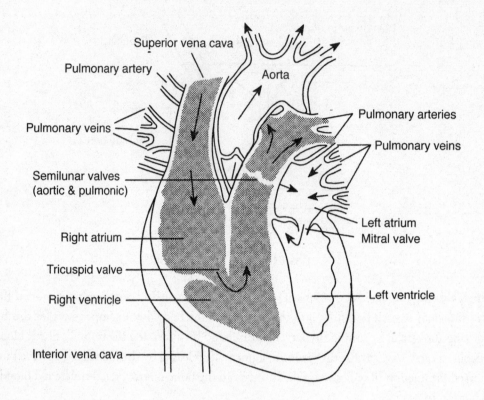

Diagram of the Heart

The atrioventricular (AV) valves, located between the atria and ventricles on either side of the heart, prevent backflow of blood into the atria when the ventricles are contracting. The semilunar valves in the pulmonary artery and the aorta serve a similar function: to prevent the backflow of blood into the ventricles from the lungs or aorta once the blood has been pumped out of the heart. The heart's pumping cycle is divided into two alternating phases: **systole** and **diastole**, which together make up the heartbeat. Systole is the period during which the ventricles contract, forcing blood into the lungs and aorta, and diastole is when the ventricles relax and fill with blood.

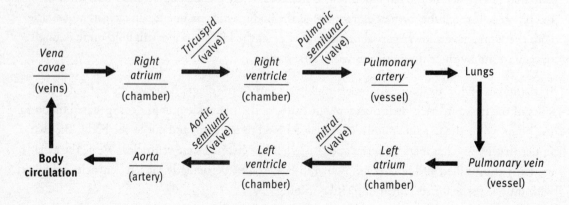

Circulation Pathway

BLOOD PRESSURE AND TEMPERATURE REGULATION ★★

Blood pressure is the force per area that blood exerts on the walls of blood vessels. It is expressed as a systolic number over a diastolic number. Blood pressure drops as blood moves farther away from the heart, partly because the distance from the pump has increased and partly because of the muscular construction of arteries versus veins.

Arteries are thick-walled, muscular, elastic vessels that transport oxygenated blood away from the heart, while veins are relatively thin-walled inelastic vessels that bring deoxygenated blood back to the heart. When veins are not filled with blood, they often collapse on themselves, only to swell open when blood is pushed through. In fact, because veins have little contractile ability of their own, blood is often pushed back to the heart from the extremities when the skeletal muscles near the veins in those areas contract. As you walk, for instance, contractions of your leg muscles propel blood sitting in leg veins back toward the heart. In contrast, arteries maintain an open **lumen** at all times and can contract against the blood within, helping to propel it away from the heart. Capillaries are the tiniest blood vessels in organisms, usually wide enough to allow blood cells to pass through only in single file. It is here, and not in arteries or veins, that diffusion of gases between the blood cells and tissue cells takes place. Keep in mind that the predominant muscle type found in all blood vessels is smooth muscle, under autonomic nervous system and hormonal control.

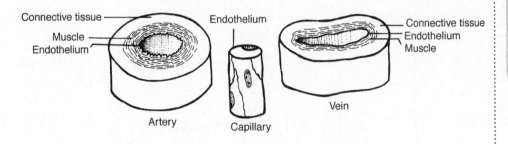

Blood Vessel and Wall Width

The path that blood takes from the heart to the rest of the body begins in the aorta, the major artery leading away from the left ventricle. The aorta splits into other arteries, smaller arterioles, and finally capillaries. As you can see from the following diagram, the increase in *overall surface area* from arteries to capillaries is huge, and this surface area is immediately decreased as capillaries coalesce into venules and then veins for the trip back to the heart.

DID YOU KNOW?

When you have your blood pressure taken, you'll get a number such as 120/80. The top number, the larger one, is your systolic blood pressure and reflects how much pressure is in your arteries when your heart contracts. The bottom number is your diastolic pressure and reflects the pressure in your arteries as the heart relaxes. As your arteries fill with fatty plaques and harden with age, the top number typically goes up.

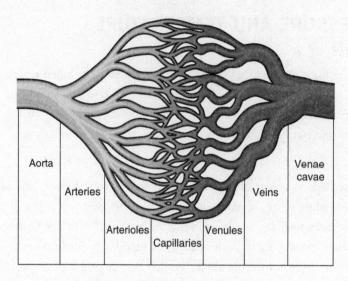

Relative Size of Blood Vessels

Because of this increase in surface area and friction between the blood cells and the blood vessel walls, average blood pressure decreases the farther away blood travels from the heart. In the veins, there is virtually no blood pressure at all.

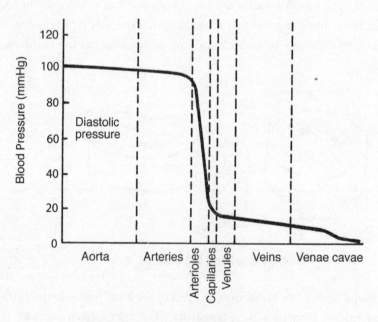

Blood Pressure and Body Location

If all the capillaries of the body had blood flowing through them at all times, your overall blood pressure would be close to zero. In other words, you do not have enough blood in your body to supply every capillary with blood all the time. Thus, the body shunts this blood from place to place, closing off some capillaries while opening others. This selective dispersal of blood is accomplished by using small bands of muscle, called **precapillary sphincters**, which exist at the base of capillaries as they branch off from nearby arterioles. When these sphincters are closed, the blood will simply bypass the capillary bed, moving from the arteriole directly into the venule on the other side of the bed. When open, these sphincters allow blood to pass into the capillary bed to nourish the tissue cells there.

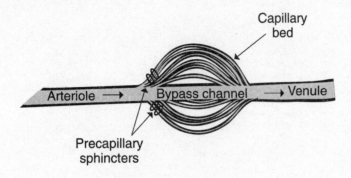

Capillary Sphincters

Another consequence of regulating blood flow using precapillary sphincters is temperature regulation. Have you ever noticed how your skin turns whitish and pale when it is cold outside, and redder when it is hot? Exercising also causes the skin to become hotter and redder in color. The reason for this is that when it is cold outside, autonomic nerves in your skin cause as many precapillary sphincters as possible to close, to conserve heat (the more blood near the surface of your body, the faster you lose heat). When it's hot or when you're exercising, many sphincters near the surface of the skin will be open, allowing more efficient cooling of the body. **Vasodilation**, or keeping lots of blood vessels open in the extremities, and sweating are the two main mechanisms the body uses to cool itself.

CHAPTER 23: DIGESTION AND EXCRETION

BIG IDEA 4: Biological systems interact, and these systems and their interactions possess complex properties.

IF YOU ONLY LEARN FIVE THINGS IN THIS CHAPTER . . .

1. Animal body systems consist of specialized cells and tissues that perform a function. You should know the structure and function of the digestive and excretory system in humans.

2. Mechanical digestion occurs in the mouth. Chemical digestion occurs in the stomach, duodenum, and small intestine. Nutrients, water, and minerals are absorbed into the body in the colon.

3. Excretion is the removal of metabolic wastes, and all metabolic processes lead to the production of mineral salts that must be excreted.

4. Kidneys regulate the concentration of water, ions, and other dissolved substances in the blood through the formation and excretion of urine.

5. The nephron is the functional unit of the kidney. It regulates the concentration of water and salts in the blood by filtration and urine production.

INTRODUCTION ★★★★

Digestion involves the degradation of large molecules into smaller molecules that are then absorbed and used directly by cells. In complex animals (such as humans), branching digestive canals and absorption into the bloodstream are necessary for the effective breakdown of nutrients and their transport to all cells of the body. Mammalian digestive tracts, which are the focus of this chapter, are organized into regions specialized for the digestion and absorption of specific nutrients.

Excretion refers to the removal of **metabolic wastes** produced in the body. It is distinguished from elimination, the removal of indigestible material. Most of the body's activities produce metabolic wastes that must be removed. For example, aerobic respiration leads to the production of carbon dioxide waste. **Deamination** of amino acids leads to the production of nitrogenous wastes, such as urea or ammonia. All metabolic processes lead to the production of mineral salts, which must also be excreted.

AMAZING BIOLOGICAL MACHINES: PHYSIOLOGICAL PROCESSES AND RESPONSES ★★

It is a wonder that organisms function at all. Most organisms are complex machines with many intricate parts and processes. It is perhaps easier to fathom the workings of a simple prokaryote cell—coordinating activity with other organisms and the liquid matrix in which it lives—than to study complex organisms. The synchronicity of the cells, tissues, and organs in a multicellular organism involves a highly complex set of processes that are not easy to simplify.

Components of the digestive tract

THE DIGESTIVE SYSTEM ★★

MOUTH

Food is ingested into the mouth. Both mechanical and chemical digestion begin in the mouth. The salivary glands produce saliva, which contains salivary amylase (ptyalin). Salivary amylase digests starch to maltose (disaccharide).

ESOPHAGUS

Food moves down the esophagus by peristaltic motion and into the stomach via the cardiac sphincter.

NOTE

All physiological processes have an underlying means of chemical or electrochemical control, which will be reviewed in this chapter as a foundation for further investigation into specific systems in different groups of organisms.

KEY POINT

In addition to synthesis of information, the College Board most strongly emphasizes form and function in the course outline and for this topic on the exam.

STOMACH

The stomach is protected from acidic material by a mucosal lining. Ulcers result when a portion of this lining is digested and a perforation in the stomach mucosa develops. Proteins are digested by pepsin. After being digested in the stomach, the mixture is called chyme.

SMALL INTESTINE

The chyme moves into the small intestine via the pyloric sphincter. There are three sections of the small intestine (duodenum, jejunum, and ileum). When food enters the duodenum (pronounced: doo-uh-dee-num), it stimulates the gallbladder and pancreas to secrete their substances into the duodenum. The duodenum is the site of the digestion of proteins, carbohydrates, and fats. The small intestine is lined with villi (which contain capillaries and lacteals) that increase the surface area for absorption of nutrients into the blood. Absorption of nutrients occurs along the jejunum and ileum.

LARGE INTESTINE

The large intestine (cecum, colon, rectum) is the major site of water reabsorption; *E. coli* inhabits the large intestine and produces vitamin K as a metabolic byproduct. Feces is stored in the rectum. Wastes are eliminated via the anus.

LIVER

The liver stores nutrients (such as glycogen), detoxifies chemicals and drugs in the bloodstream, forms urea, forms glycogen from glucose or breaks down glycogen to release glucose, and produces bile.

GALLBLADDER

The gallbladder stores and concentrates bile, which is released to emulsify fats. When the gallbladder is removed, the liver is still producing bile. But bile runs into the small intestine as it is produced, rather than dumped when you need it. People without gallbladders will be fine as long as they change their diets to match their bile production, because their bodies can no longer store and release the bile to match the diet.

PANCREAS

The pancreas performs an endocrine function, affecting blood glucose concentration via the secretion of insulin and glucagon. It also secretes pancreatic juices into the small intestine (exocrine function). These secreted juices contain hydrolytic enzymes, such as amylase, trypsinogen, chymotrypsinogen, and pancreatic lipase. In addition, the pancreas also secretes bicarbonate ions into the small intestine to neutralize the acidic chyme from the stomach. The zymogens are converted to their active enzyme form in the duodenum; most pancreatic enzymes function optimally at an alkaline pH.

THE DIGESTIVE TRACT ★★★★

The digestive tract is simply a long tube that travels through a mammal, allowing the exchange of nutrients. The lumen, the space inside the tube, is filled with a liquid matrix that carries particles along like a conveyor belt. By the time the digestive tract leaves the animal through the anus, the tissues have taken what they want and eliminated what they don't, leaving only waste. The different tissues of the digestive tract are localized in organs such as the stomach and intestines that coordinate with each other in the uptake of carbohydrates, proteins, fats, vitamins, and any other nutrients the body needs. Just as with gas exchange and the circulatory system, the digestive tract simply provides a mode of transportation for these specialized tissues to transport goods. Again, don't think of the digestive tract as being inside the body. Think of it as an external surface that has been moved inside so it can imitate the liquid environment from which all mammals originated.

NUTRIENTS

After a mammal regulates its body with hormones and nervous coordination and supplies its cells with the gases necessary for cellular respiration, the next most immediate concern is getting nutrients. Nutrients include not only the energy required to fuel cellular activity, but the micronutrients (vitamins and minerals) necessary for chemical function and, most importantly, water.

KEY STRATEGY

As you prepare for the AP Biology exam, make sure you know the differences in structure and function between herbivore and carnivore digestive tracts.

As in respiration, animals gradually evolved from single-celled organisms that exchanged molecules with the outside environment across their entire bodies to complex organisms with tissue and organ specialization. Even though the exchange of molecules occurs inside the body cavity of these organisms, the actual surface where this exchange takes place in the respiratory and digestive systems is the interface between the inside and outside of the animal. This means that the contents of the stomach, although inside the abdomen of an animal, are actually part of the outside environment (think of the entire digestive system as being a hole that passes through an organism). The internal compartments of the alveoli are also part of the outside environment. These systems occur externally in more primitive animals.

Ingested food is first subjected to mechanical shearing and grinding in the mouth to reduce the size of food particles. Saliva begins digestion of carbohydrates and lubricates the ball of food that is known as a bolus. Stretch receptors in the stomach cause chief cells and parietal cells to secrete gastric juices, a highly acidic ionic solution that denatures proteins by disrupting hydrogen bonds and secondary structure. Gastric juice thereby causes connective tissues of animal meat to break apart. Unfolding of proteins provides increased surface area for proteases active **only** at low pH levels.

As the food mass breaks up, it stratifies in the stomach according to size, density, and water solubility. Fat cells and lipids float to the top of the stomach, carbohydrates and smaller proteins remain in the middle, and the larger proteins sink to the bottom. This is very important for the next step, because the contractions of the stomach force proteins and carbohydrates out first and the lipids next. Evolutionarily speaking, this design favors early digestion and absorption of the most readily available sources of chemical energy (polysaccharides); protein and fat digestion come afterward.

The partially digested mass of food, known as **chyme**, enters the small intestine. Pancreatic fluid, rich in bicarbonate ions to neutralize stomach acid and full of enzymes to break down proteins and lipids, is secreted. Bile is secreted by the liver to help fat absorption by emulsifying the chyme. The proper mixture (depending on what's in the meal that was eaten) of chyme, pancreatic fluid, and bile is obtained by adjusting the rates of gastric emptying, pancreatic secretion, bile secretion, and intestinal motility (muscle movement). The smooth muscle of the small intestine, like an elevator, continuously moves the chyme down the intestine. Chemical breakdown continues, and products are absorbed by cells lining the inside of the small intestine. By the time the chyme gets to the end of the small intestine—the terminal **ileum**—all of the digestible food will have been digested and absorbed. **Bile salts** are secreted by the liver to be used again for the next meal.

Unabsorbed material is pushed into the large intestine where the remaining water and salt are reabsorbed. The leftover material is the **feces**, which is stored in the terminal colon (large intestine) until enough material accumulates that the defecation reflex is activated.

Overall, you should consider that the digestive process involves the secretion of over 7 liters of fluids and enzymes *per day* to the inside of the digestive tract (which is technically outside of the body). Humans only have 5 or 6 liters of blood; thus, an enormous amount of fluid is used for digestion, and the fluid needs to be restored quickly if dehydration and eventual death are to be avoided.

The variations seen among vertebrate digestive systems are a result of dietary differences. These variations involve the shape of teeth, the length of the alimentary canal (how long the digestive tract is), and the presence of enlarged,

AP TEST STRATEGY

The secreted fluids and enzymes used in digestion are ultimately reabsorbed back into the body. The process in mammals is not all that different from the way a fly digests and absorbs its food—just more discreet.

IMPORTANT

Accumulation of nitrogenous waste is dangerous; it can cause changes in mental alertness and bleeding, particularly in the stomach and small intestine.

multichambered stomachs as seen in mammals called **ruminants**. Whereas carnivores possess sharp and pointed incisors to kill prey and tear flesh, herbivores have round flat teeth for the crushing and grinding of stems and leaves. In addition, herbivores and animals that eat mainly vegetation have much longer digestive systems than those that are primarily meat eaters. The reason for this is that longer periods of digestion are needed to break down and absorb the biomolecules found in vegetation, especially cellulose, which must be broken down by symbiotic bacteria living within the gut. Animals such as cattle and horses have enlarged cecums; the cecum is the pouch where the small intestine joins the large intestine. It is here that hordes of bacteria sit and digest cellulose present in the plant material that is consumed in large quantities. In fact, cattle and sheep also have multichambered stomachs, divided into three or four sections (chambers) that allow slow and repeated digestion of food before the food actually proceeds into the small intestine for absorption.

EXCRETORY SYSTEM ★★

The principal organs of excretion in humans are the lungs, liver, skin, and kidneys. In the lungs, carbon dioxide and water vapor diffuse from the blood and are continually exhaled. Sweat glands in the skin excrete water and dissolved salts (and small quantities of urea). Perspiration serves to regulate body temperature because the evaporation of sweat produces a cooling effect. The liver processes blood pigment wastes and other chemicals for excretion. **Urea** is produced by the deamination of amino acids in the liver, and it diffuses into the blood for excretion in the kidneys. The kidneys function to maintain the osmolarity of the blood and to conserve glucose, salts, and water.

On an average day-to-day basis, an adult human maintains more or less the same weight, a relatively constant blood pressure, and a certain amount of salts, metabolic wastes, and water volume in his tissues. This is due to the fact that **renal** (kidney) **output** varies around a set point for each individual, whereby renal output of salt, water, and urea goes up as dietary intake increases and goes down as dietary intake decreases. The mechanism by which the kidney achieves this is through filtration of the blood through many small capillary bundles called **glomeruli** (singular: glomerulus) and reabsorption or rejection of this filtrate as it passes down long, twisting tubes known as nephrons. Within each kidney, there are perhaps 1,000,000 glomeruli, each attached to a nephron tube. The filtrate that makes it through the nephron tubes and is reabsorbed back into the bloodstream is called the final filtrate, or urine. It empties from the ends of the nephrons into the renal pelvis and is then carried to the bladder for storage and eventual release during urination.

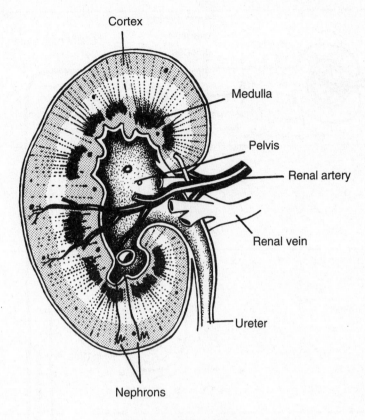

Structure of the Kidney

Blood enters the kidney via the renal artery. From there, it goes into the afferent arterioles and into the glomerulus. Blood leaves the kidney via the efferent arteriole and out the renal vein.

Approximately 200–300 liters of blood pass through the two kidneys every day, entering via the renal artery and exiting via the **renal vein**. The average human body has only 6 or 7 liters of blood, which means that *the entire blood volume is swept through the kidneys at least 20 times every day*. As blood enters, it is quickly disseminated into tiny capillaries that end at a glomerulus, a tiny ball of capillaries contained within a pouch called **Bowman's capsule**. Bowman's capsule leads into a long and coiled nephron that is divided into functionally separate units: the **proximal convoluted tubule**, the **loop of Henle**, the **distal convoluted tubule**, and the **collecting duct**. The loop of Henle of most nephrons dips down through the medulla of the kidney, while the glomeruli and convoluted tubules remain positioned in the cortex.

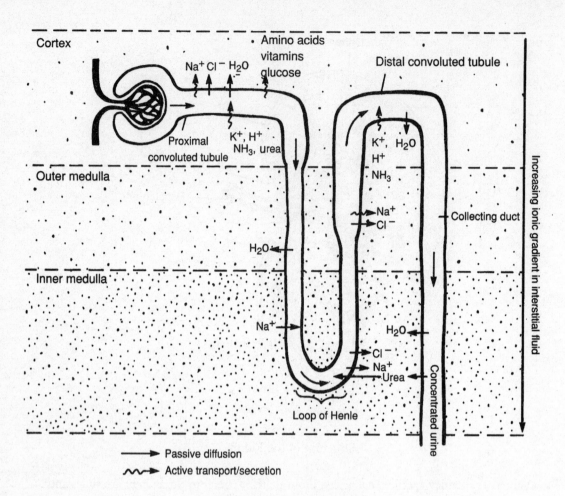

Structure of the Nephron

The kidney filters, secretes, and reabsorbs to maintain the body's homeostasis.

FILTRATION

Filtration occurs at the level of the glomerulus. The glomerulus has small holes to allow small molecules and water through, but excludes cells and large proteins.

SECRETION

Some solutes are actively secreted into the urine by transporters in the membranes of epithelial cells. Examples include organic molecules such as *p*-aminohippuric acid and penicillin.

REABSORPTION

Some molecules are reabsorbed from the urine into the capillary. Examples are water, electrolytes, glucose, and amino acids. These may require a transporter or may occur by passive diffusion.

In the glomerulus, blood is filtered out of capillaries into Bowman's capsule and the nephron. Many substances do not pass through the glomerulus into the nephrons; instead, they simply exit the kidneys through the renal vein without ever passing through the nephron tubes. Large proteins, such as **albumin**, and proteins with lots of negative charge, are repelled due to the structure of the glomerulus, with its triple-layered capillary cells. Cells remain within the bloodstream. Waters, salts, sugars, and urea do enter the nephron, however, and this fluid is called the filtrate. The bloodstream drops off material at the glomerulus and then doubles back to wind very closely around the proximal and distal tubules and the loop of Henle. The closeness of the bloodstream to the filtrate in the nephron tube sets up a concentration gradient whereby much of the material in the filtrate will diffuse back into the nearby blood or be moved there by active transport. After filtration, the next step is the selective reabsorption of most of the filtrate.

The filtrate finds itself within the proximal tubule. On the surface of the cells that line this tubule are Na^+/H^+ ion exchange pumps. While kicking out hydrogen ions into the tube, these membrane proteins absorb sodium from the filtrate so that it travels into the cells that line the tubule. This sodium moves through to the opposing end of the cells where it is extruded into the bloodstream by a Na^+/K^+ ATPase pump. The end result is the movement of sodium ions from the filtrate into the bloodstream. Water naturally follows due to the hypertonic conditions that occur as salt concentration goes up in the nearby bloodstream, and it is here in the proximal tubule that most of the water and salt that were originally filtered out of the blood are returned back into the blood.

The proximal tubule is also responsible for the reabsorption of glucose using active transport. In a healthy individual, there is no glucose excreted in the urine because all glucose is efficiently reabsorbed in the proximal tubule. In the loop of Henle, either water or salt is absorbed independently. The descending part of the loop absorbs water only, and the ascending loop absorbs salt ions only. More water and salt reabsorption takes place in the distal tubule, and in the collecting duct, certain hormones regulate the concentration of the urine.

The selective permeability of the tubules establishes an **osmolarity gradient** in the surrounding interstitial fluid. By exiting and reentering at different segments of the nephron tube, solutes create a situation in which tissue osmolarity increases from the cortex into the inner medulla. While most of the urea that ends up in the collecting duct is excreted, some diffuses into the nearby interstitial fluid and reenters the ascending loop of Henle. Overall, the gradient that is established by the concentration of solutes in the kidney medulla helps to maximize water conservation and make the urine as hypertonic (concentrated) as possible.

HORMONAL REGULATION ★★★

Principal cells in the collecting duct respond to various hormones, such as aldosterone, to absorb more water or salt depending on how concentrated the filtrate is at that point. **Aldosterone** is produced in the adrenal cortex and stimulates the reabsorption of Na^+ from the collecting duct. Na⁺ stimulates water reabsorption as well, so aldosterone is released when blood pressure falls in the body. People with Addison's disease produce insufficient aldosterone and have urine that has too high Na⁺ concentrations. This causes dehydration and a drop in blood pressure, as water is pulled out of the bloodstream to follow the excess salt in the urine. Another hormone, **angiotensin**, is also produced when blood volume falls, and it works in a similar fashion to aldosterone: retaining water to increase blood volume.

ADH, or **antidiuretic hormone**, also controls the concentration of the urine and acts on cells lining the collecting duct of the nephron. As its name implies, release of this hormone, which takes place in the posterior pituitary, causes water reabsorption and a decrease in urine output. ADH is also known as **vasopressin**, and it works mainly by opening up active transport water channels called **aquaporins** on the luminal surfaces of principal cells in the collecting duct. This causes water to leave the filtrate and move into the medulla, where it is eventually reabsorbed into the bloodstream, leaving the filtrate more concentrated and the eventual urine much less diluted.

EVENTS THAT MAKE BLOOD PRESSURE FALL ★★★

Blood is mostly water, so anything that causes water loss, such as vomiting, diarrhea, blood loss, or heavy sweating, can cause a drop in blood pressure. The body responds by releasing aldosterone and angiotensin, which allow the cells in the collecting duct to reabsorb more salt and water. This extra fluid adds volume, and therefore pressure, back to the blood.

CHAPTER 24: THE MUSCULOSKELETAL SYSTEM

BIG IDEA 4: Biological systems interact, and these systems and their interactions possess complex properties.

IF YOU ONLY LEARN FOUR THINGS IN THIS CHAPTER . . .

1. The musculoskeletal system gives humans the ability to move using muscles and the skeletal system (bones).

2. Sarcomeres combine to make a myofibril. Many myofibrils bundle together to make a muscle fiber. Several muscle fibers combine to make a bundle. Many bundles make up a muscle.

3. Muscle fibers create tension for contraction by the cross-bridge cycling of actin and myosin. Action potentials travel down the axon toward the neuromuscular junction where voltage-gated calcium channels are activated, releasing acetylcholine across the synapse.

4. Bones have many functions, including protection, movement, structure, blood production, and mineral storage. Bones are composed of many osteons, each containing a central or Haversian canal surrounded by lacunae containing osteocytes, lamellae, and canaliculi.

SKELETAL MUSCLE CONTRACTION ★★

Vertebrate skeletal muscles are bundles of parallel fibers. Each fiber is a multinucleated cell created by the fusion of mononucleate embryonic cells. The nuclei are usually found at the periphery of a muscle cell.

Embedded in the fibers are filaments called **myofibrils**, which are further divided into contractile units known as **sarcomeres**. The myofibrils are enveloped by a modified ER (endoplasmic reticulum) that holds concentrated stores of calcium (Ca^{2+}) ions and is called the sarcoplasmic reticulum. The membrane surrounding a muscle fiber, the sarcolemma, is folded into a series of transverse tubules (T tubules) that can propagate an action potential almost instantly through the entire fiber.

The sarcomere is composed of thin and thick filaments. The thin filaments are chains of globular actin molecules associated with two other proteins, troponin and tropomyosin. The thick filaments are composed of organized bundles of myosin molecules; each myosin molecule has a head region and a tail region.

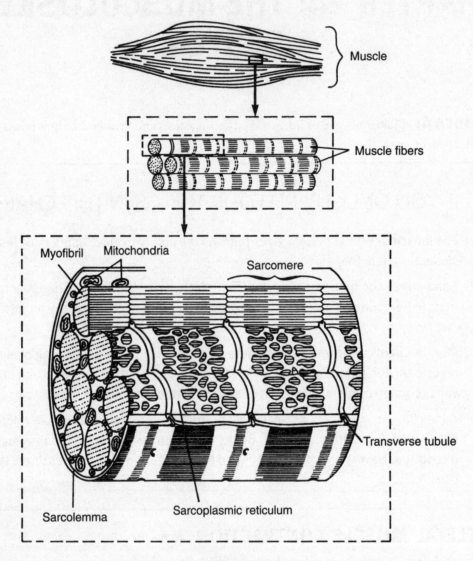

Skeletal Muscle

Electron microscopy reveals that the sarcomere is organized as follows: Z lines define the boundaries of a single sarcomere and anchor the thin filaments. The M line runs down the center of the sarcomere. The I band is the region containing thin filaments only. The H zone is the region containing thick filaments only. The A band spans the entire length of the thick filaments and any overlapping portions of the thin filaments. Note that during contraction, the A band is not reduced in size, while the H zone and I band are.

Sarcomere → myofibril → fiber → bundle → muscle

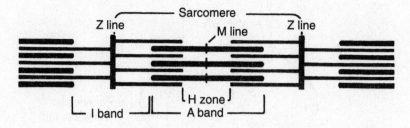

Muscle before Contraction

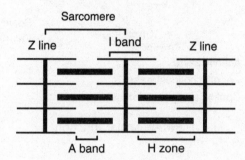

Muscle after Contraction

ACETYLCHOLINE AT THE NEUROMUSCULAR JUNCTION ★★

A single muscle fiber is usually innervated by only one motor nerve axon. The neurotransmitter that is released by this motor axon at the synapse with skeletal muscle is ACh, **acetylcholine**, and the receptors on the surface of the muscle cells are ACh receptors. The special region where the motor nerve synapses on the muscle is called the **motor end plate**. As the motor axon approaches the end plate, it loses its myelin sheath and splits into branches. At the ends of these branches are presynaptic terminals that contain vesicles filled with ACh. Calcium influx causes the ACh vesicles to release their contents into the synapse between the neuron and muscle fiber. The binding of the ACh to receptors on the muscle fibers causes depolarization of the muscle fiber as Na^+ ions flood across into the muscle cell. This depolarization triggers muscle contraction.

The synthesis of ACh in motor neurons requires choline, acetyl-CoA, and the enzyme choline acetyl transferase (CAT). The availability of choline is the rate-limiting factor for the synthesis of the ACh. Removal and breakdown of ACh, once it is released into the synapse, uses the enzyme **acetylcholinesterase** (AChE), which breaks the ACh into choline and acetate. The removal of ACh is required to terminate the nervous signal that leads to muscle contraction. The choline can then be taken back into the presynaptic cell via high-affinity active transport pumps.

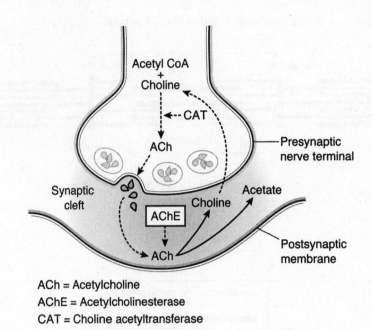

ACh = Acetylcholine
AChE = Acetylcholinesterase
CAT = Choline acetyltransferase

Neuromuscular Junction

MECHANISM OF MUSCLE CONTRACTION AND MUSCLE RESPONSE ★★

Once an action potential is generated, it is conducted along the sarcolemma and the T system and into the interior of the muscle fiber. This causes the sarcoplasmic reticulum to release Ca^{2+} into the sarcoplasm. The Ca^{2+} binds to the troponin molecules, causing the tropomyosin strands to shift, thereby exposing the myosin-binding sites on the actin filaments.

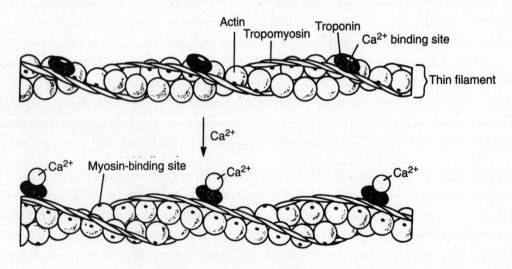

Muscle Fiber

The free globular heads of the myosin molecules move toward and then bind to the exposed binding sites on the actin molecules, forming actin-myosin cross-bridges. In creating these cross-bridges, the myosin pulls on the actin molecules, drawing the thin filaments toward the center of the H zone and shortening the sarcomere. ATPase activity in the myosin head provides the energy for the power stroke that results in the dissociation of the myosin head from the actin. (An ATPase is an enzyme that hydrolyzes ATP.) The myosin returns to its original position and is now free to bind to another actin molecule and repeat the process, thus further pulling the thin filaments toward the center of the H zone.

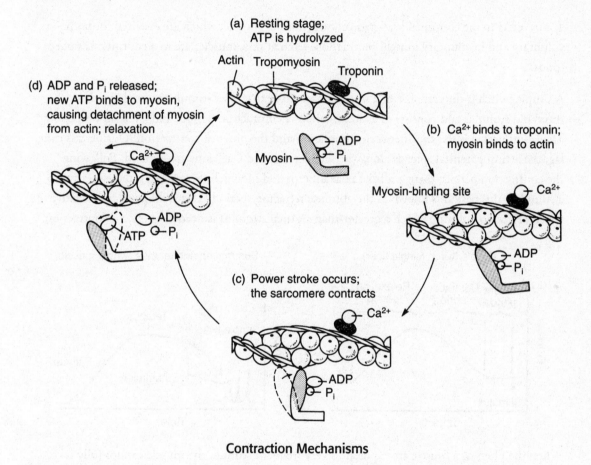

Contraction Mechanisms

When the sarcolemmic receptors are no longer stimulated, the Ca^{2+} is pumped back into the sarcoplasmic reticulum. The products of ATP hydrolysis are released from the myosin head, and a new ATP binds to the head. This results in the dissociation of the myosin from the thin filament, and the sarcomere returns to its original width. In the absence of Ca^{2+}, the myosin-binding sites on the actin are again covered by tropomyosin molecules, thereby preventing further contraction. After death, ATP is no longer produced, and the myosin heads can't detach from actin, and therefore the muscle cannot relax. This condition is known as rigor mortis.

STIMULUS AND MUSCLE RESPONSE ★★

Individual muscle fibers generally exhibit an all-or-none response; only a stimulus above a minimal value, called the **threshold value**, can elicit contraction. The strength of the contraction of a single muscle fiber cannot be increased, regardless of the strength of the stimulus. Whole muscle, on the other hand, does not exhibit an all-or-none response. Although there is a minimal threshold value needed to elicit a muscle contraction, the strength of the contraction can increase as stimulus strength is increased by involving more fibers. A maximal response is reached when all of the fibers have reached the threshold value and the muscle contracts as a whole.

Tonus refers to the continual low-grade contractions of muscle, which are essential for both voluntary and involuntary muscle contraction. Even at rest, muscles are in a continuous state of tonus.

A simple twitch is the response of a single muscle fiber to a brief stimulus at or above the threshold stimulus and consists of a latent period, a contraction period, and a relaxation period. The latent period is the time between stimulation and the onset of contraction. During this time lag, the action potential spreads along the sarcolemma and Ca^{2+} ions are released. Following the contraction period, there is a brief relaxation period in which the muscle is unresponsive to a stimulus; this period is known as the absolute refractory period. This is followed by a relative refractory period, during which a greater-than-normal stimulus is needed to elicit a contraction.

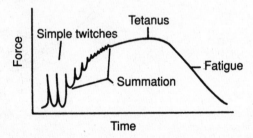

When the fibers of a muscle are exposed to very frequent stimuli, the muscle cannot fully relax. The contractions begin to combine, becoming stronger and more prolonged. This is known as frequency summation. The contractions become continuous when the stimuli are so frequent that the muscle cannot relax. This type of contraction is known as tetanus and is stronger than a simple twitch of a single fiber. If tetanization is prolonged, the muscle will begin to fatigue.

SKELETAL SYSTEMS ★★

There are three main functions of any skeleton: support, aid in movement, and protection. Humans have an endoskeleton or internal system of bones and cartilage that support surrounding soft tissues. **Cartilage** is a type of connective tissue that is softer and more flexible than bone. It is composed of an elastic-form matrix, called **chondrin**, that is secreted by specialized cells known as chondrocytes. Embryonic skeletons in mammals are made mostly of cartilage before turning into bone due to calcification and hardening. Most cartilage is avascular (lacking blood vessels) and devoid of nerves but receives nourishment from nearby capillaries.

Bone is a mineralized connective tissue that comes in two basic types: compact and spongy. It is in the spongy bone that we find bone marrow, where the production of red and white blood cells takes place. The bones of the appendages, the long bones, are characterized by a cylindrical shaft called a **diaphysis** and dilated ends called epiphyses. The diaphysis is composed primarily of compact bone surrounding a cavity containing bone marrow. The epiphyses are made of spongy bone surrounding a layer of compact bone. The **epiphyseal plate**, or growth plate, is a disk of cartilaginous cells separating the diaphysis from the epiphysis, and it is the site of bone growth longitudinally (in length). The epiphyseal plate remains cartilaginous as long as growth continues. When the plate seals and calcifies, growth ceases. A fibrous sheath known as the periosteum protects the bone.

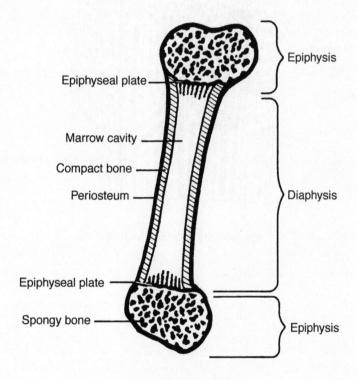

Basic Structure of Bone

Bone is much more dynamic than you might think. Bones exist in a dynamic equilibrium between being broken down by osteoclasts and being built up by osteoblasts. Osteoclasts eat away at the bony matrix, adding calcium to the blood, and osteoblasts take calcium out of the blood to deposit it onto the bones. This constant bone remodeling takes place throughout one's entire life.

The bony matrix that builds bone is deposited in structural units called osteons, each of which consists of a tiny channel known as a **Haversian canal** that is surrounded by concentric circles called **lamellae**. Blood vessels, nerves, and lymph vessels travel through these Haversian canals to supply the bone tissue with nutrients, gases, and innervation. Also within the bone are mature bone cells, osteocytes, which are involved in the maintenance of the bone's structure. **Canaliculi** that radiate from gaps (lacunae) in the bony matrix serve to indirectly connect all the Haversian canals for exchange of nutrients and waste.

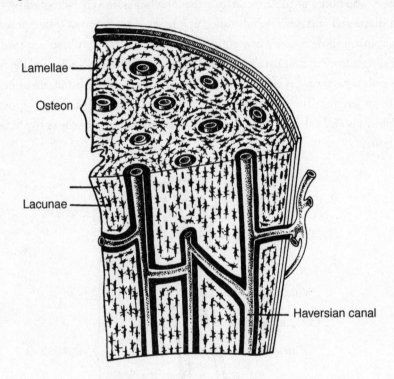

Microscopic Bone Structure

CHAPTER 25: ECOLOGY

BIG IDEA 4: Biological systems interact, and these systems and their interactions possess complex properties.

IF YOU ONLY LEARN SIX THINGS IN THIS CHAPTER . . .

1. Populations with infinite resources increase exponentially in a J-shaped curve. Usually the environment has a carrying capacity, the maximum number of individuals it will support. Factors affecting population size can be density-dependent (overcrowding) or density-independent (a storm wipes out part of a population).

2. Species that are *r*-selected have many offspring and provide almost no parental care (fish, insects). They succeed best in new habitats with many resources. *K*-selected species have few offspring and care for them for an extended period (elephants, humans). Those species do best in stable environments near carrying capacity.

3. A biome is a climatic zone with associated animals and vegetation (tundra, desert, etc.). An ecosystem comprises a community of living organisms and its habitat. Different populations of organisms make up the community. Each organism is adapted to a specific niche or role.

4. Food webs trace energy flow in a community. Different trophic levels consist of primary producers (plants), primary consumers (herbivores), secondary consumers (carnivores), and decomposers (bacteria).

5. Materials such as water, nitrogen, carbon, and phosphorus travel through the environment and are recycled in biogeochemical cycles involving different life forms.

6. Human activity has affected the environment significantly, e.g., ozone depletion, habitat destruction, and global warming.

IT'S NOT A VACUUM OUT THERE: HOW ORGANISMS INTERACT WITH EACH OTHER AND THEIR ENVIRONMENT

The natural world is a harsh place. Organisms face the daily challenge of finding nutrients (often in the form of other organisms), safe havens, and mates. Organisms must not only adjust to the abiotic (nonliving) factors presented by the environment, but also compete for resources with other organisms while avoiding being eaten. Some scientists cite examples of cooperative exchanges among and between species, but others suggest that these are selfish behaviors that help an individual survive and pass on its DNA. Fortunately, the College Board has concentrated their questions about this hypothesis-dense discipline on just a few concepts.

PRIMARY CONCEPTS

Primarily, four ecology concepts are found on the exam:

1. The basics of population biology, including abiotic and biotic factors that affect population size, growth, and decline

2. The diagnostic features of the different levels of environmental systems, or the "classification" of the environment

3. Nutrient cycling and energy exchange

4. Human impact on the global environment

POPULATION DYNAMICS ★★★★

What do you get if you put one bacterium in a lot of space and give it an unlimited amount of resources and time to reproduce? You get a whole lot of bacteria, that's what! As a matter of fact, you get an exponential rise in the number of bacteria according to the equation:

$N_t = N_0 e^{rt}$, where

N_t = the number of bacteria at time t,

N_0 = the number of bacteria at the beginning,

r = the rate of population growth (a difference of the reproduction and death rates),

t = the number of chronological steps of reproduction (seconds, minutes, years, etc.), and

e = the constant 2.71828…

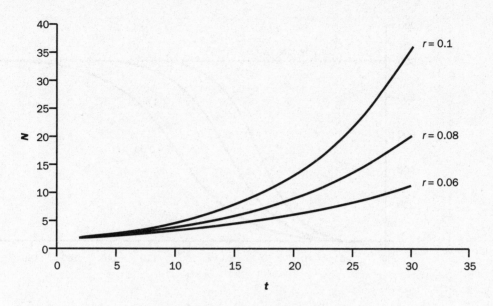

Exponential Population Increase

The three lines on the graph show the exponential increase in population size with no limits according to the equation $N_t = N_0 e^{rt}$. Each line represents a population increasing at the rate indicated, having started with two individuals.

This graph shows how rapidly populations can grow when the increase is exponential. Remember that t is a measure of the population's rate of producing another generation, which is very short for bacteria. Some bacteria can reproduce every 10 to 30 minutes, so 30 time steps would only take between 5 and 15 hours.

DID YOU KNOW?

What is really incredible is that a population of bacteria can grow to numbers in the millions in only a matter of hours!

The reality is that populations do not grow exponentially without limits. Most reach a size at which they plateau, called the **carrying capacity**. The rate of change in size of natural populations can be estimated by the equation $\dfrac{\Delta N}{\Delta t} = rN\left(\dfrac{K - N}{K}\right)$ where

$\dfrac{\Delta N}{\Delta t}$ = the change in the population size over the given time,

N = the size of the population at the beginning of time t,

r = the rate of population growth (again, a difference of the reproduction and death rates), and

K = the carrying capacity.

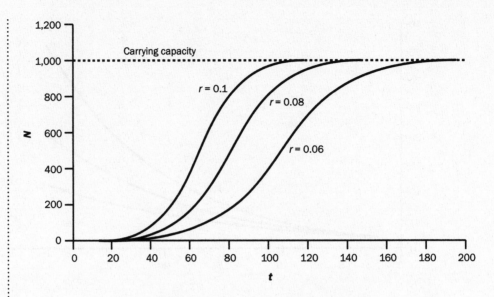

Population Increase to Carrying Capacity

These three lines show the increases in population size with the same rates of population growth as in the previous graph, but with a limited carrying capacity of 1,000 individuals. This graph shows that carrying capacity is the ultimate limit to population size, while the rate of population growth affects how quickly the population increases. A change in carrying capacity will change where the plateau occurs.

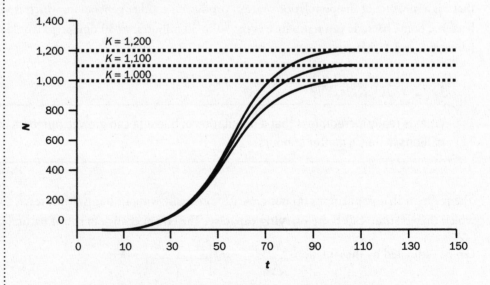

Populations with Different Carrying Capacities

The three lines in the previous graph show the increases in population size with a constant rate of population growth ($r = 0.10$) for three populations whose carrying capacities are not the same. Carrying capacities are indicated on the graph. When the carrying capacity increases for a population with a constant growth rate, the shape of the *growth curve* hardly changes. It is the ultimate population size that is really affected.

POPULATION GROWTH FACTORS

Factors that affect population growth fall into two categories, density-independent and density-dependent.

- **Density-independent** factors affect populations in the same way regardless of how many individuals are in the population at that given time. These factors tend to be more catastrophic, abiotic factors, such as flood, drought, hurricane, or fire.

- The impact of **density-dependent** factors on populations increases as the number of individuals in the population, and therefore the density in a given area, increases. These factors include variables such as competition among or between species, predation, or emigration.

Different **biotic** and **abiotic** factors in the environment have contributed toward adaptive strategies that allow organisms to maximize their reproduction. Species that monopolize rapidly changing environments, such as disturbed habitats, produce many offspring quickly. These species are called *r*-selected, or *r*-strategists, because their strategy is to increase their *r*-value. Species that are *r*-strategists generally have short life spans, begin breeding early in life, and produce large numbers of offspring.

Species in more stable environments tend to produce fewer offspring and invest more resources into the success of each offspring. These species are called *K*-selected, or *K*-strategists. *K*-strategists have longer life spans, begin breeding later in life, have longer generation times, and produce fewer offspring. They take better care of their young than *r*-strategists; they have also evolved to be better at exploiting limited parts of their environments. There is a continuum between *r*-selected and *K*-selected species, as few communities are made up of strictly *r*-strategists or *K*-strategists.

Every species has a set of conditions that are optimal for its reproductive strategies; in any ecosystem, the most abundant species will be the one for which the environment most closely approximates this set of optimal conditions. If these conditions remain constant, all other things being equal, the distribution of organisms in an ecosystem should remain roughly the same.

ENVIRONMENTAL CHANGE

Without important changes to the environment or available resources, there is little reason to expect a population to boom or crash. However, these kinds of significant changes are inevitable over time, and they determine which species can thrive or dominate in an ecosystem. This change in the species makeup of an ecosystem is called **ecological succession.**

A HIERARCHICAL CLASSIFICATION OF THE ENVIRONMENT ★★

Just as every organism needs an identifier to allow effective discourse among the scientific community, the environment is also made up of named divisible units with diagnostic features. The largest division within the Earth is the biome. There is no consensus among the scientific community as to which regions constitute a biome. Deserts, grasslands, taigas, and oceans are all types of biomes. They are defined by large geographic regions with homogenous abiotic and biotic characteristics such as low rainfall, high altitude, domination by short grasses, and so on. Biomes are divided into ecosystems, which defy standardizations of scale. An **ecosystem** can be an entire mountain range or a coral reef, but can also be an ephemeral pool of water in the middle of the desert. An ecosystem is a collection of communities and their abiotic environment, grouped around a geographic feature. The key is that ecosystems include processes such as nutrient flow.

TERRESTRIAL BIOMES OF THE WORLD

Biome	Climate	Flora and Fauna
Tropical rain forest	Constant temperatures, high rainfall	Most diverse biome with many species from all domains
Deciduous forest	Temperate, warm summers, cold winters with snow	Deciduous trees, herbs, and grasses, forest-dependent fauna, migratory birds
Taiga	Long, cold winters and short, cool summers	Conifers, short grasses, low diversity, mammals and birds, few reptiles and insects
Desert	Extremely low rainfall, temperature extremes	Shrubs, cacti, and succulents, reptiles, specialized adaptations to deal with extremes in temperature and drought
Grassland	Semi-arid to temperate, hot summers and cold winters	Long or short grasses, very low diversity
Tundra	Extremely short growing season; long, cold winters	Lichens, mosses, sedges, dwarf shrubs, simplest biome with lowest diversity
Savanna	Tropical rainy season, arid dry season	Perennial grasses, shrubs, sparse trees, diverse in large mammals

There are three other terms with which you should be familiar. An ecosystem is composed of a habitat, populations, and communities. The **habitat** is the total of the various abiotic factors, such as terrain, soil type, and water. **Populations** are all the organisms that occur in a specific habitat. **Communities** are all the populations of organisms that live in the habitat. The fauna, flora, and abiotic factors collectively make up an **ecosystem**.

A **niche** is a theoretical construct that is defined by a multidimensional space made up of all of the environmental factors, both abiotic and biotic, that define where and how a population of organisms lives. For example, if a population of adult pelagic fish feed at a depth of 10–20 m, that range of depths is part of the fish's niche. If a population of thermophiles thrives in liquid clay at 100°C, both the clay soil and the high temperature are part of its niche.

ENVIRONMENTAL DYNAMICS ★★★★

Environmental dynamics deals with the flow of energy and other nutrients through the environment. **Energy flow** is traditionally shown in the form of a food web. Arrows are drawn between different species of organisms, with the direction of the arrow indicating the direction of energy flow.

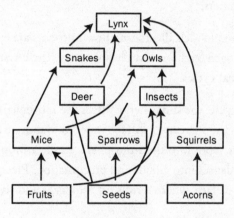

Food Web

In the illustration of a terrestrial food web, the arrows indicate the direction of energy flow between the organisms. The fruits, oaks, and grasses only have arrows pointing away from them, so they are primary producers. The lynx is the top predator (quaternary consumer) and only has arrows pointing toward it. There are no decomposers in this food web.

> **REMEMBER**
>
> Energy is lost as it moves through the various food webs. Only about 10 percent of the energy received is passed from one trophic level to the next.

Food webs can quickly become quite complicated because there are so many different organisms interacting in any given ecosystem. Food webs are rarely entirely comprehensive for the same reason. As one kind of organism feeds another, nutrition moves through a food web from one trophic level to the next. Organisms included in a food web may be **primary producers** (**autotrophs**), which produce energy from sunlight, **primary** and **secondary consumers**, and

decomposers. Because they make their own food, primary producers only have arrows pointing away from them in a food web. Organisms that get energy by consuming other living things (**heterotrophs**) have arrows pointing toward them, indicating the flow of energy from another trophic level. If decomposers are part of the food web, there will also be arrows leading from the various levels of heterotrophs to the decomposers. Primary consumers eat plants, while secondary consumers, tertiary consumers, and quaternary consumers feed on the levels below them, such as the primary consumers. **Herbivores** eat only plants (thus, they are always primary consumers). However, some secondary and **tertiary** consumers are **omnivores**, which eat both animals and plants. Omnivores in a food web are indicated by arrows pointing toward them from both primary producers and primary consumers.

Other simplified models illustrating how nutrients are cycled include those for water, nitrogen, carbon, and phosphorous. These systems are known as **biogeochemical cycles**.

In the **water cycle**, the sun's energy drives the movement of water through the environment. The sun causes surface water to evaporate and plants to transpire, releasing water vapor into the atmosphere. In the atmosphere, the water vapor cools and condenses into clouds and precipitation. Precipitation returns water to the Earth's surface, where it runs off into lakes and rivers and percolates into groundwater.

There is more nitrogen in the air than oxygen: Nitrogen gas makes up over two-thirds of Earth's air. However, most living things cannot get the nitrogen they need from nitrogen gas. Certain groups of bacteria in the soil, called **nitrogen-fixing bacteria**, convert nitrogen gas into nitrates and nitrites; plants then take up the nitrites and nitrates from the soil and from fertilizers. Animals, including humans, get nitrogen by eating plants and other living or dead organic matter. When animals and plants die, groups of bacteria and fungi break down these organic remains and release nitrogen back into the soil. Animal waste contains nitrogen as well. **Denitrifying bacteria** in the soil then change the nitrates and nitrites back into nitrogen gas, which returns to the air. Together, all of these steps make up the **nitrogen cycle**.

KEY POINT

It is important to understand productivity as it relates to energy flow in food webs—many AP Biology exam questions will deal with differences between gross and net primary productivity and factors that influence productivity within a food chain or food web.

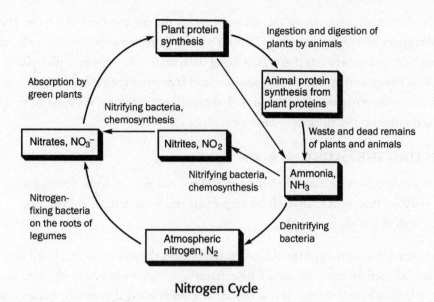

Nitrogen Cycle

Photosynthesis and respiration are the complementary reactions of the **carbon cycle**. In photosynthesis, carbon dioxide and water combine to form sugars and oxygen; the sun's energy is stored in the bonds among carbon atoms in the sugars. Respiration uses carbohydrates and oxygen to produce carbon dioxide, water, and energy; it releases the energy that is stored by photosynthesis. As animals eat the plants and one another and then breathe, carbon moves through the food web; the carbon that plants and animals contain is returned to the soil when they decompose. The carbon in the soil enters the cycle again, perhaps as part of a microorganism. A great deal of carbon is stored in the oceans and in minerals. Combustion is another way that carbon enters the atmosphere, in the form of carbon dioxide. The widespread and heavy use of fossil fuels has created concerns about how all of this combustion is affecting the environment.

KEY POINT

The biogeochemical cycles of carbon and nitrogen are good examples of areas where you can relate several concepts in biology. Examples of processes that are involved in the recycling of these elements are photosynthesis, cellular respiration, hydrolysis, and condensation.

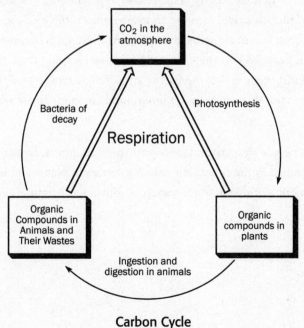

Carbon Cycle

Phosphorous is another important nutrient that moves through the environment in a cycle. The phosphorus cycle begins when phosphate (PO_4^{3-}) from weathered rock moves into soils. Plants then take up phosphate, and it becomes part of the living ecosystem. Like nitrogen, phosphate moves through living things as they feed on one another, and it reenters the ecosystem through the decomposers' action on waste and dead remains. The phosphorus cycle stands apart from the water, nitrogen, and carbon cycles because it has no gas phase.

ISSUES AFFECTING THE GLOBE ★★★★

Other than the major extinction events recorded every 100 million years or so in Earth's geological history, the only variables that seem to have global impact are related to human activity. You should be familiar with two of these variables.

The Earth is surrounded by a layer of ozone (O_3). The layer lies in the stratosphere, 10–17 km from the Earth's surface, and protects the surface from harmful UV rays emitted by the sun. Over the past 20 years, scientists have determined that the ozone layer is being depleted by compounds containing chlorine, fluorine, bromine, carbon, and hydrogen (halocarbons) that seem to be produced by human activity. One particular group of chemicals, CFCs (chlorofluorocarbons), has received a considerable amount of the blame for depletion of the ozone layer.

OZONE LAYER ALERT

The ozone layer may be depleted by as much as 60 percent over the Antarctic in the spring, causing the notorious hole in the ozone layer. Scientists worry that without changes in human activity, the ozone layer will be depleted beyond recovery, preventing certain forms of life from surviving on Earth.

The other major global impact of human activity is the release of greenhouse gases (water vapor, carbon dioxide, methane, and nitrous oxide) into the atmosphere from the inefficient burning of fossil fuels (there are sources from manmade aerosols as well). As these gases become more dense in the Earth's atmosphere, it is assumed that they prevent heat from escaping the Earth's surface, causing global warming. Scientists believe that the Earth's atmospheric temperature is steadily rising and that this will cause major shifts in seawater levels, local climates, and world weather patterns.

Habitat destruction occurs as people clear natural areas for natural resources, housing, and recreational areas. This has resulted in the extinction of many species of plants and animals and has an adverse impact on our worldwide water resources, global temperatures, and food availability.

AP BIOLOGY LAB 10: ENERGY DYNAMICS
INVESTIGATION ★★★★

Using a model ecosystem, this investigation explores how matter and energy flows, the roles of producers and consumers, and the complex interactions between organisms. First, estimate the net primary productivity of Wisconsin Fast Plants growing under lights, and then determine the flow of energy from the plants to cabbage white butterflies as the larvae consume cabbage-family plants.

Remember that the source of almost all energy on Earth is the sun. Free energy from sunlight is captured by producers that convert the energy to oxygen and carbohydrates through photosynthesis. The net amount of energy captured and stored by the producers in a system is the system's net productivity. Net productivity is calculated by assuming the change in biomass of a plant is due to uptake and use of energy.

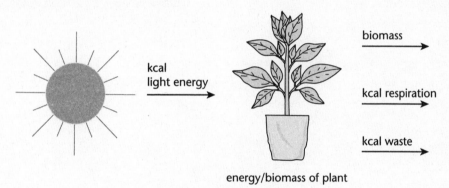

Energy/Biomass of Plants

To determine the net primary productivity of your plants, convert the difference in biomass over the growing time to energy according to the following equation:

$$\text{Energy} = \text{Biomass in grams} \times 4.35 \text{ kcal}$$

Let's look at a sample set of data for 10 plants grown over seven days. In seven days, the 10 plants gain 4.2 grams of dry mass. Using the equation, you can determine that 4.2 grams of dry mass is equivalent to 18.27 kcal of energy captured from the sun. To determine the net primary productivity per plant per day, you must divide that total amount of energy by 10 plants and by seven days. In this example, the net primary productivity is 0.26 kcal per plant per day.

Age	Dry Mass Gained/10 Plants	Energy (g biomass × 4.35 kcal)	Net Primary Productivity per Plant per Day
7 days	4.2 grams	18.27 kcal	0.26 kcal/day

The efficiency of energy transfer from producer to primary consumers varies with the type of organism and with the characteristics of the ecosystem. In the second part of the investigation, determine the biomass change in butterfly larvae that eat your plants to evaluate the energy use of primary consumers and ultimately the energy efficiency of the relationships.

APPLYING THE CONCEPTS: BIG IDEA FOUR REVIEW QUESTIONS

ENZYMES

This Review Quiz is designed to assess your content knowledge, and is not necessarily representative of the actual AP Biology exam. So don't panic if you see a new question type or a question asking about a relatively low-yield detail from the chapter.

1. Consider a biochemical reaction A → B, which is catalyzed by the human enzyme AB dehydrogenase. Which of the following statements is true of this reaction?

   (A) The reaction will proceed until the enzyme concentration decreases.

   (B) The reaction will be more favorable at 0°C.

   (C) A component of the enzyme is transferred from A to B.

   (D) The free energy change (ΔG) of the catalyzed reaction is the same as the free energy change for the uncatalyzed reaction.

2. The covalent binding of a molecule other than the substrate to the active site of an enzyme often results in

   (A) noncompetitive inhibition.

   (B) feedback inhibition.

   (C) irreversible inhibition.

   (D) increasing the energy of activation.

3. Substrates may bind into the active sites of enzymes by all of the following EXCEPT

   (A) hydrogen bonds.

   (B) peptide bonds.

   (C) polar covalent bonds.

   (D) dipole-dipole interactions.

Questions 4–7 refer to the following reaction profile.

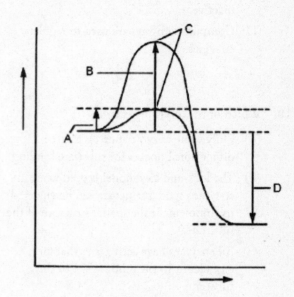

4. This is also known as $\triangle G$.

5. This is the activation energy of catalyzed reaction.

6. Activation energy of uncatalyzed reaction is this.

7. This is the transition state.

8. Enzyme specificity means

   (A) an enzyme can only catalyze a reaction once.

   (B) specific enzymes catalyze specific reactions.

   (C) enzymes can only catalyze in one direction.

   (D) enzymes specifically catalyze biological reactions.

9. Which of the following is NOT true?

   (A) When the end product works to block the original enzyme, it is called negative feedback.

   (B) Cells use inhibition to control enzymes.

   (C) Enzymes spend most of their time inactive.

   (D) Complex pathways are used to regulate enzymes.

10. Which of the following is true?

    (A) Only some enzymes possess a three-dimensional pocket for substrate binding.

    (B) The lock-and-key model is used when an enzyme's spatial structure can be changed to complement the spatial structure of the substrate.

    (C) All enzymes have active sites that are an exact fit to the binding site of the substrate.

    (D) Enzymes can be classified into two different groups based on how they bind to their substrates.

11. V_{max} is

    (A) the maximum velocity of enzyme movement.

    (B) when enzymes are saturated with substrate.

    (C) usually represented by a curved line.

    (D) increased by noncompetitive inhibitors.

12. To become active, most but not all enzymes require

    (A) proper pH.

    (B) binding sites.

    (C) cofactors.

    (D) a thermal window.

Questions 13–16 refer to the following list.

   (A) Competitive inhibition

   (B) Noncompetitive inhibition

   (C) Feedback inhibition

   (D) Coenzyme inhibition

13. This may be irreversible.

14. This is used as a form of enzyme regulation.

15. This can be overcome by adding more substrate.

16. This does not exist for enzymes.

ANSWERS AND EXPLANATIONS

1. D
Although enzymes may decrease the activation energy needed to start a reaction, the overall free energy change between products and reactants does not change. In other words, products and reactants maintain the same potential energy difference in uncatalyzed reactions as they do in catalyzed ones.

2. C
In irreversible inhibition, a molecule other than the substrate covalently bonds to the active site of the enzyme, preventing substrate molecules from accessing the active site. Although competitive inhibition, where chemicals compete with substrate molecules for a spot within the active site, may slow down an enzymatic reaction, irreversible inhibition may stop a reaction altogether and permanently. Noncompetitive inhibition takes place when a chemical decreases the affinity of an enzyme for a substrate through binding to a location on the enzyme other than the active site, inducing a conformational change in the active site without ever entering that area.

3. B
Substrates will not bind into active sites via peptide bonds. The amino acids that make up the active site are already bonded to one another via peptide bonds and there are no more places on their structure for another bond of that type to form. The bonds that do form occur between R groups sticking out into the active site and are of an ionic (hydrophobic, hydrophilic) nature or a covalent nature that does not involve amino and carboxyl groups (i.e., a peptide bond).

4. D
G is the overall energy change of the reaction. Therefore, it is measured as the difference between the initial state and the final state. That is labeled as D on this diagram.

5. A
Activation energy is the energy required to start a reaction. An enzyme lowers the activation energy of a reaction, thus allowing the reaction to proceed

from a different potential energy. This is shown as A on this diagram.

6. B
An uncatalyzed reaction will have a different, higher activation energy than a catalyzed reaction. Remember, an enzyme speeds up a reaction by lowering the activation energy.

7. C
The transition state is an intermediate, unstable molecule that forms during a reaction when the reactants collide. It occurs at the highest energy point in a reaction. This is labeled as C on the diagram.

8. B
Read carefully; even though some answers may seem like true statements, only (B) is the definition of enzyme specificity. Enzyme specificity means that an enzyme will only catalyze one specific reaction.

9. A
Don't get confused with terms from other chapters. Feedback inhibition may sound similar to negative feedback, but feedback inhibition is specific to enzymes while negative feedback is usually discussed in relation to endocrinology. All the other statements are true.

10. D
Enzymes can be grouped into two models of substrate binding. The lock-and-key model is for enzymes that have active sites that are perfectly matched to their binding sites. Induced-fit enzymes can change their active site to allow for the enzyme to fit into the binding site. Read these answers carefully.

11. B
V_{max} occurs when the enzymes are saturated with substrate and cannot work any faster. This is usually represented as the flat part of the rate versus substrate concentration curve.

12. C
Most enzymes require cofactors to become active. Proper pH, binding sites, and temperatures are necessary for all enzymes, not just most enzymes.

13 B
Noncompetitive inhibition occurs when the inhibitor causes a conformation change in the enzyme. Therefore, noncompetitive inhibition may or may not be irreversible.

14. C
Feedback inhibition is when the end product blocks the original enzyme. This is one of the many ways that cells can regulate enzyme action to keep the reaction in check.

15. A
Competitive inhibition occurs when the inhibitor and the substrate compete for the active site. This type of inhibition can be overcome by increasing the concentration of substrate.

16. D
There is no such thing as coenzyme inhibition. This is a combination of coenzymes and cofactors designed to distract the student. Be wary of such fake answers on the AP Biology exam.

PLANT VASCULAR AND LEAF SYSTEMS

This Review Quiz is designed to assess your content knowledge, and is not necessarily representative of the actual AP Biology exam. So don't panic if you see a new question type or a question asking about a relatively low-yield detail from the chapter.

1. Which of the following terms describes a plant's ability to grow toward or away from an environmental stimulus?

   (A) Taxis

   (B) Kinesthesis

   (C) Tropism

   (D) Vining

2. Placing a ripe banana in a bag with an unripe avocado will quicken the ripening of the avocado. This is caused by the positive feedback mechanism initiated by which of the following plant compounds?

   (A) Auxins

   (B) Cytokinins

   (C) Ethylene

   (D) Gibberellins

3. The creosote bush, *Larrea tridentata*, is found in circular stands of cloned individuals in the southwestern deserts of the United States. This is an example of propagation called

   (A) wind dispersal.

   (B) hybrid fitness.

   (C) polyploidy.

   (D) vegetative reproduction.

4. Nutrients that are produced as a result of photosynthesis in a leaf are delivered to the roots of a tree by the pull of

   (A) gravity.

   (B) osmotic potential between the source and the sink.

   (C) root pressure.

   (D) transpiration of water through the stomata in the leaf.

5. Deciduous trees have programmed death of their leaves in the fall and initiate this response with which hormone(s)?

   (A) Abscisic acid

   (B) Auxins

   (C) Cytokinins

   (D) Ethylene

6. Which of the following compounds is used by a plant to adjust to changing light levels as the seasons change?

   (A) Chlorophyll *a*

   (B) Chlorophyll *b*

   (C) Cytokinins

   (D) Phytochrome

7. Guttation, the accumulation of water on the leaf surface of plants due to root pressure, would most likely occur under which of the following conditions?

   (A) Midday heat

   (B) Morning humidity

   (C) Afternoon wind

   (D) Overnight freeze

8. Which of the following is not a characteristic of monocots?

   (A) One cotyledon in each seed
   (B) Parallel leaf veins
   (C) Fibrous root system
   (D) Petals in multiples of four or five

Questions 9–12 refer to the following list.

   (A) Epicotyl
   (B) Cotyledons/seed leaves
   (C) Seed coat
   (D) Endosperm

9. This develops into leaves and the upper part of the stem.

10. This stores food for the developing embryo.

11. Outer covering; protection/dormancy is this.

12. This feeds the embryo.

Questions 13–16 refer to the diagram shown.

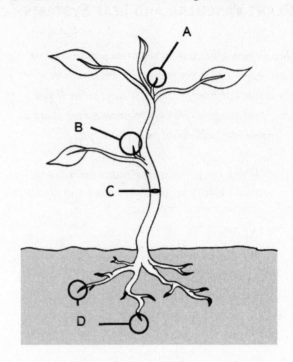

13. This is also known as cambium.

14. This letter represents the location of primary growth.

15. This is apical meristem location.

16. Location of secondary growth

ANSWERS AND EXPLANATIONS

1. C

The term *tropism* comes from the Greek *trope*, which means "to turn." Tropism is an automatic reaction in orientation or a positive or negative response to a stimulus.

2. C

This is a common trick used at home, but the reason why it works isn't usually understood. There are two good clues to the answer of this question. First, the question is asking for a compound that causes ripening. Second, the question is looking for a compound that creates a positive feedback loop. Most biological reactions are negative feedback mechanisms. The correct answer is ethylene, (C). Ethylene is the plant hormone associated with ripening and is an excellent example of a positive feedback loop.

3. D

The key issue in this question is that the individuals are clones. Cloned organisms are genetic duplicates that arise from mitosis in some form of asexual reproduction. Creosote bushes spread by vegetative reproduction through a massive root system. Scientists believe that some of these clonal groups may be 9,000 years old.

4. B

You may automatically choose (A) because gravity is included in the common phrases "the pull of gravity" or "gravity's pull," and (A) may make some intuitive sense because the nutrients in question are moving down the plant; however, this is not the correct answer. Root pressure pushes water away from the roots instead of pulling it, so (C) can't be correct. Transpiration affects xylem flow, (D). The correct answer is (B). Phloem material is pulled from sources to sinks due to osmotic potential.

5. D

The one answer that may be misleading is abscisic acid, (A). This plant hormone slows down growth, especially during periods of stress such as water loss. It also earned its name because scientists once thought it contributed to the abscission of leaves

in the fall. The correct answer is (D), however. It is productive to think of ethylene as the ripening hormone and that ripening is a form of aging and death. The role of fruits is to die and fortify the seeds, after all.

6. D

Intuitively, one would think the compound must have some kind of sensitivity to light because it is the stimulus for the reaction. Photopigments, (A) and (B), are certainly light sensitive, but the chlorophylls have not been discussed in connection with plant regulation other than photosynthesis. Cytokinins are plant hormones, so you may be tempted to choose (C). The correct answer, however, is phytochrome. Phytochrome is not a plant hormone; instead, it is the light sensor that triggers hormone response.

7. B

Guttation occurs under conditions that decrease transpiration. Transpiration is just like evaporation and, therefore, is increased by heat and wind, things that make the air dry. Guttation would be most likely in conditions that would make the air wet and cool.

8. D

Dicots, not monocots, have parts of their flowers (petals) in multiples of four or five. Monocots, however, have petals in multiples of three. All other characteristics in the answer choices are traits of monocots.

9. A

The epicotyl develops into leaves and upper portions of the stem.

10. D

The endosperm stores food for the developing embryo.

11. C

The seed coat is the outer covering for the seed and protects it, especially during dormancy.

12. B

The cotyledon feeds the developing seed embryo.

13. C

The lateral meristems (also known as cambium) are found within the stem between layers of xylem and phloem on the sides of the plant.

14. A

Growth upward that occurs as a result of cell division within apical meristems is called primary growth.

15. A

Apical meristems exist at the tips of roots and stems.

16. C

Secondary growth is outward or lateral growth. This growth happens around the stems and is the reason we see growth rings on trees.

RESPIRATION AND CIRCULATION

This Review Quiz is designed to assess your content knowledge, and is not necessarily representative of the actual AP Biology exam. So don't panic if you see a new question type or a question asking about a relatively low-yield detail from the chapter.

1. Carbon dioxide waste is transported from working cells in the body to the lungs in the form of

   (A) bicarbonate ions.

   (B) oxygen.

   (C) dissociated carbon and oxygen.

   (D) glucose and water.

2. Which of the following contribute(s) to gas exchange in the alveoli?

   I. Low partial pressure of O_2 in the pulmonary capillaries when compared with inhaled air

   II. Low partial pressure of CO_2 in the pulmonary capillaries when compared with inhaled air

   III. Presence of surfactant

   (A) I only

   (B) II only

   (C) II and III only

   (D) I and III only

3. Which of the following does NOT characterize the human respiratory system?

   (A) The trachea branches into two bronchi.

   (B) Gas exchange takes place across the alveolar capillary membrane.

   (C) Contraction of the diaphragm muscle expands the thoracic cavity and permits the lungs to fill.

   (D) The rate of breathing can be increased by the parasympathetic nervous system.

4. Which of the following regulates ventilation?

   (A) Right hemisphere

   (B) Basal ganglia

   (C) Occipital lobe

   (D) Brain stem

5. How many subunits are on each hemoglobin molecule?

   (A) 1

   (B) 2

   (C) 4

   (D) 16

6. Which of the following is the correct sequence of the passages through which air travels during inhalation?

   (A) Pharynx → trachea → lungs → bronchi → alveoli

   (B) Larynx → pharynx → bronchi → lungs → alveoli

   (C) Larynx → pharynx → trachea → bronchi → alveoli

   (D) Pharynx → larynx → trachea → bronchi → alveoli

7. Which of the following is a key characteristic of maximal gas exchange for almost all species?

   (A) Humidity is more important than surface area.

   (B) Lungs are the most efficient organ for gas exchange.

   (C) Gas exchange is based on pressure.

   (D) It is a passive process.

8. Which of the following is true?

   (A) Deoxygenated blood is found in all veins.

   (B) All arteries contain oxygenated blood.

   (C) Pulmonary arteries carry deoxygenated blood.

   (D) The aorta contains deoxygenated blood.

9. Which of the following is NOT true?

   (A) At high altitudes, the partial pressure of oxygen in the atmosphere decreases, making it easier to get oxygen.

   (B) Pulmonary capillaries have a high partial pressure of carbon dioxide.

   (C) Inhaled air has a high partial pressure of oxygen.

   (D) The partial pressure of carbon dioxide in the capillaries is higher than that of inhaled air.

10. There are _____ hemoglobin molecules in one erythrocyte.

    (A) 10

    (B) 1,000

    (C) 100,000

    (D) 1,000,000

11. Which of the following places the organs and vessels in the correct order for circulation?

    (A) Left ventricle, pulmonary vein, aorta, right atrium

    (B) Right ventricle, pulmonary artery, lungs, left atrium

    (C) Aorta, vena cava, right ventricle, tricuspid valve

    (D) Pulmonary vein, lungs, left ventricle, aorta

12. Systolic blood pressure

    (A) measures artery pressure under heart contraction.

    (B) is usually the smallest number.

    (C) measures vein pressure as the heart relaxes.

    (D) decreases with age.

13. Average blood pressure

    (A) increases close to the heart because of friction between blood and vessel walls.

    (B) is absent in veins.

    (C) decreases with age.

    (D) is most dependent upon oxygen partial pressure.

14. Which of the following is NOT true of precapillary sphincters?

    (A) They help with selective dispersal of blood.

    (B) They are small bands of muscle.

    (C) They exist at the base of capillaries as they branch off from bronchioles.

    (D) They allow blood to bypass a capillary bed.

15. Which of the following is NOT true of thermoregulation and circulation?

    (A) When exposed to cold, precapillary sphincters close to conserve heat.

    (B) When exercising, skin capillaries vasodilate to change skin tone.

    (C) Precapillary sphincters help with temperature regulation.

    (D) Vasodilation is regulated by the autonomic nervous system.

ANSWERS AND EXPLANATIONS

1. A
Choice (B) might be tempting because oxygen is converted to carbon dioxide in some processes, but this is not correct. If the amount of CO_2 in the blood is too high, the pH drops too much, causing acidosis. The answer to this problem is to convert the CO_2 into bicarbonate until it reaches the lungs, where it is liberated again as gas.

2. D
Compared with inhaled air, the pulmonary capillaries have a higher partial pressure of CO_2 and a lower partial pressure of O_2. Thus, in the alveoli, gas exchange occurs when CO_2 flows down its concentration gradient from the capillaries to the alveoli, and O_2 flows down its gradient from the alveoli into the capillaries. The presence of surfactant lowers the surface tension of the alveoli and keeps the membranes moist to facilitate gas exchange. Therefore, statements I and III contribute to gas exchange, (D).

3. D
Choice (A) is true—the trachea branches into two bronchi. Choice (B) is true—carbon dioxide and oxygen are exchanged across the alveolar membranes. Choice (C) is true—contraction of the diaphragm expands the thoracic cavity, creating a partial vacuum in the interpleural space, which allows the lungs to expand to fill that space. Choice (D) is not true—breathing cannot be increased by the parasympathetic nervous system.

4. D
Ventilation is controlled by two parts of the brain stem: the medulla oblongata and the pons. Because pons is not among the answer choices, the answer is (D).

5. D
Each hemoglobin molecule has four subunits. Each subunit is capable of binding to four different oxygen molecules. Therefore, each hemoglobin molecule can transport 16 oxygen molecules.

6. D
The pharynx is the cavity that channels air from the mouth and nose into the larynx, which sits atop the trachea. The trachea then branches into the left and right bronchi, which lead to the left and right lungs and their alveoli. Note that alveoli appear at the end of each list, so it can be disregarded. Choices (B) and (C) start with the larynx; therefore, they can be eliminated first.

7. D
Humidity is not more important than surface area. Together, they are both important. Lungs are not the most efficient organ. That statement is too bold. There are other equally good organs, such as fish gills, that are good at gas exchange. Gas exchange may be influenced by pressure, but exchange is not dependent upon pressure. Gas exchange is based on simple diffusion across cell membranes. The correct answer is (D).

8. C
You need to have a good understanding of arteries and veins, and it is incorrect to state that all arteries contain oxygenated blood and all veins contain deoxygenated blood. The pulmonary arteries carry blood from the heart to the lungs; therefore, they carry deoxygenated blood. The pulmonary veins carry blood from the lungs to the heart, so they carry oxygenated blood. A helpful hint is: Veins visit, arteries away!

9. A
At high altitudes, the partial pressure of O_2 in the atmosphere declines, making it more difficult to get sufficient oxygen to diffuse into the capillaries. Be sure you understand the differences between the partial pressures of O_2 and CO_2, and don't get confused about when each is higher or lower.

10. D
Each red blood cell or erythrocyte contains approximately 1 million hemoglobin molecules.

11. B

Blood flows from the body into the venae cavae, then the right atrium, through the tricuspid valve into the right ventricle. Then blood travels through the pulmonic semilunar valve into the pulmonary artery and into the lungs. Blood leaves the lungs through the pulmonary veins and enters the heart in the left atrium. It goes through the mitral valve and into the left ventricle. It leaves the heart from the aortic semilunar valve, through the aorta and into general circulation.

12. A

Your systolic blood pressure (the top larger number) reflects how much pressure is in your arteries when your heart contracts. The lower, bottom number is your diastolic pressure. It measures the pressure in your arteries as the heart relaxes.

13. B

Average blood pressure decreases the farther away blood travels from the heart due to an increase in surface area and friction between the blood cells and the blood vessel walls. There is virtually no blood pressure in the veins. Choice (C) is incorrect because many health factors are responsible for changes in blood pressure with age.

14. C

Precapillary sphincters are found at the base of capillaries as they branch off from nearby arterioles, so (C) is the correct answer. All other statements are true.

15. B

Vasodilation occurs to open blood vessels and allow for more efficient cooling of the body during heat or exercise. Any change in skin color is a result of the vasodilation, not a purpose of it. All other statements are true.

DIGESTION AND EXCRETION

This Review Quiz is designed to assess your content knowledge, and is not necessarily representative of the actual AP Biology exam. So don't panic if you see a new question type or a question asking about a relatively low-yield detail from the chapter.

1. A doctor analyzes the urine of a patient, and the results suggest the kidneys may not be functioning normally. Which of the following compounds present in urine indicates possible kidney dysfunction in a human?

   (A) Water

   (B) Urea

   (C) Salt

   (D) Protein

2. Which of the following terms describes the synchronized muscular contractions that force chyme through the digestive tract?

   (A) Peristalsis

   (B) Arrhythmia

   (C) Gastric reflux

   (D) Colitis

3. Which of the following nutrients are absorbed in the stomach?

   (A) Carbohydrates, lipids, and protein are equally subject to absorption in the stomach, as it is the main organ used for nutrient absorption.

   (B) Only lipids and proteins are absorbed in the stomach, as the acid from HCl is able to emulsify the fats, and the proteins are digested by pepsin.

   (C) Only proteins are absorbed in the stomach, as the peptidase pepsin is released to aid in the absorption of the resulting amino acids.

   (D) No nutrients are absorbed in the stomach.

4. Aspirin (acetylsalicylic acid) is a common analgesic and anti-inflammatory drug that can cause damage to the gastroduodenal mucosa, in part by deactivating growth factors involved in mucosal defense and repair. This damage to the mucosa is most likely to first result in

   (A) malnutrition.

   (B) diarrhea.

   (C) ulceration.

   (D) death.

5. The absence or reduction of gut flora, due to a cause such as broad-spectrum antibiotic use, may lead to what complication?

   (A) Invasion of opportunistic species

   (B) Reduced fermentation of carbohydrates

   (C) Diarrhea

   (D) All of the above

6. Physical trauma to the kidneys may cause glomerular capillaries to become permeable enough to allow plasma proteins to enter the renal tubule. What effect would this have on the urine?

   (A) Increased urine output

   (B) Decreased urine output

   (C) Increased urine urea concentration

   (D) Decreased urine osmolarity

7. Considering again the physical trauma to the kidneys described in the previous question, what effect would this have on systemic fluid levels?

    (A) Increased blood pressure within the arteries

    (B) Increased osmotic pressure within the veins

    (C) Increased fluid retention within the interstitium

    (D) No effect except to the kidneys

8. Renal failure is characterized by an insufficient rate of glomerular filtration. This is most likely to first result in

    (A) dehydration due to lack of water.

    (B) lethargy due to buildup of urea.

    (C) pain due to high concentrations of endorphins.

    (D) hyperactivity due to excess glucose.

9. _____ of amino acids leads to the production of urea or ammonia.

    (A) Deamination

    (B) Filtration

    (C) Catabolism

    (D) Metabolism

10. Which of the following is the correct order of organs that food will encounter during digestion?

    (A) Esophagus, duodenum, large intestine, ileum

    (B) Stomach, pancreas, dudoenum, large intestine

    (C) Stomach, duodenum, jejunum, rectum

    (D) Esophagus, stomach, liver, small intestine

11. ADH

    (A) controls the concentration of the blood.

    (B) is released by the pituitary.

    (C) closes aquaporins.

    (D) increases urine output.

12. Which of the following would not cause a decrease in blood pressure?

    (A) Heavy sweating

    (B) Diarrhea

    (C) Aldosterone

    (D) Vomiting

13. Renal output depends upon

    (A) metabolic wastes.

    (B) weight.

    (C) water volume.

    (D) All of the above

14. Which of the following is NOT a function of the kidney?

    (A) Maintain blood pressure

    (B) Maintain blood osmolarity

    (C) Conserve glucose

    (D) Conserve salts

15. The pancreas

    (A) is strictly endocrine.

    (B) is strictly exocrine.

    (C) is both endocrine and exocrine.

    (D) is neither endocrine nor exocrine.

ANSWERS AND EXPLANATIONS

1. D

Urine is a waste product of normal body function that contains large amounts of water, urea (a byproduct of protein metabolism), and salts. This excludes (A), (B), and (C). The kidneys filter out large protein molecules in the nephron, so protein, (D), in the urine would indicate a problem with the filtration process in the kidney.

2. A

It takes more than gravity and the pressure of more food coming down the gullet to get chyme moving through the digestive tract. The digestive organs are lined with smooth muscle that contracts under the control of the sympathetic nervous system in rhythmic contractions that form waves. This process is called peristalsis, (A).

3. D

Mechanical and chemical digestion is the function of the stomach. HCl and pepsin are produced by the stomach during digestion. The small intestine is the first site of absorption, so the stomach does not absorb any nutrients.

4. C

The gastroduodenal mucosa is the innermost layer of the GI tract and surrounds the lumen of the duodenum, the upper part of the small intestine. This layer of mucosa is in direct contact with chyme, partially digested food that passes into the duodenum from the stomach. The mucosa protects the epithelial cells of the small intestine from the acidic nature of the chyme. Any damage to the gastroduodenal mucosa would first affect the epithelial cells of the duodenum, resulting in ulcer formation. Choices (A), (B), and (D) are all distortions that could occur if the damage to the duodenal mucosa persisted for a long period of time.

5. D

Certain types of gut bacteria have enzymes that break down certain polysaccharides, such as lactose and maltose. Human cells lack these enzymes and mostly absorb carbohydrates as monosaccharides in the small intestine. Without flora in the gut, undigested carbohydrates cannot be utilized by the human body. Gut flora also promote a healthy immune system. The bacteria that comprise gut flora promote the development of the gut's mucosal immune system and stimulate lymphoid tissue associated with the gut mucosa to synthesize antibodies to pathogens. The immune system is trained to fight harmful bacteria and not harm the bacteria naturally present in the gut. With diminished beneficial bacteria, invasive bacteria can multiply and produce toxic substances, which can lead to diarrhea.

6. A

The healthy glomerulus filters small molecules and water into the nephron while keeping large proteins and cells in the blood. If plasma proteins are filtered into the renal tubule through the glomerulus, they will not be able to return back into the plasma because the rest of the nephron is still impermeable to plasma proteins. The plasma proteins in the filtrate will attract water due to osmosis, and this will cause an increase in urine output. Choices (B), (C), and (D) are all opposite to what occurs because the plasma proteins would partially replace urea in urine and attract more water while still slightly increasing urine osmolarity.

7.   C

This question seems very similar to the previous question. Be sure to read carefully. On this one, you need to think about what would happen to the systemic fluids when plasma proteins are filtered through the glomerulus and into the renal tubule. The plasma proteins would pass from the collecting duct of the nephron into the ureter into the bladder and out the urethra in the urine. The loss of plasma proteins will lead to a drop in osmotic pressure of the blood (blood pressure drop). As a result, water that leaves the blood at the arteriole end of a capillary will not be reabsorbed, and excess water will remain in the interstitial fluids. Choices (A) and (B) are incorrect because they state the opposite of what would happen to systemic fluid levels.

8.   B

Filtration occurs at the level of the glomerulus, which has small pores that allow small molecules such as urea and water through while preventing cells and large proteins from diffusing. If glomerular filtration is occurring at an insufficient rate, there will be a decrease of urine production. Waste products will remain in the blood because they will not be filtered into the renal tubules to be excreted in the urine. Choice (B) is correct—an increase of urea and other waste products in the blood can lead to tiredness. Dehydration is the opposite because extra water is retained. Endorphins inhibit pain and would likely remain unaffected. Glucose is not normally present in the urine; therefore, it would remain unaffected as well.

9.   A

Deamination of amino acids leads to the production of nitrogenous wastes such as urea or ammonia.

10.   C

Be careful: The question asked for organs the food will encounter. Choice (B) is incorrect because food will not pass through the pancreas. Choice (D) is also incorrect for similar reasons—food will not pass through the liver. Choice (C) is correct.

11.   B

ADH, antidiuretic hormone, controls the concentration of the urine, so (A) is correct. It opens aquaporins on the luminal surfaces of principal cells in the collecting duct and decreases urine output; thus, (C) and (D) are incorrect.

12.   C

Anything that causes water loss will cause a drop in blood pressure. Aldosterone is released in response to low blood pressure. It causes the cells in the collecting duct to reabsorb more salt and water and, thus, increases blood pressure.

13.   D

Renal output varies around a set point. This is because weight, blood pressure, and amounts of salt, metabolic wastes, and tissue water volume remain relatively constant. All of these things are responsible for renal output, so the answer is (D).

14.   A

The kidneys function to maintain the osmolarity of the blood and to conserve glucose, salts, and water. Even though blood pressure is directly related to water and salts, it is not directly maintained by the kidneys.

15.   C

The pancreas performs an endocrine function of affecting blood glucose concentration via the secretion of insulin and glucagon. It performs an exocrine function of secreting pancreatic juices into the small intestine. Therefore, it is both endocrine and exocrine.

MUSCULOSKELETAL SYSTEM

This Review Quiz is designed to assess your content knowledge, and is not necessarily representative of the actual AP Biology exam. So don't panic if you see a new question type or a question asking about a relatively low-yield detail from the chapter.

1. Sarcomere shortening in muscle fibrils requires

   (A) the rapid influx of sodium ions into the cytoplasm after release from the sarcoplasmic reticulum.

   (B) T-tubule shortening after calcium ion release by the sarcoplasmic reticulum.

   (C) ATP release from the sarcoplasmic reticulum and subsequent myosin attachment to actin filaments.

   (D) conformational modifications of the tropomyosin-troponin complex within muscle fibers.

2. When a muscle fiber is subjected to very frequent stimuli,

   (A) an oxygen debt is incurred.

   (B) a muscle tonus is generated.

   (C) the threshold value is reached.

   (D) the contractions combine in a process known as summation.

3. Restoration of the resting state in muscles begins when neural stimulation stops and calcium ions are transported back into the

   (A) postsynaptic terminal of the nearby axon.

   (B) sarcoplasmic reticulum.

   (C) presynaptic axon terminal.

   (D) neuromuscular junction.

4. Which of the following statements about the musculoskeletal system is NOT true?

   (A) Tendons join muscle to muscle while ligaments join muscle to bone.

   (B) Contraction is initiated in muscle tissue by a cascade of Ca^{2+} ions.

   (C) Ca^{2+} levels for the muscle are regulated by hormones secreted from the thyroid.

   (D) Specialized cells react to changing body conditions to maintain serum calcium levels.

5. What is generally not required for a muscle to elongate?

   (A) Influx of Ca^{2+} into the sarcoplasmic reticulum

   (B) Consumption of ATP

   (C) Contraction of an antagonist muscle

   (D) Elongation of actin filaments

6. Myotoxins, potentially lethal proteins found in snake bites, cause the calcium gradients of the sarcoplasmic reticuli to dissipate. The most likely immediate effect of such a snake bite is

   (A) vomiting.

   (B) hyperventilation.

   (C) paralysis.

   (D) relaxation.

Questions 7–10 refer to the following list.

    (A) Myofibril

    (B) Muscle fiber

    (C) Sarcomeres

    (D) Sarcolemma

7. This is a multinucleated cell.

8. These are contractile units.

9. These are the filaments embedded in muscle fibers.

10. These contain I bands and Z lines.

11. Which of the following is NOT a function of the skeletal system?

    (A) Movement

    (B) Protection

    (C) Immunity

    (D) Support

Questions 12–15 refer to the following list.

    (A) Epiphyseal plate

    (B) Periosteum

    (C) Epiphysis

    (D) Diaphysis

12. This is composed primarily of compact bone.

13. This is the site of production of red and white blood cells.

14. This is the site of longitudinal bone growth.

15. This is the fibrous sheath.

ANSWERS AND EXPLANATIONS

1. D

Recall that a sarcomere is a unit of layered actin and myosin filaments, and within a given muscle fiber there are repeating sarcomere units that traverse the length of the muscle cells. The shortening of the sarcomeres is regulated by a nerve impulse outside the fiber, resulting in calcium ion release from the sarcoplasmic reticulum, a specialized ER found in muscle cells. Calcium allows a conformational change to take place within the sarcomere such that tropomyosin is pulled off of the binding site (troponin) for the myosin heads on the actin filaments. With the myosin binding sites uncovered, myosin can attach to actin and slide past it, shrinking the length of the sarcomere.

2. D

When fibers of a muscle are exposed to very frequent stimuli, the muscle cannot fully relax. The contractions begin to combine, becoming stronger and more prolonged. This is known as frequency summation. If the stimuli become so frequent that the muscle cannot relax, the contractions become continuous, which is known as tetanus.

3. B

Calcium ions are stored in muscle cells within the sarcoplasmic reticulum (SR), and they are pulled back into the SR after muscle contraction for use in the next contraction. All other answer choices deal with structures that are involved in the neuronal stimulation of the muscle fiber (using acetylcholine), not in the contraction itself.

4. A

Choice (A) is not true because tendons join muscle to bone while ligaments join bone to bone. Choice (B) is true because contraction is initiated by a release of calcium ions from the sarcoplasmic reticulum. Choice (C) is true since the thyroid gland secretes calcitonin, which decreases the calcium concentration in blood. Choice (D) is also true because the specialized cells that maintain blood serum calcium levels are osteoblasts and osteoclasts.

5. D

The sarcoplasmic reticulum ATPase pumps in calcium when a muscle is at rest. When excited, it releases the calcium, which binds with myosin and causes the muscle to contract. To stop contracting, the myosin needs to exchange the calcium with an ATP and the calcium needs to be pumped back into the sarcoplasmic reticulum, also using ATP. Once contraction has stopped, the action of an antagonist group pulls the muscle to elongate it. Choice (D) is correct, as it is the only false statement. Actin filaments themselves do not elongate during elongation, but rather move farther apart from one another.

6. C

Motor neurons stimulate muscle cells via the neurotransmitter acetylcholine. When a muscle cell is stimulated, its sarcoplasmic reticulum becomes more permeable to calcium. Calcium then diffuses into the cytoplasm and binds to troponin on actin. This binding triggers a series of events to occur, which leads to muscle contraction. Without a concentration gradient, muscles cannot contract. Therefore, if this gradient is dissipated, the muscles go into a state of paralysis. Choices (A) and (B) require coordinated motion, and (D) is the opposite of what would happen.

7. B

Muscle fibers are multinucleated cells created from the fusion of mononucleated embryonic cells.

8. C

Sarcomeres are the contractile units of a muscle.

9. A

Hopefully the word *fibril* clued you in that myofibrils are the filaments that are embedded in the muscle fibers.

10. C

As the contractile units of muscles, sarcomeres contain all the contractile elements, including the I band, A band, H zone, Z line, and M line.

11. C

There are three main functions of any skeleton: support, aid in movement, and protection. Technically, there are some types of immunity that come from the bone marrow, but it is regulated by other systems and, thus, is not the main function of the skeletal system.

12. D

The diaphysis is composed primarily of compact bone surrounding a cavity containing bone marrow.

13. C

Bone marrow is found in the spongy bone, where the production of red and white blood cells takes place. The epiphyses are made of spongy bone surrounding a layer of compact bone.

14. A

The epiphyseal plate, or growth plate, is a disk of cartilaginous cells separating the diaphysis from the epiphysis. It is the site of bone growth longitudinally.

15. B

The periosteum is a fibrous sheath that protects the bone.

ECOLOGY

This Review Quiz is designed to assess your content knowledge, and is not necessarily representative of the actual AP Biology exam. So don't panic if you see a new question type or a question asking about a relatively low-yield detail from the chapter.

1. The maximum population size for a species based on the environmental factors is called

   (A) ecological footprint.

   (B) carrying capacity.

   (C) population growth.

   (D) environmental capacity.

2. Which of the following is not a density-dependent factor?

   (A) Emigration

   (B) Predation

   (C) Competition

   (D) Geographic isolation

3. Which of the following groups of organisms is most likely the primary producer in a tundra biome?

   (A) Conifer forests

   (B) Small shrubs

   (C) Fleshy succulents

   (D) Subterranean fungi

4. All of the following could be a top predator in a food chain except a

   (A) tiger.

   (B) reticulated python.

   (C) great white shark.

   (D) squid.

5. According to scientists, the Earth's ozone layer is being depleted primarily by which of the following?

   (A) Greenhouse gases

   (B) Acid rain

   (C) UV rays from the sun

   (D) CFC (chlorofluorocarbon) molecules

6. Which of the following is not an important part of the nitrogen cycle?

   (A) Nitrogen-fixing bacteria

   (B) Herbivores

   (C) Oxygen

   (D) Denitrifying bacteria

7. A biome that experiences all four seasons— winter, spring, summer, and fall—and is characterized by warm, moist springs and summers is the

   (A) tropical deciduous forest.

   (B) taiga.

   (C) temperate deciduous forest.

   (D) chapparal.

8. Photoperiod, or day length, is a critical factor for many plants and animals, as it regulates metabolic and growth processes. In northern alpine forests, many species show a variation in their metabolic response to changes in day length. Northern species are genetically programmed to slow down their metabolism more quickly than southern species as day length shortens and winter approaches. This geographic variation is known as

   (A) a cline.
   (B) gametic disequilibrium.
   (C) genetic drift.
   (D) selective photoperiodism.

9. Which of the following is the complementary reaction of the carbon cycle?

   (A) The citric acid cycle
   (B) Fermentation
   (C) Respiration
   (D) The electron transport chain

Questions 10–13 refer to the following list.

   (A) Clade
   (B) Hybrid zone
   (C) Phylogeny
   (D) Ecological niche

10. What is the history of the descent of a group of organisms from common ancestors?

11. This is a region in which genetically distinct populations come into contact with each other and produce some offspring of mixed ancestry.

12. This is the combination of all relevant environmental variables in which a species or population lives.

13. This is the set of species descended from a particular ancestral species.

14. The total of all the abiotic factors is called a(n)

   (A) population.
   (B) habitat.
   (C) ecosystem.
   (D) community.

15. As it moves through the various food webs, energy

   (A) is lost.
   (B) is gained.
   (C) stays the same.
   (D) None of the above

ANSWERS AND EXPLANATIONS

1. B

The maximum population size for any species is called carrying capacity, (B). This is based on the different environmental factors that control for important resources that limit population growth. The measure for a species demand on an environment is called ecological footprint, (A), and this is not correct. Population growth, (C), describes how a population grows based on the availability of resources. Choice (D), environmental capacity, is not a recognized term for the AP Biology exam.

2. D

Density-dependent factors can change as the number of individuals in the population increases or decreases. These factors include things such as competition, predation, and emigration. Choice (D), geographic isolation, or the development of new species due to population isolation resulting from geographic changes, does not depend on population size.

3. B

You might initially confuse the tundra biome with the taiga biome because they are both cold regions with little biodiversity, but the important distinction is that taiga is dominated by gymnosperms. Watch out for conifer forests, (A), which are characteristic of taiga, not tundra. Succulents, (C), are parts of arid zones with high sunlight, and fungi, (D), are not the primary producers of any biome. The tundra is dominated by lichens, shrubs, and sedges, so (B) is correct.

4. D

Top predators have few or no predators of their own. These animals are at the top of the food chain and prey on other animals in the food chain. Tigers are a common apex predator so (A) is incorrect. Although it is not often to see snakes at the top of a food chain, reticulated pythons have no natural predators and so (B) is incorrect. Similar to tigers, the great white shark has no known predators, therefore (C) is incorrect. Squid are a common food item for many marine creatures throughout the food web, so the correct answer is (D).

5. D

All of the possible answers are connected with conservation issues, so you have to go beyond general familiarity. The ozone layer blocks UV rays, so you might be tempted to consider (C), but this is incorrect. Ultimately, barring being familiar with intricate atmospheric chemistry, you will simply need to remember that CFCs are harmful to the ozone layer, (D).

6. C

Nitrogen-fixing bacteria convert nitrogen gas into nitrates and nitrites which are then taken up by plants. Herbivores eat these plants. After death, the nitrogen is broken down from the herbivores' bodies and released back into the soil. Denitrifying bacteria in the soil converts nitrates and nitrites back into nitrogen gas that is released into the air. While oxygen is a requirement for most herbivores, it is not a requirement for the nitrogen cycle. Herbivores living in a low oxygen environment would still be able to eat plants containing nitrogen, so oxygen is not a requirement for the nitrogen cycle.

7. C

The characteristics mentioned in the question stem describe a temperate deciduous forest, such as that which exists in the northern United States. These biomes experience a spring and summer growing season, cold winters, and plenty of moisture. All other answer choices are incorrect as they include biomes with climates and seasonal variability (markedly different) from the biome described in the question.

8. A

A cline can be defined as a gradual change in the expression of a trait over a particular geographic range. In this case, northernmost species slow down their metabolisms more rapidly than southernmost species as day length shortens. This change as one moves from the south to the north is defined as a cline.

9. C

Photosynthesis and respiration are the complementary reactions of the carbon cycle. Carbon dioxide and water combine to form sugar and oxygen in photosynthesis. Therefore the answer is (C).

10. C

The evolutionary history of an organism or species is its phylogeny. The phylogeny of a species traces its ancestry back in time.

11. B

Offspring of mixed ancestry are known as hybrids, and a hybrid zone is the area of overlapping geographic ranges between two subspecies where viable offspring can sometimes be produced from interbreeding populations.

12. D

Also known as the role each organism plays in its habitat, an organism's niche is the combination of how the organism interacts with the important environmental factors surrounding the organism (e.g., predators, prey, climate, resource availability).

13. A

A cladogram is an evolutionary tree that groups related organisms by shared features, so a clade is a set of species that descends from the same common ancestor and shares similar features and body structures.

14. B

The habitat is the total of the abiotic factors such as terrain, soil type, and water. Populations are all the organisms that occur in a specific habitat. Communities are all the populations of organisms that live in the habitat. The fauna, flora, and abiotic factors collectively make up an ecosystem.

15. A

Energy is lost as it moves through various parts of the food webs. Ten percent of all the energy is passed from one trophic level to another.

Part Three

PRACTICE
TESTS

HOW TO TAKE PRACTICE TESTS

The next section of this book consists of practice tests. Taking a practice AP exam gives you an idea of what it's like to answer these test questions for a longer period of time, one that approximates the real test. You'll find out which areas you're strong in and where additional review may be required. Any mistakes you make now are ones you won't make on the actual exam, as long as you take the time to learn where you went wrong.

For the most accurate results you should approximate real test conditions as closely as possible. Before taking a practice test, find a quiet place where you can work uninterrupted for three hours. Time yourself according to the time limit at the beginning of each section. The full-length diagnostic test includes 63 multiple-choice questions, six grid-in items, and two long and six short free-response questions. You are allotted 90 minutes for the multiple-choice questions and grid-in items, a ten-minute break, and 90 minutes to answer the free-response questions, which begins with a recommended ten-minute reading period. Use the reading period to plan your answers for the free-response questions, although you may begin writing your responses before the ten minutes are up.

As you take the practice tests, remember to pace yourself. Train yourself to be aware of the time you are spending on each question. Try to be aware of the general types of questions you encounter, and alert to certain strategies or approaches that help you to handle the various question types more effectively.

After taking a practice exam, be sure to read the detailed answer explanations that follow. These will help you identify areas that could use additional review. Even when you've answered a question correctly, you can learn additional information by looking at the answer explanation.

Finally, it's important to approach the test with the right attitude. You're going to get a great score because you've reviewed the material and learned the strategies in this book.

Good luck!

HOW TO COMPUTE YOUR SCORE

SCORING THE MULTIPLE-CHOICE QUESTIONS AND GRID-IN ITEMS

To compute your score on this portion of the two sample tests, calculate the number of questions you got right on each test, then divide by 69 to get the percentage score for the multiple-choice portion of that test.

SCORING THE FREE-RESPONSE QUESTIONS

Reviewers will have key bits of information that they are looking for in response to each free-response question. Each piece of information that they are able to check off in your essay is a point toward a better score.

To figure out your approximate score for the free-response questions, look at the key points found in the sample response for each question. For each key point you included, add a point. Figure out the number of key points there are in each question, then add up the number of key points you did include for each question. Divide by the total number of points available for all the free-response questions to get the percentage score for the free-response portion of that test.

CALCULATING YOUR COMPOSITE SCORE

Your score on the AP Biology exam is a combination of your scores on the multiple-choice portion of the exam and the free-response section. The free-response section is worth 50 percent of the exam score, and the multiple-choice is worth 50 percent.

To determine your score, multiply the percentage score for the free-response section by 0.5, and for the multiple-choice section multiply by 0.5. Add these numbers together and multiply by 100 to get your final score. Remember, however, that much of this depends on how well all of those taking the AP test do. If you do better than average, your score would be higher. The numbers here are just approximations.

The approximate score range is as follows:

5 = 65–100 (extremely well qualified)

4 = 55–64 (well qualified)

3 = 45–54 (qualified)

2 = 40–43 (possibly qualified)

1 = 0–39 (no recommendation)

If your score falls between 55 and 100, you're doing great, keep up the good work! If your score is lower than 54, there's still hope—keep studying and you will be able to obtain a much better score on the exam before you know it.

Practice Test 1 Answer Grid

1. Ⓐ Ⓑ Ⓒ Ⓓ
2. Ⓐ Ⓑ Ⓒ Ⓓ
3. Ⓐ Ⓑ Ⓒ Ⓓ
4. Ⓐ Ⓑ Ⓒ Ⓓ
5. Ⓐ Ⓑ Ⓒ Ⓓ
6. Ⓐ Ⓑ Ⓒ Ⓓ
7. Ⓐ Ⓑ Ⓒ Ⓓ
8. Ⓐ Ⓑ Ⓒ Ⓓ
9. Ⓐ Ⓑ Ⓒ Ⓓ
10. Ⓐ Ⓑ Ⓒ Ⓓ
11. Ⓐ Ⓑ Ⓒ Ⓓ
12. Ⓐ Ⓑ Ⓒ Ⓓ
13. Ⓐ Ⓑ Ⓒ Ⓓ
14. Ⓐ Ⓑ Ⓒ Ⓓ
15. Ⓐ Ⓑ Ⓒ Ⓓ
16. Ⓐ Ⓑ Ⓒ Ⓓ
17. Ⓐ Ⓑ Ⓒ Ⓓ
18. Ⓐ Ⓑ Ⓒ Ⓓ
19. Ⓐ Ⓑ Ⓒ Ⓓ
20. Ⓐ Ⓑ Ⓒ Ⓓ
21. Ⓐ Ⓑ Ⓒ Ⓓ

22. Ⓐ Ⓑ Ⓒ Ⓓ
23. Ⓐ Ⓑ Ⓒ Ⓓ
24. Ⓐ Ⓑ Ⓒ Ⓓ
25. Ⓐ Ⓑ Ⓒ Ⓓ
26. Ⓐ Ⓑ Ⓒ Ⓓ
27. Ⓐ Ⓑ Ⓒ Ⓓ
28. Ⓐ Ⓑ Ⓒ Ⓓ
29. Ⓐ Ⓑ Ⓒ Ⓓ
30. Ⓐ Ⓑ Ⓒ Ⓓ
31. Ⓐ Ⓑ Ⓒ Ⓓ
32. Ⓐ Ⓑ Ⓒ Ⓓ
33. Ⓐ Ⓑ Ⓒ Ⓓ
34. Ⓐ Ⓑ Ⓒ Ⓓ
35. Ⓐ Ⓑ Ⓒ Ⓓ
36. Ⓐ Ⓑ Ⓒ Ⓓ
37. Ⓐ Ⓑ Ⓒ Ⓓ
38. Ⓐ Ⓑ Ⓒ Ⓓ
39. Ⓐ Ⓑ Ⓒ Ⓓ
40. Ⓐ Ⓑ Ⓒ Ⓓ
41. Ⓐ Ⓑ Ⓒ Ⓓ
42. Ⓐ Ⓑ Ⓒ Ⓓ

43. Ⓐ Ⓑ Ⓒ Ⓓ
44. Ⓐ Ⓑ Ⓒ Ⓓ
45. Ⓐ Ⓑ Ⓒ Ⓓ
46. Ⓐ Ⓑ Ⓒ Ⓓ
47. Ⓐ Ⓑ Ⓒ Ⓓ
48. Ⓐ Ⓑ Ⓒ Ⓓ
49. Ⓐ Ⓑ Ⓒ Ⓓ
50. Ⓐ Ⓑ Ⓒ Ⓓ
51. Ⓐ Ⓑ Ⓒ Ⓓ
52. Ⓐ Ⓑ Ⓒ Ⓓ
53. Ⓐ Ⓑ Ⓒ Ⓓ
54. Ⓐ Ⓑ Ⓒ Ⓓ
55. Ⓐ Ⓑ Ⓒ Ⓓ
56. Ⓐ Ⓑ Ⓒ Ⓓ
57. Ⓐ Ⓑ Ⓒ Ⓓ
58. Ⓐ Ⓑ Ⓒ Ⓓ
59. Ⓐ Ⓑ Ⓒ Ⓓ
60. Ⓐ Ⓑ Ⓒ Ⓓ
61. Ⓐ Ⓑ Ⓒ Ⓓ
62. Ⓐ Ⓑ Ⓒ Ⓓ
63. Ⓐ Ⓑ Ⓒ Ⓓ

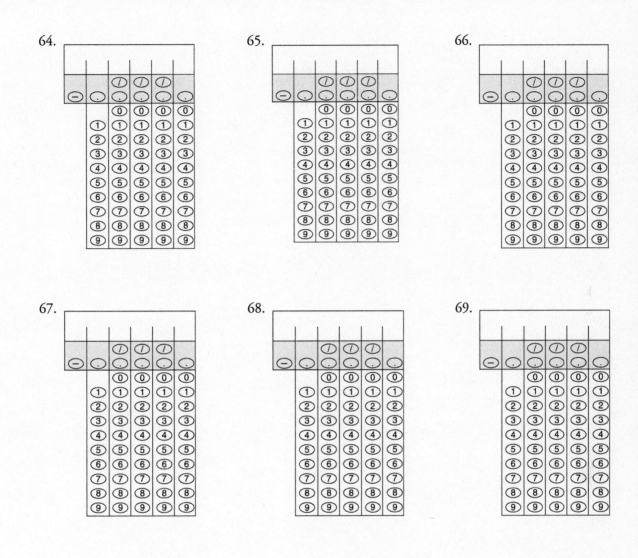

PRACTICE TEST 1

Section I: Multiple-Choice Questions and Grid-In Items

Time: 90 Minutes
63 Multiple-Choice Questions and 6 Grid-In Items

Directions: Choose the best answer choice for the following questions.

1. Why is the muscle of the left ventricle of the heart thicker than the muscle of the right ventricle?

   (A) It only appears to be more muscular because the ventricular space is smaller.

   (B) It has to contain all of the blood collected from the rest of the body.

   (C) It has to pump blood against the pressure of the lungs.

   (D) It has to pump blood to the rest of body.

2. Self-pollination in plants is an example of

   (A) asexual reproduction because a single parent is involved.

   (B) asexual reproduction because offspring are genetically identical to the parent plant.

   (C) asexual reproduction because offspring are genetically unique.

   (D) sexual reproduction because offspring are produced via fusion of gametes.

3. How are bacteria used to produce human insulin?

   (A) They are grown on media rich in sugar, which stimulates insulin production in bacteria.

   (B) The DNA sequence that codes for human insulin production is inserted into the bacterial genome.

   (C) Human pancreas cells are grown in culture with bacteria and transformation occurs.

   (D) Specific bacteriophage viruses are used to produce the correct mutation in the bacterial genome.

GO ON TO THE NEXT PAGE

Site	Plants Species	Plants Individuals	Amphibians Species	Amphibians Individuals	Reptiles Species	Reptiles Individuals	Mammals Species	Mammals Individuals	Total Species	Total Individuals
Bluewater Swamp	15	113	2	8	3	8	5	7	25	136
Papago Buttes	5	27	0	0	2	4	2	3	9	34
Beaver's Bend	8	121	2	2	0	0	3	18	13	141
Sherwood Forest	4	159	1	1	0	0	6	24	11	184
Tortilla Flats	4	63	0	0	3	24	1	5	8	92

4. Based on the table, which site has the greatest species diversity?

(A) Bluewater Swamp

(B) Papago Buttes

(C) Beaver's Bend

(D) Sherwood Forest

5. Allosteric regulation of enzyme activity involves

(A) competitive binding at the enzyme active site.

(B) turning off genes that code for enzyme production.

(C) conformational change in the enzyme due to binding at the allosteric site.

(D) binding of allosteric inhibitors to the substrate.

6. Which of the following are classified in the phylum *Arthropoda*, animals with hard exoskeletons and jointed appendages?

(A) Lobsters

(B) Oysters

(C) Snails

(D) Clams

7. Transpiration against the force of gravity is possible in trees 100 m tall because

(A) water is actively transported from roots to leaves.

(B) evaporation from stomata pulls water up through the tree.

(C) high pressure in the soil pushes the water up.

(D) gravity creates pressure in the xylem, squeezing water out of the stomata.

8. Chimpanzees demonstrate insight learning when presented with a problem. Stacking boxes and climbing on them to bat down a suspended banana with a stick demonstrates the ability to

(A) combine separate experiences to solve novel problems.

(B) associate a new stimulus with a particular reward or punishment.

(C) retain memories with no immediate consequence and use them later.

(D) learn to ignore stimuli that have no positive or negative associations or consequences.

GO ON TO THE NEXT PAGE

9. What are the products of double fertilization in angiosperms?

    (A) An embryo and endosperm

    (B) Two seeds

    (C) A seed and a fruit

    (D) A fruit and a flower

10. Plants form close associations with mycorrhizae, fungi that colonize plant roots. The plant benefits because the fungus makes soil phosphorus available to the plant. The fungus benefits because the plant provides it with sugars. This is an example of

    (A) commensalism.

    (B) competition.

    (C) parasitism.

    (D) mutualism.

11. Internal fertilization probably evolved to allow vertebrates to

    (A) develop two different sexes.

    (B) conserve energy as they produce larger eggs.

    (C) identify their own species.

    (D) rely less on water as they invaded the land.

12. In the very early evolution of animals, bilateral symmetry led to the formation of left- and right-sidedness, dorsal and ventral surfaces, and anterior and posterior ends. This trend was followed by

    (A) appendages.

    (B) cephalization.

    (C) gills.

    (D) pseudopodia.

13. Why does acid rain affect evergreen trees (e.g., conifers) more adversely than it does deciduous trees (e.g., oaks)?

    (A) Evergreen leaves have thinner cuticles and are more sensitive to acid.

    (B) Acid rain occurs more frequently in habitats dominated by evergreens.

    (C) Deciduous trees are adapted to very acidic conditions.

    (D) Loss of leaves each year reduces long-term effects in deciduous trees.

14. What kind of organic macromolecules are formed when amino acids undergo dehydration synthesis and the resulting polypeptide chains twist and fold into three-dimensional structures?

    (A) Proteins

    (B) Carbohydrates

    (C) Lipids

    (D) Nucleic acids

15. A cell from an unknown organism is examined under a microscope. It has a membrane-bound nucleus, a large central vacuole, chloroplasts, and a cell wall. What type of cell is it?

    (A) Prokaryotic

    (B) Bacterial

    (C) Fungal

    (D) Plant

16. An enzyme speeds up chemical reactions by

    (A) giving up electrons in reduction reactions.

    (B) gaining electrons in oxidation reactions.

    (C) acting as a reactant to form new products.

    (D) acting as a catalyst by loosening chemical bonds.

GO ON TO THE NEXT PAGE ▷

17. A man who has a sex-linked recessive disorder carries the gene for the condition on his X chromosome. If he marries a woman who does not have the gene on either of her X chromosomes, what are the chances that their first son will have the disease?

   (A) 100 percent

   (B) 75 percent

   (C) 50 percent

   (D) 0 percent

18. Which of the following relationships is NOT an example of symbiosis?

   (A) A tick sucking blood from a dog

   (B) A clownfish living in the tentacles of a sea anemone for protection

   (C) Ants farming aphids to feed on their honeydew

   (D) Cows eating grass and leaving manure behind as fertilizer

19. DNA replication is described as being semi-conservative, which means that

   (A) one strand is identical to the parent molecule and one is unique.

   (B) one strand of DNA in a chromosome comes from the male gamete and the other from the female gamete.

   (C) half of the genetic code is conserved when the molecule replicates.

   (D) a replicated DNA molecule is composed of one old strand and one new strand.

20. Water has extraordinary properties, not the least of which is that it is less dense in its solid form than in its liquid form. This allows ice to float, so the greatest volume of water

   (A) freezes from the top down.

   (B) freezes from the bottom up.

   (C) remains unfrozen.

   (D) freezes more quickly.

21. The sequence of nucleotides in DNA codes for production of proteins, and that code is carried to the ribosome as mRNA. Which of the following protein structures represents the longest strand of nucleotide bases?

   (A) The gene for hemoglobin, a protein, which consists of 20 amino acids

   (B) The normal chloride channel protein gene associated with cystic fibrosis, which consists of 63 nucleotide bases

   (C) The mutated sickle cell form of the hemoglobin gene, which has a single substituted base

   (D) The mutated cystic fibrosis mRNA strand, which has a base insertion

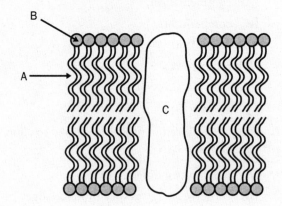

22. Which of the following statements is true?

   (A) The region of the membrane labeled A is nonpolar.

   (B) The region of the membrane labeled B is hydrophobic.

   (C) The structure labeled C is a complex carbohydrate.

   (D) Charged ions such as Na⁺ diffuse directly through the membrane bilayer.

GO ON TO THE NEXT PAGE

23. Human white blood cells can "crawl" to damaged tissues to contact and phagocytize bacteria at the site of the damage. Which of the following processes facilitates this type of movement?

    (A) Addition of phospholipids to the cell plasma membrane

    (B) Rapid formation and deformation of actin filaments in the cytoskeleton

    (C) Beating of cilia against vessel and tissue surfaces

    (D) Whip-like motion of the cell's flagellum

24. In humans, short-term energy supplies are stored as glycogen and long-term energy supplies are stored as fat. Which of the following statements explains this phenomenon?

    (A) Lipids are more readily soluble in water than carbohydrates.

    (B) Lipids are more easily broken down by mitochondria than carbohydrates.

    (C) Lipids are more oxidized and produce less energy per gram.

    (D) Glycogen is more readily converted to glucose than fat.

25. Which of the following statements is NOT true about DNA and RNA in eukaryotes?

    (A) DNA is double-stranded and RNA is single-stranded.

    (B) Adenine is a nucleotide common to both DNA and RNA.

    (C) Guanine is a nucleotide common to both DNA and RNA.

    (D) Both DNA and RNA can be found in the cell cytoplasm.

26. During complete aerobic cellular respiration, each molecule of glucose broken down in the mitochondria can yield 36 molecules of ATP. What conditions might lead to a decrease in the amount of ATP produced in a given system?

    (A) An increase in the amount of glucose added to the system

    (B) A decrease in the amount of light the system is exposed to

    (C) A decrease in the amount of oxygen available in the system

    (D) A decrease in the amount of carbon dioxide available in the system

27. Which of the following are most closely related to humans?

    (A) Lobsters and crabs

    (B) Clams and mussels

    (C) Bees and wasps

    (D) Sea stars and sea urchins

28. While exploring a tropical ecosystem, a student encountered an unidentified plant that had xylem and phloem, a free-living haploid generation, and spore-producing structures called sori on the underside of compound leaves. What type of plant did the student find?

    (A) A bryophyte

    (B) A fern

    (C) A gymnosperm

    (D) An angiosperm

29. Junipers are conifers that produce seeds with fleshy coats that resemble blueberries. Therefore, it is NOT true that junipers

    (A) produce fruits.

    (B) reproduce sexually.

    (C) are gymnosperms.

    (D) are naked-seeded plants.

GO ON TO THE NEXT PAGE

30. If an animal's jaw contains teeth with broad, rigid surfaces, you can conclude that the animal is probably

(A) an herbivore.

(B) a producer.

(C) a top consumer.

(D) a carnivore.

31. Amphibians require standing water for successful reproduction, while reptiles do not. Which of the following factors contribute to this phenomenon?

(A) Internal fertilization in amphibians

(B) Amniotic eggs of reptiles

(C) Metamorphosis in reptiles

(D) Scales in amphibians

32. Birds are thought to be very closely related to crocodiles. Which of the following is a characteristic shared by birds and crocodiles that is NOT shared by birds and turtles?

(A) A four-chambered heart

(B) Amniotic eggs

(C) Homeothermy (warm-bloodedness)

(D) Deuterostome development

33. Grasslike plants such as wheat and rice are monocots, while deciduous trees and most woody shrubs are dicots. Based on this information, which of the following statements is FALSE?

(A) The leaves of grasses have parallel venation.

(B) Deciduous trees are flowering plants.

(C) The seeds of woody shrubs have a single cotyledon.

(D) Wheat's flower parts occur in multiples of three.

34. An unknown animal is observed to have bilateral symmetry. What other characteristic would you expect this animal to have?

(A) Multicellularity

(B) Diploblastic development

(C) A notochord

(D) Deuterostome development

35. Tapeworms are endoparasites, which means that they

(A) live attached to the outside of their hosts.

(B) cannot live outside the body of a host.

(C) live inside the body of a host.

(D) must reproduce within the body of a host.

36. A defective protein has been produced due to an error in transcription. The molecule that contains the error immediately after transcription is

(A) DNA.

(B) rDNA.

(C) rRNA.

(D) mRNA.

37. How do macrophages contribute to the body's defense?

(A) Macrophages ingest invading organisms.

(B) Macrophages reduce inflammation.

(C) Macrophages secrete histamine.

(D) Macrophages secrete antibodies.

GO ON TO THE NEXT PAGE

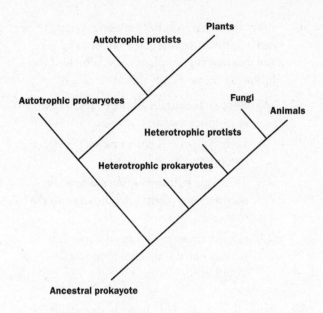

38. The phylogenic analysis shown in the diagram suggests that

(A) animals are descended from fungi.

(B) fungi are more similar to animals than to plants.

(C) protists are a monophyletic group.

(D) the first organisms were eukaryotes.

39. Which of the following is an example of postzygotic reproductive isolation?

(A) Species in the same area occupying incompatible habitats

(B) Species separated geographically by a physical barrier such as an ocean

(C) Prevention of gamete fusion

(D) Production of sterile hybrid offspring

40. Evolution may be defined as changes over time in the allele frequency of a population or species. All of the following are examples of evolutionary processes EXCEPT

(A) artificial selection in domestic dogs.

(B) populations in Hardy-Weinberg equilibrium.

(C) mutations that decrease reproductive fitness.

(D) natural selection between different colors of insects.

41. Archaea are unique prokaryotes that are thought to be more closely related to eukaryotes than they are to bacteria. Which characteristic listed supports this idea?

(A) Many archaea are adapted to extreme environments, such as deep-sea thermal vents.

(B) The cell walls of archaea lack peptidoglycans.

(C) Archaea have introns in some genes.

(D) Archaea lack a membrane-bound nucleus and organelles.

42. Organisms that reproduce sexually exhibit zygotic, gametic, or sporic meiosis. One way to determine the type of life cycle an organism has is by

(A) observing embryonic development.

(B) comparing the diploid and haploid forms of the organism.

(C) determining when in the life cycle fertilization occurs.

(D) determining if gametes are multicellular or unicellular.

GO ON TO THE NEXT PAGE ⟩

43. Which of the following animals is most closely related to an octopus?

   (A) Whale
   (B) Trout
   (C) Snake
   (D) Clam

44. Which shows the correct order of hierarchy from simple to complex?

   (A) Molecule → tissue → cell → organism
   (B) Cell → tissue → organ → organism
   (C) Organism → species → biosphere → ecosystem
   (D) Molecule → cell → organism → tissue

45. Organisms transform energy when they ingest food, breaking it down into nutrients used to build tissues and make repairs. Each energy transfer increases the universe's level of

   (A) order.
   (B) stability.
   (C) disorder.
   (D) energy.

46. What is the primary danger of a population relying on an agricultural monoculture for its dietary staple?

   (A) People get tired of eating the same thing.
   (B) Children develop aversion to foods that appear too often in their diet.
   (C) Food allergies develop after repeated exposure.
   (D) Genetic similarity makes an entire crop vulnerable to a single pest or pathogen.

47. A botanist observes that a mature root cell can dedifferentiate in tissue culture and give rise to a diversity of plant cells. Which of the following best explains this observation?

   (A) Root cells contain all the genes necessary to produce a variety of cells.
   (B) Each type of cell has a unique genetic blueprint.
   (C) The tissue culture transferred proteins necessary for plant differentiation to the root cell.
   (D) mRNA transcripts from the root cell are translated only after appropriate stimulation.

48. According to the ABC hypothesis for the functioning of organ identity genes, three classes of genes (A, B and C) are responsible for the spatial pattern of floral parts. Sepals develop from the region where only A genes are active. Petals develop where both A and B genes are expressed. Stamens arise where B and C genes are active and carpels arise where only C genes are expressed. Furthermore, it is observed that if A gene or C gene activity is missing, then the activity of the other spreads throughout.

   Which of the following could be the floral morphology for a mutant lacking C gene activity?

   (A) Sepal-petal-stamen-carpel-stamen-petal-sepal
   (B) Carpel-stamen-carpel-stamen-carpel
   (C) Sepal-carpel-sepal
   (D) Sepal-petal-sepal-petal-sepal

GO ON TO THE NEXT PAGE

49. In a certain terrestrial ecosystem, over a given period of time, primary producers had 1.1 MJ (megajoules) of energy available in the form of sunlight, whereas the primary consumers had approximately 12 kJ and secondary consumers, 1.1 kJ. Which of the following is the most reasonable conclusion from this data?

(A) The majority of the energy in a trophic level is transferred to the next higher level.

(B) The transfer of energy from producers to primary consumers is more efficient than the transfer between consumers.

(C) Energy within an ecosystem is constantly cycled through the various levels.

(D) Approximately 90% of the energy available to consumers is not transferred to the next highest level.

50. Which of the following lack vascular tissue?

(A) Mosses

(B) Ferns

(C) Gymnosperms

(D) Monocots

51. The concept of gradualism was initially used to explain the formation of geologic features, however, certain tenets of gradualism were later incorporated into Darwin's theory of evolution. Which of the following best describes those tenets?

(A) Change occurs mainly through catastrophic events.

(B) Slow and continuous processes can lead to drastic changes.

(C) Certain heritable traits are gradually favored over others.

(D) Resources are limited and there is a struggle for existence among individuals.

Questions 52–54 refer to the following image.

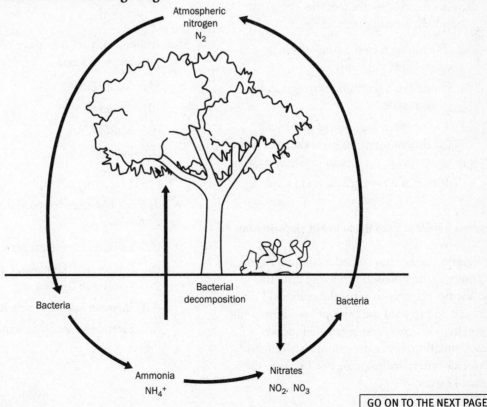

GO ON TO THE NEXT PAGE

52. Which of the following is a form of nitrogen that animals are capable of utilizing?

(A) N_2

(B) NH_4^+

(C) Amino acids

(D) NO_3^-

53. The conversion of ammonia (NH_3) to NO_3^- is most analogous to

(A) the aerobic metabolism of glucose to CO_2.

(B) the reduction of carbon dioxide in photosynthesis.

(C) the synthesis of polypeptides in translation.

(D) the hydrolysis of ATP in muscle cells.

54. Nitrogen is the mineral nutrient that contributes the most to plant growth and crop yields. Which of the following best explains why plants can suffer from nitrogen deficiencies, despite the fact that the atmosphere is nearly 80% nitrogen?

(A) The nitrogen in the atmosphere is unavailable to plants.

(B) Plants can only utilize nitrogen in its gaseous state.

(C) The nitrogen available to plants is dependent upon the breakdown of rock.

(D) 80% of the atmosphere is only a fraction of the total nitrogen demand of plants.

Questions 55–59 refer to the following information

Your biology teacher has created a computer game that simulates the function of a living cell. The object for the player is to act as the command center for the cell and to send the appropriate information for protein production and organelle function. For each situation, select the appropriate cellular structure or functional response needed from the command center.

55. More ATP is needed.

(A) Increase protein synthesis

(B) Increase mitochondrial activity

(C) Produce more cytoskeletal elements

(D) Initiate cell division

56. The cell is in the S-stage of interphase.

(A) Replicate DNA

(B) Align chromosomes at the cell equator

(C) Rest

(D) Pull sister chromatids to opposite poles of the cell

57. The cell has been dropped into a hypertonic solution.

(A) Take up water through aquaporins

(B) Close all membrane-bound proteins

(C) Pump out solutes

(D) Lose water through aquaporins

58. Amino acids are needed at ribosomes for protein production.

(A) Send rDNA

(B) Send mRNA

(C) Send rRNA

(D) Send tRNA

59. The cell has engulfed particles that need to be disposed of.

(A) Initiate chloroplast activity

(B) Stimulate endoplasmic reticulum to synthesize lipids

(C) Increase lysosome activity

(D) Shut down enzyme activity

GO ON TO THE NEXT PAGE

Questions 60–63

The following diagram shows energy transformations within a cell. Each form of energy is represented by the symbols E I–E IV. Two cellular organelles are represented by the letters A and B. Answer the following questions about the various processes depicted in the diagram and about the cell in which they are occurring.

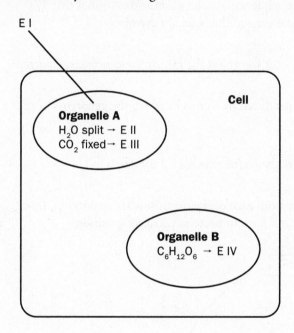

60. What form of energy is represented by E II?

(A) Radiant energy in the form of photons

(B) Chemical energy being stored in the bonds of glucose

(C) Chemical energy in the form of ATP

(D) Chemical energy released by glycolysis

61. If the transformation depicted in organelle B requires oxygen, what form of energy is represented by E IV?

(A) Radiant energy in the form of photons

(B) Chemical energy being stored as glycogen

(C) Chemical energy in the form of ATP

(D) Chemical energy released by glycolysis

62. What cellular organelles are represented as A and B, respectively?

(A) The nucleus and the ribosome

(B) The mitochondrion and the chloroplast

(C) The mitochondrion and the ribosome

(D) The chloroplast and the mitochondrion

63. Which kind of organism could the cell shown belong to?

(A) A photosynthetic bacteria

(B) A photosynthetic protist

(C) A heterotroph

(D) A fungi

GO ON TO THE NEXT PAGE

GRID-IN ITEMS

The following questions require a numeric answer. Calculate the correct answer for the question, and then enter the answer on the provided grids. You may use a simple four-function calculator and the included formula sheet.

64. If a population has 500 individuals and 127 have the *bb* genotype, assuming simple dominance of the *B* allele, what is the frequency of the *Bb* genotype?

65. A scientist extracts DNA from the nucleus of cells and sequences it. The scientist determines that 27 percent of the nucleotide bases are guanine. What percentage of the bases are thymine?

66. If the pH of a solution has decreased from 6 to 2, by what factor has the H^+ ion concentration changed?

67. What is the probability that the genotype *rrss* will be produced by a cross in which the genotypes of the parents are both *RrSs*?

68. How many possible combinations of chromosomes are there after meiosis I if $2n = 6$?

69. The ocean food web shown depicts the flow of carbon from each of the organisms via respiration, from producer to consumer to consumer via consumption and back to bacteria via decomposition.

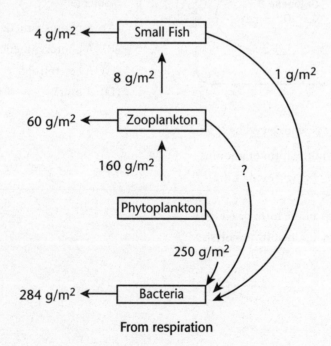

How much carbon (in g/m^2) from zooplankton is used by decomposers? Give your answer to the nearest whole number.

IF YOU FINISH BEFORE TIME IS CALLED, YOU MAY CHECK YOUR WORK ON THIS SECTION ONLY. DO NOT TURN TO ANY OTHER SECTION IN THE TEST.   STOP

10-MINUTE READING PERIOD

Take the next 10 minutes to glance over the four questions that comprise Section II of this test. You can take notes in the margins, and begin writing your responses, but it is advisable that you use this time to develop your responses.

When the 10-minute period is over, you will have 80 minutes to complete the eight free-response questions.

Section II: Free-Response Questions

Time: 90 Minutes
2 Long Questions
6 Short Questions

Part A Directions: Answer each part of the following free-response questions with complete sentences. Answers should be in essay form. Make sure to provide a detailed response to each part of each question. Diagrams and other figures may be included to demonstrate knowledge of a topic, but they should not be the only answer you provide.

1. The human body manifests physiological changes in every organ system throughout the process of eating, from the time a person first anticipates food to the elimination of waste.

   (A) **Design** a controlled experiment where you test the hypothesis that the body's reaction to eating begins before food enters the body.

   (B) **Describe** physiological changes that occur in three different organ systems when food is ingested and how these systems coordinate a response in the human body.

2. Human impact on the Earth's environment is poignantly demonstrated in the Mississippi River drainage into the Gulf of Mexico. Every spring, high levels of nitrogen and phosphorous wash into the river from the midwestern United States farming region and into the northern Gulf. These nutrients supply a huge algal bloom at the water surface, creating a dangerously destructive hypoxia that all but destroys the marine ecosystem.

   (A) **Discuss** the cycling of oxygen in the natural marine ecosystem and how this cycle is upset during an algal bloom.

   (B) Pick two organisms from two different trophic levels in the marine ecosystem, and **explain** the impact of the algal bloom on their ecology.

   (C) It has been suggested that the hypoxic condition in the Gulf might be contributing to an increased number of shark attacks along the Texas coast. **Explain** the thinking behind this suggestion.

GO ON TO THE NEXT PAGE ⟩

Part B Directions: Answer each of the following short free-response questions with complete sentences. Answers should be in paragraph form. Diagrams and other figures may be included to demonstrate knowledge of a topic, but they should not be the only answer you provide.

3. All living organisms contain genetic information that provides several functions inherent to the individual organism and the perpetuation of its species. Discuss how the nature of genetic material both perpetuates the identity of an individual and provides for high biodiversity.

4. In a long-term project studying the interactions of several species of animals on an isolated island, scientists counted the number of individuals of each species visiting a site on the island over the course of several days, every summer for 100 years. The results from that study are shown in the following graph.

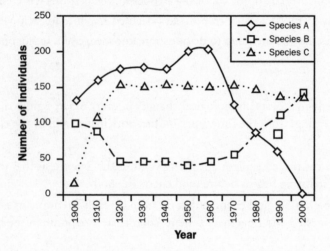

Analyze the relationship between Populations A, B, and C, and make conclusions about the state of the populations in the year 2000.

GO ON TO THE NEXT PAGE

5. A man is found unconscious on the sidewalk, and a passerby rushes him to a nearby hospital emergency room. The patient has no identification and no medical history is available. On initial examination, the ER physician finds that the man's temperature is elevated and that his blood pressure is slightly decreased. His EKG is normal and there are no obvious signs of trauma on his body. A laboratory workup is ordered and the following test results are obtained.

	Patient Value	Normal Value
Hematology		
Hemoglobin	15	13–18 gm/dl
White blood cell count	36,000	5,000–10,000 µl/mm^3
% Neutrophils	97	48–73%
% Lymphocytes	3	18–48%
Platelet count	175,000	150,000–350,000/ml
Blood Chemistry		
Glucose	444	70–110 mg/dl
Blood urea	87	7–18 mg/dl
Ethanol	0.1	0–0.1 mg/dl
Blood pH	7.0	7.35–7.45

What is the likely diagnosis for this patient? What additional information would the doctor likely want to confirm the diagnosis?

6. The following pedigree shows the occurrence of a genetic disorder through several generations and several lineages of a family. The squares represent males and the circles represent females; the semicircles represent females heterozygous for, but unaffected by, the defective gene. None of the males are heterozygous for the condition. The symbols G1, G2, and G3 represent the first, second, and third generations subsequent to the initial mating pair. The symbols L1, L2, and L3 represent the three lineages of descent from the initial generation.

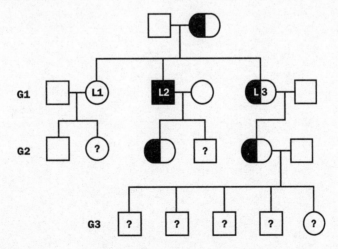

What type of disorder is most likely shown by this pedigree?

GO ON TO THE NEXT PAGE

7. In one type of enzyme regulation, the presence of the end product of a metabolic pathway inhibits an enzyme that functions in an early step in the pathway. What is this type of enzyme regulation?

8. Achondroplasia is a form of dwarfism among humans. A male and female who both have achondroplasia and are under 3.5 ft tall produce an offspring of normal size. What are the alleles for the father, mother, and their child, respectively?

PRACTICE TEST I ANSWER KEY

1.	D	24.	D	47.	A
2.	D	25.	D	48.	D
3.	B	26.	C	49.	D
4.	A	27.	D	50.	A
5.	C	28.	B	51.	B
6.	A	29.	A	52.	C
7.	B	30.	A	53.	A
8.	A	31.	B	54.	A
9.	A	32.	A	55.	B
10.	D	33.	C	56.	A
11.	D	34.	A	57.	D
12.	B	35.	C	58.	D
13.	D	36.	D	59.	C
14.	A	37.	A	60.	C
15.	D	38.	B	61.	C
16.	D	39.	D	62.	D
17.	D	40.	B	63.	B
18.	D	41.	C	64.	0.5
19.	D	42.	B	65.	23
20.	C	43.	D	66.	10,000
21.	D	44.	B	67.	1/16 or 0.06
22.	A	45.	C	68.	8
23.	B	46.	D	69.	92

ANSWERS AND EXPLANATIONS

SECTION I: MULTIPLE-CHOICE QUESTIONS AND GRID-IN ITEMS

1. D
The left ventricle has to pump blood to the most distal parts of the body. The right atrium is the collecting site of blood from the body, (B), and the right ventricle pumps blood to the lungs, (C), necessarily at far less pressure. Choice (A) is untrue.

2. D
By definition, asexual reproduction is production of offspring genetically identical to the single parent, primarily via cell division. Sexual reproduction involves fusion of unique haploid gametes to produce genetically unique diploid offspring. Self-pollination is a form of sexual reproduction. A single parent plant produces both male and female gametes, but each is the product of meiosis and genetically unique. Fusion of two gametes produces a unique diploid zygote, which will develop into a unique adult.

3. B
The only way a bacterium can be induced to produce a human protein such as insulin is to have the gene that codes for production of that protein inserted into the bacterial genome. Choice (A) is clearly incorrect; bacteria are cultured on sugar-rich media routinely, and it does not confer human abilities upon them. Both (C) and (D) suggest technically feasible possibilities, except that the chances of such a beneficial transfer of genetic information occurring in either case would be virtually zero.

4. A
Species diversity is a measure of the number of different kinds of organisms present at a site. Papago Buttes, (B), does not have very many species or many individuals. Beaver's Bend, (C), has a lot of different species, but the overwhelming majority of animals there are mammals. Although Sherwood Forest, (D), has the most individual plants and animals, there are few species.

5. C
Allosteric regulators may be inhibitors or activators, but by definition they always change the shape of the protein due to binding at an allosteric site, and not by binding at the enzyme active site, (A). Allosteric inhibitors bind to the enzyme allosteric site and not the substrate, (D). They change the shape of the enzyme so that the substrate cannot be bound at the enzyme active site. Allosteric regulation is unrelated to genetic regulation, ruling out (B).

6. A
Lobsters are arthropods. Oysters, snails, and clams are all mollusks. Their shells are not exoskeletons, and they lack jointed appendages.

7. B
Transpiration is a passive process, one that does not require energy expenditure by the plant, ruling out (A). Although positive pressure in the roots can cause some water loss, it is not the typical case and does not occur in very tall trees, so (C) is incorrect. Gravity, (D), could only exert pressure downward. Because water is cohesive and adhesive, evaporation literally pulls water through the tree, much like sucking up liquid through a straw.

8. A
Insight learning is the ability to exhibit a useful or productive behavior in a situation with which the animal has no experience. Choice (B) is classical conditioning, (C) is observational learning, and (D) is habituation.

9. A
Angiosperms, or flowering plants, are considered a monophyletic group (a group arising from one common ancestor) based on the characteristic of double fertilization. One sperm fertilizes the egg of a mature ovule, and the other one fertilizes the polar nuclei. The ovule develops into an embryo,

and the polar nuclei develop as endosperm, food reserves for the embryo. Although many seeds may be produced, (B), and a fruit surrounds the seed or seeds, (C), each seed with its embryo is a product of a double fertilization event. Flowers, (D), are where fertilization occurs.

10. D
Commensalism, (A), is a form of symbiosis in which one organism benefits and the other is neither benefited nor harmed. Competition, (B), occurs when two organisms both use the same limited resources, such as food or sunlight. Parasitism, (C), is another form of symbiosis, one in which one organism benefits while the other is harmed. Mutualism, (D), is the form of symbiosis in which both organisms involved benefit.

11. D
The two primary forms of fertilization are external and internal. External fertilization requires that eggs be deposited in safe environments, protected from heat and desiccation, so that eggs are invariably deposited in water or areas that are reliably damp. Internal fertilization frees animals from these constraints.

12. B
Cephalization was the evolutionary development in animals that created an anterior end in which sensory equipment (e.g., for taste) was localized. Appendages, (A), and gills, (C), came after the head. Single cells (such as amoebae and macrophages) move by way of pseudopodia, (D).

13. D
Acid rain is harder on evergreen trees because their leaves have to endure many years of cumulative effects. Evergreens typically have leaves with thicker cuticles than deciduous trees, so (A) is incorrect. Choice (B) is incorrect because acid rain is not a localized phenomenon. Although (C) might be a reasonable guess, one must consider that acid rain is composed of sulfuric acid, and very few living things can tolerate regular sulfuric acid showers.

14. A
Amino acids are the monomers of proteins. All organic macromolecules are formed via dehydration synthesis of monomers. For carbohydrates, (B), these monomers are monosaccharides and for lipids, (C), fatty acids and glycerol. Nucleotides are the monomers of DNA and RNA, (D). Only proteins are made of amino acids, and polypeptides fold and twist into distinct three-dimensional shapes.

15. D
Choices (A) and (B) are incorrect because the organism has a membrane-bound nucleus, making it a eukaryote. Fungi, (C), lack chloroplasts. Only a plant cell would have all of the structures mentioned.

16. D
Enzymes catalyze chemical reactions but are not used up or changed by the reactions they catalyze. Either reduction, (A), or oxidation, (B), would change the enzyme, as would its being chemically altered into a new product, (C).

17. D
This disorder is sex-linked and X-linked, passed on the X chromosome. The father can pass the gene on to his daughters, but not to his sons. Because boys have one X and one Y chromosome and mothers lack Y chromosomes, a boy always gets his Y chromosome from his father. Sons would therefore never get the X-linked gene. In this instance, no daughters would have the disorder either, because they would all get one healthy X chromosome from their mother. Because it is a recessive disorder, the condition would not affect the daughters.

18. D
Symbiosis is simply defined as organisms living in close association with each other. Choice (A) is an example of parasitism, (B) of commensalism, and (C) of mutualism, all forms of symbiosis. Choice (D) is an example of predation, one organism feeding on another, which is not a form of symbiosis.

19. D

DNA replication produces a double-stranded molecule identical to its parent because after unzipping the double helix, a complementary strand is produced for each strand. This results in a new molecule composed of one old strand and one new complementary strand. All other answer choices have implications for a replicated DNA molecule different from the original.

20. C

Ice acts as an insulating cap on a body of water, allowing the bulk of water to remain unfrozen, which protects water-dwelling organisms. If ice sank and bodies of water froze from the bottom up, then they would become solid, unable to support much life. Choice (D) is a variation on the theme of overall freezing.

21. D

Both the nonmutated hemoglobin and cystic fibrosis gene sequences consist of 20 amino acids plus the stop codon. Because each amino acid has a three-nucleotide codon, both are composed of 63 bases, so (A) and (B) would be the same length. Choice (C) is incorrect because the mutation associated with sickle cell is a substitution; one nucleotide replaces another, so the length of the molecule is the same. The cystic fibrosis mutation involves an insertion—the addition of a nucleotide—making it the longest molecule.

22. A

The region labeled A is composed of long fatty acid tails of the phospholipids that make up the cell plasma membrane. The lipid end of the molecule is nonpolar and thus hydrophobic, which prevents charged ions such as Na^+ from crossing the membrane, so (D) is incorrect. Region B is the polar, hydrophilic region of the molecule. The structure labeled C is a membrane-bound protein, making (C) incorrect.

23. B

Most cell movement is associated with the cytoskeleton. Changes in the plasma membrane, (A), are associated with endocytosis and exocytosis. Cilia, (C), and flagella, (D), are not associated with the amoeboid movement of human white blood cells.

24. D

Carbohydrates serve as quick short-term energy reserves and are more soluble in water relative to insoluble fats. Glycogen is a type of complex carbohydrate that is converted to glucose as needed. The mitochondria act primarily on carbohydrates; using fat as an energy source is less efficient.

25. D

Double-stranded DNA is confined to the nucleus, while single-stranded RNA, (A), can be found in the cell's cytoplasm, (D). Adenine, (B), and guanine, (C), are monomers of both molecules; thymine is unique to DNA and uracil is unique to RNA.

26. C

Complete cellular respiration is an aerobic (oxygen-requiring) process. In the absence of oxygen, fermentation occurs and the amount of ATP yielded decreases. Adding glucose to the system, (A), would not decrease the amount of ATP produced, nor would it increase the amount produced per glucose molecule. Light, (B), and carbon dioxide, (D), are important to photosynthesis, not cellular respiration.

27. D

Sea stars and sea urchins are the only deuterostomes given in the answer choices. Despite the radial symmetry of the adults, their larvae are bilaterally symmetrical (making them members of Bilateria). All of the others in the list are protostomes.

28. B

Bryophytes, (A), are nonvascular plants and so lack xylem and phloem. Gymnosperms, (C), and angiosperms, (D), are both seed plants, and the haploid forms are not free living. Ferns are the seedless vascular plants that bear spores in sori.

29. A

Junipers are conifers and therefore gymnosperms, (C), or naked-seeded plants, (D). This tells us that they do not produce fruits; only angiosperms produce fruits. Junipers reproduce sexually, (B).

30. A

Herbivores tend to have broad teeth, which are good for grinding up plant material; they are primary consumers feeding on producers. Producers, (B), are plants, while top consumers, (C), and carnivores, (D), tend to have pointed teeth for tearing meat.

31. B

Amphibians must have water to successfully reproduce because eggs are fertilized outside the body in water. Reptiles have scales made of keratin, have internal fertilization, and can lay eggs away from water because the embryo is enclosed in a watertight structure (amniotic egg). Metamorphosis is associated with amphibians and not with reptiles.

32. A

Both birds and crocodiles have four-chambered hearts, while most other reptiles, such as turtles, have three-chambered hearts. Birds, crocodiles, and turtles all lay amniotic eggs, (B), and have deuterostome development, (D). Of the three, only birds are warm-blooded, (C).

33. C

Monocots such as grasses, (A), and wheat, (D), tend to have a single cotyledon, parallel leaf venation, flower parts in threes or multiples of three, and diffuse vascular bundles. Dicots such as woody shrubs, (C), have two cotyledons. Both monocots and dicots (such as deciduous trees) can be flowering plants.

34. A

By definition, all animals are multicellular. Diploblastic development, (B), is associated with Cnidarians, which have radial symmetry. Both a notochord, (C), and deuterostome development, (D), are associated with only two phyla, the echinoderms and chordates. There are many phyla that have bilateral symmetry but are not necessarily echinoderms or chordates (such as arthropods and mollusks).

35. C

Ectoparasites such as ticks live attached to the outside of a host body, (A). Endoparasites can live outside the body of a host, (B), and often reproduce, (D), outside the host's body; however, they live much of their lives and feed attached to the inside of the host's body, making (C) the best choice.

36. D

DNA is used to make mRNA during transcription, so (A), DNA, is incorrect. rDNA is recombinant DNA, associated with genetic engineering, so (B) is incorrect. Errors in rRNA or tRNA would lead to translational errors, ruling out (C).

37. A

Macrophages are phagocytic. They ingest invading organisms.

38. B

Cladograms are intended to show the phylogenic relationships among organisms, not to imply ancestral descent, so (A) is incorrect. This cladogram implies that protists are polyphyletic, ruling out (C), and that the first organisms were prokaryotes, which are also polyphyletic, ruling out (D).

39. D

Postzygotic reproductive isolation is that which precludes successful reproduction at some point of development in a diploid offspring, after the fertilization event. Geographic separation, (A) and (B), and prevention of gamete formation, (C), are both prezygotic factors. Only sterile hybrids (such as mules) represent postzygotic examples of reproductive isolation.

40. B

Artificial selection, (A), is a process that can act on gene mutations, (C), both factors contributing to evolution. Populations in Hardy-Weinberg equilibrium, (B), are the antithesis of evolutionary processes; they represent stable allele frequencies and change little over time. Natural selection, (D), is a mechanism of evolution.

41. C
All of the answer choices state something true about archaea, but only (C) relates them to eukaryotes. The presence of introns is an advanced eukaryotic characteristic.

42. B
Life cycles are characterized by the timing of meiosis and the characteristics of the diploid and haploid generations. Embryonic development, (A), does not reveal either of these characteristics, nor does fertilization, (C). All gametes are unicellular, ruling out (D).

43. D
An octopus is classified in the phylum *Mollusca*, as are bivalves such as clams. Whales, (A), trout, (B), and snakes, (C), are all vertebrates, a subgrouping of the phylum *Chordata*.

44. B
Choice (A) is incorrect because tissues are composed of cells. In (C), biosphere should follow ecosystem, as it encompasses all other levels. (D) shows tissues being made up of organisms, and the opposite is true.

45. C
The second law of thermodynamics says that every change in energy or energy transfer contributes to the entropy of the universe. Entropy is disorder. Choice (D) is incorrect because it violates the first law of thermodynamics: energy can be neither created nor destroyed.

46. D
Potato blight caused the Irish famine in the 1800s, destroying the population's dietary staple. All potato plants were susceptible to the blight because they were so genetically uniform. Choice (A) may be true but is irrelevant. Choices (B) and (C) are incorrect.

47. A
Based on the observation and fundamental knowledge of genetics, the best explanation is that the mature root cell contains the necessary genetic information to produce the various plant cells. This genetic information is DNA. The best answer is (A).

48. D
The mutant lacking the C gene will be incapable of producing stamens and carpels, so rule out all answer choices with those organs. According to the observation mentioned in the last sentence, if a mutant lacks C gene activity, then A gene activity will spread throughout. So expect only sepals and petals. The correct answer is (D).

49. D
Approximation reveals that about 1% of the available energy is transferred from primary producers to primary consumers and that about 10% is transferred from consumer to consumer. Therefore, the best answer is (D).

50. A
Mosses lack xylem and phloem, and so are nonvascular plants.

51. B
Gradualism was defined by a slow, gradual change over time. The answer that best matches this tenet is (B). (A) describes a contrasting viewpoint known as catastrophism. (C) and (D) are tenets of Darwin's theory, but not ones clearly influenced by gradualism.

52. C
Animals can only utilize organic forms of nitrogen. The only organic form of nitrogen is given in (C).

53. A
The conversion of NH_3 to NO_3^- is an oxidation reaction. The most analogous reaction is the oxidation of glucose to CO_2.

54. A
From the figure it can be seen that nitrogen does not flow directly from the atmosphere into plants. This makes (A) a viable explanation. Plants obtain nitrogen through absorption of ammonium and nitrate, ruling out (B). Unlike other minerals,

ammonium and nitrate are not derived from the breakdown of rock, so (C) is wrong as well. (D) is untrue, as the atmosphere is the main reservoir of nitrogen (although it is inaccessible to plants).

55. B

When cells need ATP, it is time for the organelles responsible for cellular respiration—the mitochondria—to go to work. ATP is produced when glucose is broken down and the energy stored in its covalent bonds is released.

56. A

Most of the cell cycle is spent in interphase, which is anything but a resting stage. One of the most important events to occur during interphase, which consists of the G_1, S, and G_2 stages, is DNA replication.

57. D

In a hypertonic solution, a cell will lose water through aquaporins via osmosis. This is because the solute concentration is higher outside the cell relative to inside the cell. Cells cannot simply close membrane protein channels to prevent this loss. They can alter their internal tonicity, a process which would involve moving solutes (not water) into the cell.

58. D

Protein production occurs at ribosomes when the genetic code recorded on mRNA is translated. In order to physically build proteins, amino acid monomers must be shuttled to the ribosome. This is accomplished by transfer RNA (tRNA), which has an anticodon complementary to the mRNA strand. The anticodon ensures that the correct amino acid will be transported to the ribosome.

59. C

Lysosomes are the garbage collectors and recyclers of the cell. They contain enzymes that help break down particles engulfed by the cells so that the molecules they contain can be recycled.

60. C

The reaction occurring in organelle A is photosynthesis, as indicated by the splitting of water and fixation of CO_2. The first reaction in this organelle represents the light reactions, in which the sun's energy is converted to chemical energy in the form of ATP (E II).

61. C

The reaction occurring in organelle B is cellular respiration, as indicated by the release of energy stored in the bonds of glucose. Because the question specifies that oxygen is required, the process depicted must show aerobic respiration rather than glycolysis. The form of energy released in aerobic respiration is chemical energy in the form of ATP (E IV).

62. D

The organelle labeled A is transforming energy via the process of photosynthesis, a process that occurs in the chloroplasts. The organelle labeled B is transforming energy via cellular respiration, which takes place in the mitochondria.

63. B

Although both bacteria and protists are capable of photosynthesis, bacteria lack true organelles such as chloroplasts (Organelle A) and mitochondria (Organelle B). Of the choices given, only photosynthetic protists can carry out photosynthesis and have true organelles.

64. 0.5

You should use the Hardy-Weinberg equilibrium to solve this problem.

$$p^2 + 2pq + q^2 = 1$$
$$p + q = 1$$

The frequency of the *bb* phenotype is 127/500 or 0.254. Therefore, $q^2 = 0.254$ and $q = 0.5$. Since $p + q = 1$, $p = 0.5$. The frequency of the *Bb* genotype is equal to $2pq$ or $2(0.5)(0.5) = 0.5$.

65. 23

In DNA, guanine pairs with cytosine and adenine pairs with thymine. If 27 percent of the nucleotides are guanine, then 27 percent of the bases are also cytosine. Therefore, 46 percent of the bases are adenine and thymine, which each occur at the same rate. Therefore, 23 percent of the nucleotide bases are thymine.

66. 10,000

The pH scale is negative and logarithmic. Therefore, for each unit that pH decreases, H^+ ion concentration increases by a factor of 10. A decrease of 4 units in pH translates to a ten-thousand-fold (10^4) increase in H^+ ion concentration. The answer is 10,000.

67. 1/16 or 0.06

We have to create a Punnett square for the following cross: *RrSs* × *RrSs*.

	RS	*Rs*	*rS*	*rs*
RS	*RRSS*	*RRSs*	*RrSS*	*RrSs*
Rs	*RRSs*	*RRss*	*RrSs*	*Rrss*
rS	*RrSS*	*RrSs*	*rrSS*	*rrSs*
rs	*RrSs*	*Rrss*	*rrSs*	*rrss*

According to the Punnett square, only 1 out of every 16 of the offspring will have the *rrss* genotype. Therefore, the probability of the *rrss* genotype is 1/16, or approximately 0.06.

68. 8

Organisms with a diploid number of six have three pairs of chromosomes ($2n = 6$), so each parent has three chromosomes to give to their offspring. When the chromosome pairs separate during the production of the haploid sperm or egg, one of each chromosome can be chosen for each of the three pairs, independently of all the other pairs, assuming no crossing over. Because there are three pairs of chromosomes, the number of possible combinations are $2^3 = 8$.

69. 92

Matter and energy are conserved in ecosystems. When consuming phytoplankton, zooplankton take in 160 g/m^2. The amount of carbon released into the atmosphere through cellular respiration is 60 g/m^2, and 8 g/m^2 are transferred to the small fish that consume zooplankton. Therefore, the amount of carbon that is transferred back to decomposers is $160 - 60 - 8 = 92$ g/m^2.

SECTION II: FREE-RESPONSE QUESTIONS

PART A

1. The first part of this question allows you some creativity in the response as long as the basics of experimental design are followed. The most important thing to include is the development of a testable hypothesis. Include a control, an independent variable, and a dependent variable to demonstrate a familiarity with the experimental process. The most likely response variables to measure in this experiment are salivation and peristalsis, but gastric secretion is also a candidate. If an unsuitable response variable is chosen, a considerable number of points may still be available for good design. It is not necessary for you to be familiar with any particular type of device or that any devices used in the design be potentially viable. It is most important that a particular variable be identified and measured. To get the most points, you will be expected to know the body's reactions to food stimulus as well as how to test a hypothesis. Part (B) should be answered with specific information about body systems. The focus should be integration so that coordination can be included in the response, but the response still should include specifics about each system. The question does not ask for all the responses of three different systems, so the response should not be comprehensive for any one system. There is likely to be one specific response for three different systems.

(A) Key points to include: testable hypothesis based on a specific example of the body's response to food, control, dependent variable, independent variable

Here is a possible response:

As Pavlov showed with his dogs and the trained response to a sound, there is a salivation reaction to food stimulus before food comes in contact with the mouth. When food is ingested, the salivary glands produce saliva that includes the enzyme amylase, which begins the process of digesting starch. The salivary response can also be initiated by the smell or sight of food or simply thinking about food. The hypothesis for this experiment is that only the smell of food will cause an increased volume of saliva to be produced in the mouth. The independent variable will be the smell of food and the dependent variable will be the amount of saliva produced. To measure the saliva, a device could be attached to a subject's mouth that would collect saliva, such as a suction tube used by a dentist. Subjects would be randomly assigned to a control group that would receive no smell stimulus and an experimental group that would receive the smell stimulus. The amount of saliva produced by the control and experimental group could be compared directly or with replicates, using a t-test.

(B) Key points to include: identify three systems, provide specific response for each, homeostasis, nutrient absorption

Here is a possible response:

As the body tries to maintain homeostasis, almost no organ system responds independently. When food is ingested, it causes acute changes in the digestive system, but also changes in the nervous and circulatory systems. The digestive system responds to the presence of nutrients in the lumen in different ways within different organs, but the majority of digestion takes place in the duodenum of the small intestine. In the duodenum, hydrolytic enzymes secreted by the pancreas after the release of chyme through the pyloric sphincter of the stomach do the majority of the work in breaking down macronutrients. The majority of absorption also takes place in the duodenum, as micronutrients are made available to the bloodstream.

The circulatory system closely follows digestive activity, controlled by nervous impulses to the smooth muscle tissue that lines arteries and capillaries. These nervous impulses are coordinated by the hindbrain and take place without conscious thought. The constriction of vessels in the appendages and skin shunts blood flow away from the periphery and into the core near the organs of absorption, such as the duodenum. At the same time that nutrients are broken down in the small intestine, blood flow is increased around the small intestine. This increases the volume and speed of nutrient delivery to the rest of the body.

In addition to the change in blood flow, the action of the digestive system imparts a change in the endocrine system at the pancreas. Glucose absorbed by the small intestine and delivered by the increased blood flow causes an elevated blood glucose volume that is detected by the beta-cells in the pancreas. These cells produce and secrete insulin in response to the glucose in the bloodstream. Insulin is delivered by the same increased blood flow throughout the body, where it is detected by receptor cells in the liver and muscles. The response is for the muscle and liver cells to absorb glucose and convert it to glycogen for storage.

2. This question is unlike many of those previously presented in its specificity, but it still requires an intricate level of synthesis. You are expected to direct your response to a particular system, but are required to pull on knowledge from several different areas to create a comprehensive answer. You should begin with the primary producers, giving specific taxa at every level if possible, and trace the path of oxygen through the ecosystem. A diagram may not be a bad idea here, but don't let the diagram stand alone. While composing this portion, you should be able to pinpoint trophic levels that would be affected greatly by hypoxia. In part (B), you should include aspects of the organisms' ecology outside of the oxygen cycle. If you are using an omnivore for your example, consider factors such as what organisms it eats, where it lives, and how it finds a mate. Part (C) expands the negative impact on the oxygen cycle into the food web since sharks are top-predator carnivores. Remember, sharks still need oxygen to breathe, but they also need nutrients to grow and for metabolism.

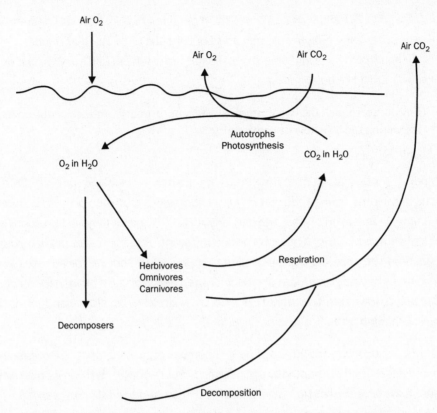

(A) Key points to include: primary producers, hypoxia, dissolved oxygen, photosynthesis, respiration, trophic levels

Here is a possible response:

Except for trace amounts of free oxygen that originate from the air-water interface, free oxygen exists as a result of photosynthesis by autotrophic organisms using energy from the sun to fix carbon from atmospheric CO_2 while releasing H_2O and O_2. All other forms of life needing oxygen use this reservoir. In the marine system, the primary producers are various forms of seaweed algae and pelagic photosynthetic plankton like diatoms. The oxygen these organisms generate is released into water where some escapes to the atmosphere depending on the atmospheric pressure. Dissolved oxygen in the water table is utilized for metabolic respiration by every other member of the ecosystem, such as carnivorous and herbivorous fish, herbivorous echinoderms, and detritivorous crustaceans. Dissolved oxygen is also necessary for certain chemical reactions involving the breakdown of organic tissue during decomposition. The processes of respiration and decomposition recycle the oxygen into CO_2, which enters the water table and the atmosphere where it is again utilized by the autotrophs.

This cycle is upset in the Gulf of Mexico because the nitrogen- and phosphorous-rich fresh water entering the marine system supports a surface-level algal bloom that blocks sunlight from reaching the natural primary producers on the marine floor and in the water table. The biomass of the algae increases beyond the system's ability to produce oxygen so that the rate of O_2 production can't meet the need for respiration and decomposition. As a result, the organisms needing oxygen and sunlight die until the algal bloom disappears. If enough of the organisms from the natural system can survive, the system begins to recover until the next spring.

(B) Key points to include: two example organisms, primary producer, consumers, respiration, energy flow in community

Here is a possible response:

The algal bloom affects the autotrophs most significantly by blocking sunlight. This prevents the organisms from producing oxygen themselves, but also prevents them from producing energy in the form of storage carbohydrates. All organisms in the ecosystem are connected somehow and autotrophs are no different. As organisms in the ecosystem die, the resources they provide disappear. Autotrophs rely on corals for substrate anchoring points. They rely on predators to decrease herbivory. As oxygen levels decrease in the water due to lower photosynthetic rates, slow-growing corals die and predators leave the area, causing an increase in herbivory.

Herbivores, like sea urchins, are affected as the number of grazing spots decreases due to the death and decreased productivity of their algal food source. Herbivores also depend on their algal forest for protection from ocean currents and predators. As oxygen levels decrease in the water, herbivores with limited mobility, like sea urchins, suffer increased death rates, as they can't move to more productive areas.

(C) Key points to include: top predator, trophic levels, oxygen deprivation, decreased food supply

Here is a possible response:

Two factors may be contributing to a relocation of a top predator like a shark. First, the oxygen content of the water may be unsuitable for a shark population. Normal function is not possible when the oxygen supply is limited. Sharks that are mobile can migrate to an area that isn't affected or is less affected by the algal bloom. The second factor is the food supply. As primary productivity decreases, it affects all levels of the ecosystem. Herbivore, omnivore, and primary carnivore species are affected so they either die or migrate to more suitable habitats. All of these trophic levels are part of the shark's nutrient reservoir. A shark would have to increase its hunting range to encounter more food items. A shark might migrate with its prey items if they move to a more suitable habitat. Finally, a shark might need to switch to a different food source if its naturally occurring sources of nutrients disappear. Unfortunately, human bathers may become an alternate food source for larger sharks that are normally not considered man-eaters.

PART B

3. Key points to include: genes, base pairs, replication, proteins, diversity

Here is a possible response:

DNA is composed of only four different base pairs, and there is only one different base pair, uracil instead of thymine, in RNA. This has created a very simple system for duplication. When a cell replicates its DNA during mitosis, it simply unzips the double helix and adds nucleic acids along its length until a duplicate set of complementary strands is created. The base pairs bond with hydrogen bonds, with cytosine bonding to guanine and adenine bonding to thymine. DNA molecules are very long, but because the code only involves four different base pairs, it can easily be duplicated. The DNA strands are actually copied in small segments and linked together to form an entire strand.

Although there are only four different base pairs, the longer the strand gets, the more combinations of base pairs can be created. These long series of base pairs code for longer and more complex proteins that can differ in minor or very significant ways. For example, sickle cell anemia is caused by only one base pair difference in the genetic code for hemoglobin, but this base pair causes an incredible difference in the function of the protein. The high number of potential combinations from only four base pairs and the impact of small changes in the sequence of base pairs contributes to the diversity of proteins in an organism. This diversity of proteins contributes to the diversity of organisms that exist because physiological function is essentially protein driven. In essence, long and diverse code leads to diverse proteins, which leads to diverse organisms.

4. Key points to include: predation, carrying capacity, mutualism

Here is a possible response:

In the year 2000, Species A is effectively out of the picture, Species B is increased, and Species C is in slight decline. It is possible that Species B and C are predators of Species A, causing its virtual extinction in the ecosystem over the last 60 years. It is also possible that Species B is a predator of Species C and the expansion of Species B's population has resulted in a decline in the population in Species C. Alternatively, the slight decline in Species C over the previous 30 years could mean that it has reached its carrying capacity.

5. Key points to include: diabetes, kidney failure, urinalysis, infection

Here is a possible response:

The lab results show that the patient has a normal value for hemoglobin, discounting anemia. All of the other results, however, point to diabetes. High glucose levels are associated with kidney failure (seen in the high blood urea levels), ketoacidosis (low blood pH), and chronic infections (sepsis, elevated white blood cell count). Based on the results of urinalysis, the physician would probably decide to order additional bacterial blood and urine cultures on the patient. High white blood cell and neutrophil counts indicate some type of bacterial infection. These tests will help the doctor identify the causative agent.

6. Key points to include: sex-linked inheritance, recessive, dominant, hemophilia (or other sex-linked recessive disorder)

Here is a possible response:

None of the males in this pedigree are heterozygous because males carry only one X chromosome. The females who are heterozygous do not have the condition because they have two X chromosomes, and the dominant form blocks expression of the defective allele, so it must be a sex-linked recessive disorder. In a sex-linked recessive disorder, about half of the children would carry the allele, but only the males would be affected by the condition. This pedigree could represent the inheritance pattern of hemophilia.

7. Key points to include: feedback inhibition, substrate, metabolic pathway

Here is a possible response:

In competitive inhibition, a molecule that resembles the normal substrate of an enzyme competes for the active site. In noncompetitive inhibition, a molecule binds to a part of the enzyme other than the active site, causing a conformational change in the enzyme. This conformational change alters the effectiveness of the enzyme. In irreversible inhibition, a molecule covalently bonds to the active site of the enzyme, preventing substrate from accessing the active site. In feedback inhibition, the presence of the end product of a metabolic pathway inhibits an enzyme that functions in an early step in the pathway. Therefore, this example is feedback inhibition.

8. Key points to include: dominance, heterozygous, recessive

Here is a possible response:

The allele that causes achondroplasia is dominant, which can be inferred from the fact that both parents are affected and the offspring is not. To produce unaffected offspring, both parents must be heterozygous Aa, and their normal-sized offspring must have the recessive genotype aa.

Practice Test 2 Answer Grid

1. Ⓐ Ⓑ Ⓒ Ⓓ
2. Ⓐ Ⓑ Ⓒ Ⓓ
3. Ⓐ Ⓑ Ⓒ Ⓓ
4. Ⓐ Ⓑ Ⓒ Ⓓ
5. Ⓐ Ⓑ Ⓒ Ⓓ
6. Ⓐ Ⓑ Ⓒ Ⓓ
7. Ⓐ Ⓑ Ⓒ Ⓓ
8. Ⓐ Ⓑ Ⓒ Ⓓ
9. Ⓐ Ⓑ Ⓒ Ⓓ
10. Ⓐ Ⓑ Ⓒ Ⓓ
11. Ⓐ Ⓑ Ⓒ Ⓓ
12. Ⓐ Ⓑ Ⓒ Ⓓ
13. Ⓐ Ⓑ Ⓒ Ⓓ
14. Ⓐ Ⓑ Ⓒ Ⓓ
15. Ⓐ Ⓑ Ⓒ Ⓓ
16. Ⓐ Ⓑ Ⓒ Ⓓ
17. Ⓐ Ⓑ Ⓒ Ⓓ
18. Ⓐ Ⓑ Ⓒ Ⓓ
19. Ⓐ Ⓑ Ⓒ Ⓓ
20. Ⓐ Ⓑ Ⓒ Ⓓ
21. Ⓐ Ⓑ Ⓒ Ⓓ

22. Ⓐ Ⓑ Ⓒ Ⓓ
23. Ⓐ Ⓑ Ⓒ Ⓓ
24. Ⓐ Ⓑ Ⓒ Ⓓ
25. Ⓐ Ⓑ Ⓒ Ⓓ
26. Ⓐ Ⓑ Ⓒ Ⓓ
27. Ⓐ Ⓑ Ⓒ Ⓓ
28. Ⓐ Ⓑ Ⓒ Ⓓ
29. Ⓐ Ⓑ Ⓒ Ⓓ
30. Ⓐ Ⓑ Ⓒ Ⓓ
31. Ⓐ Ⓑ Ⓒ Ⓓ
32. Ⓐ Ⓑ Ⓒ Ⓓ
33. Ⓐ Ⓑ Ⓒ Ⓓ
34. Ⓐ Ⓑ Ⓒ Ⓓ
35. Ⓐ Ⓑ Ⓒ Ⓓ
36. Ⓐ Ⓑ Ⓒ Ⓓ
37. Ⓐ Ⓑ Ⓒ Ⓓ
38. Ⓐ Ⓑ Ⓒ Ⓓ
39. Ⓐ Ⓑ Ⓒ Ⓓ
40. Ⓐ Ⓑ Ⓒ Ⓓ
41. Ⓐ Ⓑ Ⓒ Ⓓ
42. Ⓐ Ⓑ Ⓒ Ⓓ

43. Ⓐ Ⓑ Ⓒ Ⓓ
44. Ⓐ Ⓑ Ⓒ Ⓓ
45. Ⓐ Ⓑ Ⓒ Ⓓ
46. Ⓐ Ⓑ Ⓒ Ⓓ
47. Ⓐ Ⓑ Ⓒ Ⓓ
48. Ⓐ Ⓑ Ⓒ Ⓓ
49. Ⓐ Ⓑ Ⓒ Ⓓ
50. Ⓐ Ⓑ Ⓒ Ⓓ
51. Ⓐ Ⓑ Ⓒ Ⓓ
52. Ⓐ Ⓑ Ⓒ Ⓓ
53. Ⓐ Ⓑ Ⓒ Ⓓ
54. Ⓐ Ⓑ Ⓒ Ⓓ
55. Ⓐ Ⓑ Ⓒ Ⓓ
56. Ⓐ Ⓑ Ⓒ Ⓓ
57. Ⓐ Ⓑ Ⓒ Ⓓ
58. Ⓐ Ⓑ Ⓒ Ⓓ
59. Ⓐ Ⓑ Ⓒ Ⓓ
60. Ⓐ Ⓑ Ⓒ Ⓓ
61. Ⓐ Ⓑ Ⓒ Ⓓ
62. Ⓐ Ⓑ Ⓒ Ⓓ
63. Ⓐ Ⓑ Ⓒ Ⓓ

64.

65.

66.

67.

68.

69.

PRACTICE TEST 2

Section I: Multiple-Choice Questions and Grid-In Items

Time: 90 Minutes

63 Multiple-Choice Questions and 6 Grid-In Items

Directions: Choose the best answer choice for the following questions.

1. Jaws and teeth provide information about how an animal lives and what it eats because

    (A) jaws evolved from gill arches, so the structure of an animal's jaws and teeth show whether it is ancient or more recently derived.

    (B) carnivores' teeth have broad, hard surfaces used for grinding, distinguishing meat eaters from plant eaters.

    (C) diet influences the way an animal's skull and dentition evolve, providing clues to its position in the food chain.

    (D) herbivores have pointed front teeth adapted for tearing that distinguish their skulls and jaws from those of carnivores.

2. Independent assortment of homologous chromosomes, crossing over, and random fertilization generate which of the following?

    (A) Haploid gametes

    (B) Diploid gametes

    (C) Genetic variability in sexually reproducing organisms

    (D) Genetic variability in asexually reproducing organisms

3. All of the following are functions of enzymes EXCEPT

    (A) they lower activation energy.

    (B) they are catalysts.

    (C) they change ΔG for a reaction.

    (D) they are affected by pH.

GO ON TO THE NEXT PAGE

4. Genetically engineered crop plants

   (A) are still very rare.

   (B) are usually more difficult to create than genetically engineered animals.

   (C) may hybridize with wild relatives, allowing novel genes to "escape" from crops.

   (D) require periodic inoculation with recombinant plasmids in order to retain bioengineered genes.

5. Three different lizard species live in sympatry. Each species specializes in feeding on a different size of prey. This would be an example of

   (A) the competitive exclusion principle.

   (B) resource partitioning.

   (C) symbiosis.

   (D) interference competition.

6. All of the following are true statements about free and bound ribosomes EXCEPT

   (A) they are structurally identical.

   (B) they are identified by where they are found.

   (C) they tend to produce proteins destined for different locations.

   (D) subunits of bound ribosomes are made in the nucleolus while subunits of free ribosomes are made in the cytosol.

7. What assumption underlies the methods used to construct a genetic map?

   (A) Recombination frequencies are directly proportional to the distance between genes on a chromosome.

   (B) Recombination frequencies are inversely proportional to the distance between genes on a chromosome.

   (C) Linked genes never cross over.

   (D) Recessive genes are less common than dominant genes.

8. Which of the following reduces weight in birds, making flight more efficient?

   (A) A gizzard instead of stomach and intestines

   (B) Reduced pectoral muscles

   (C) Air sacs instead of lungs

   (D) Bones with a honeycomb internal structure

9. Indigenous rain forest people have used toxins from the brightly colored poison arrow frogs for hunting large animals. These types of tree frogs show

   (A) Müllerian mimicry.

   (B) Batesian mimicry.

   (C) aposematic coloration.

   (D) cryptic coloration.

10. Which of the following is NOT true of plasmodesmata?

    (A) They allow water to pass from cell to cell.

    (B) They allow cytosol to pass from cell to cell.

    (C) They allow cells to be organized into tissues.

    (D) They are found in animal cells.

GO ON TO THE NEXT PAGE ▷

11. What component of a plant's photosystem loses electrons to the primary electron acceptor?

(A) The thylakoid membrane

(B) Antenna pigment molecules

(C) The photon

(D) Chlorophyll *a* at the reaction center

12. Which of the following does NOT occur in meiosis II?

(A) Synapsis

(B) Cytokinesis

(C) Separation of sister chromatids

(D) Formation of the spindle apparatus

13. A plant cell's central vacuole may store all of the following substances EXCEPT

(A) cytosol.

(B) proteins.

(C) pigments.

(D) inorganic ions.

14. Competitive inhibitors of enzymes can be reversed by

(A) increasing the pH above the enzyme's optimal range.

(B) increasing the concentration of substrate.

(C) adding noncompetitive inhibitors.

(D) lowering the temperature below the enzyme's optimal range.

15. If a reaction results in $\Delta G = -625$ kcal/mol, one can conclude that

(A) free energy was absorbed.

(B) the system is closed.

(C) the system is at equilibrium.

(D) the reaction is exergonic.

16. Which of the following best corresponds to the four levels of protein structure, moving from primary to quaternary structure?

(A) Polypeptide folds, aggregation of polypeptide subunits, side-chain bonding, sequence of amino acids

(B) Polypeptide folds, side-chain bonding, sequence of amino acids, aggregation of polypeptide subunits

(C) Sequence of amino acids, side-chain bonding, aggregation of polypeptide subunits, polypeptide folds

(D) Sequence of amino acids, polypeptide folds, side-chain bonding, aggregation of polypeptide subunits

17. Clumps of overlapping mammalian cells would indicate a problem with cell cycle regulation because

(A) normal cells use cyclin-dependent kinases for regulation.

(B) normal cell division is stimulated by active anaphase-promoting complex.

(C) normal cells show density-dependent inhibition and anchorage dependence.

(D) normal cell division is stimulated by the presence of growth factor.

GO ON TO THE NEXT PAGE

18. When ATP is used for transport, mechanical, or chemical work,

   (A) it is the adenosine that performs the work.

   (B) it is the phosphorylates that perform the work.

   (C) it is the precursor, ADP, that performs the work.

   (D) it is the energy released when ATP leaves the mitochondria that performs the work.

19. Cell A and Cell B differ in the following ways: Cell A contains a cell wall. Cell B contains centrioles. Based only on this information, what can be concluded about these cells?

   (A) Cell A is prokaryotic, and Cell B is eukaryotic.

   (B) Cell A is eukaryotic, and Cell B is prokaryotic.

   (C) Cell A is from a plant or prokaryote, and Cell B is from an animal.

   (D) Cell B is from a plant or prokaryote, and Cell A is from an animal.

20. Deep-diving air breathers (e.g., seals, whales, penguins) have numerous adaptations that allow them to dive to great depths and remain underwater for long periods of time. Which of the following would be LEAST likely to be found in a deep diver?

   (A) Decreased O_2 storage in the lungs and increased O_2 storage in the blood

   (B) High concentrations of myoglobin

   (C) A large volume of blood per body mass ratio

   (D) Efficient use of buoyancy to aid locomotion

21. What is the function of the electron transport chain in cellular respiration?

   (A) To break a large free-energy drop into smaller energy-releasing steps

   (B) To make ATP

   (C) To store electrons for use in the Krebs cycle

   (D) To convert pyruvate to acetyl-CoA

22. Most biomass pyramids show a rapid decrease in biomass as trophic level increases. In aquatic systems, however, this pattern may be reversed so that one observes a larger standing crop of consumers compared with producers. What explains this pattern?

   (A) Aquatic producers tend to have larger body sizes than terrestrial producers.

   (B) Water is an easier medium to live in, and aquatic organisms require less food.

   (C) Biomass in aquatic systems cannot be measured accurately.

   (D) Phytoplankton is rapidly consumed, but it has a high turnover rate.

23. Sweat cools the human body because

   (A) water has a high heat of vaporization.

   (B) water has a low heat of vaporization.

   (C) water dissolves salts.

   (D) water adheres to heat receptors.

24. Which property of water allows ice to float?

   (A) Relatively low surface tension

   (B) Relatively high heat of vaporization

   (C) Attraction to hydrophilic materials

   (D) Hydrogen bonding

GO ON TO THE NEXT PAGE

25. Which sequence correctly depicts the order of events in the cell cycle? Assume that these events are continuous (i.e., after the last phase in each list, the cell returns to the first phase listed and again proceeds through the order of events).

    (A) $G_1 \rightarrow$ Cytokinesis $\rightarrow G_2 \rightarrow$ Mitosis $\rightarrow$ S

    (B) $G_1 \rightarrow G_2 \rightarrow$ S $\rightarrow$ Mitosis $\rightarrow$ Cytokinesis

    (C) S $\rightarrow G_2 \rightarrow$ Cytokinesis $\rightarrow$ Mitosis $\rightarrow G_1$

    (D) S $\rightarrow G_2 \rightarrow$ Mitosis $\rightarrow$ Cytokinesis $\rightarrow G_1$

26. The main difference between the lytic and lysogenic cycles of phage reproduction is that

    (A) the lysogenic cycle alters the bacteria while the lytic cycle does not.

    (B) the lytic cycle alters the bacteria while the lysogenic cycle does not.

    (C) the lysogenic cycle kills the host cell while the lytic cycle does not.

    (D) the lytic cycle kills the host cell while the lysogenic cycle does not.

27. During photosynthesis

    (A) light reactions produce sugar, while the Calvin cycle produces O_2.

    (B) light reactions produce NADPH and ATP, while the Calvin cycle produces sugar.

    (C) light reactions photophosphorylate ADP, while the Calvin cycle produces ATP.

    (D) the Calvin cycle produces both sugar and O_2.

28. Primary succession will occur after

    (A) the climax community is established.

    (B) a mudslide covers and kills the existing community of plants.

    (C) *K*-selected species out-compete *r*-selected species.

    (D) the formation of a new volcanic island.

29. Processes that can alter chromosome structure include all EXCEPT which of the following?

    (A) Inversion

    (B) Reduction

    (C) Deletion

    (D) Duplication

30. The oldest fossils observed are prokaryotes, then fish, amphibians, reptiles, and finally mammals and birds (youngest fossils). This pattern

    (A) is an example of how comparative embryology supports Darwinian evolution.

    (B) supports the idea of macroevolution.

    (C) supports the idea of microevolution.

    (D) is an example of how Lamarckian evolution has modified species over time.

31. If the intracellular release of calcium within a fertilized egg cell were blocked or inhibited, what effect would this have on reproduction?

    (A) The fertilization envelope would not be formed.

    (B) The sperm and egg nuclei will not fuse.

    (C) The zygote would not contain 46 chromosomes.

    (D) The acrosomal reaction would be blocked.

GO ON TO THE NEXT PAGE

32. A potential disadvantage of asexual reproduction when compared with sexual reproduction is that

    (A) asexual reproduction requires more resources than sexual reproduction.

    (B) asexual reproduction requires that an organism remain immobile.

    (C) it tends to be easier to produce numerous offspring quickly with asexual reproduction.

    (D) asexual reproduction tends to limit the amount of possible genetic variation.

33. Which sequence best describes what happens when a retrovirus infects a host cell?

    (A) RNA-DNA hybrid → reverse transcriptase → viral proteins → provirus

    (B) Reverse transcriptase → RNA-DNA hybrid → provirus → viral proteins

    (C) Provirus → viral proteins → RNA-DNA hybrid → reverse transcriptase

    (D) Viral proteins → RNA-DNA hybrid → reverse transcriptase → provirus

34. Which of the following does NOT have the potential to regulate gene expression in eukaryotes?

    (A) Alternative splicing during RNA processing

    (B) Repressor activity on operons

    (C) Chromatin structure

    (D) Translation initiation factors

35. Members of a species provide parental care and rear few offspring during a given reproductive cycle. Which type of survivorship curve would be most consistent with these characteristics?

    (A) Type I

    (B) Type II

    (C) Type III

    (D) Cohort specific

36. The effects of a hurricane on an island population of butterflies would be

    (A) an example of both a density-dependent and a density-independent factor because the population cycles over time.

    (B) an example of a density-dependent factor because the storm will be more severe when the population is near carrying capacity.

    (C) an example of a density-dependent factor because populations at higher density will be better able to recover from the storm.

    (D) an example of a density-independent factor because the population size has nothing to do with the occurrence or intensity of the storm.

37. What source of energy does ATP synthase use to generate ATP from ADP and inorganic phosphate in cellular respiration?

    (A) Light energy

    (B) Fermentation

    (C) Chemiosmosis of H^+

    (D) Enzymes

GO ON TO THE NEXT PAGE

38. Cellular respiration produces much more ATP per glucose molecule than fermentation because

    (A) respiration uses glycolysis to oxidize glucose.

    (B) oxygen is necessary to release energy stored in pyruvate.

    (C) fermentation uses NAD^+ as the oxidizing agent in glycolysis.

    (D) respiration produces 2 ATP via glycolysis.

39. Which organelle modifies products from the endoplasmic reticulum (ER) and then transports these products to other parts of the cell?

    (A) The Golgi apparatus

    (B) The lysosomes

    (C) The mitochondria

    (D) The ribosomes

40. When NAD^+ is reduced to NADH by dehydrogenase, what is released into the surrounding solution?

    (A) An electron

    (B) Hydrogenase

    (C) A proton

    (D) Hydrogen

41. In their respective natural habitats, saltwater fish continuously lose water and gain salts, while freshwater fish gain water and lose salts. What causes this to happen?

    (A) Excretion

    (B) Homeostasis

    (C) Active transport

    (D) Osmosis

42. What ion channel(s) is/are required for an action potential to travel along the axon of a neuron?

    (A) Na^+, K^+, Ca^{2+}

    (B) K^+

    (C) Na^+

    (D) Na^+, K^+

43. In an alternation of generations life cycle, the sporophyte refers to the individual that is

    (A) diploid.

    (B) haploid.

    (C) fruiting.

    (D) germinating.

44. The trade-off between water loss and CO_2 uptake has resulted in the evolution of what adaptation in many plants that experience a hot, dry environment?

    (A) C4 and CAM photosynthesis

    (B) C3 and CAM photosynthesis

    (C) C3 and C4 photosynthesis

    (D) Photorespiration that generates ATP

45. Which situation illustrates a reaction norm?

    (A) Adopted and biological children in a single family make the same grammatical errors.

    (B) Skin color depends on multiple inherited genes.

    (C) Soil acidity affects whether hydrangea flowers of the same genotype will be violet-blue or pink.

    (D) Boldness in mice is correlated with tendency to explore new environments.

GO ON TO THE NEXT PAGE

46. If an egg were to be fertilized by multiple sperm cells, one would hypothesize that

    (A) there was an abnormality in the acrosomal or cortical reaction.

    (B) the egg's metabolism was not activated by the increase in Ca^{2+}.

    (C) the nucleus of the first sperm did not fuse with the egg's nucleus rapidly enough.

    (D) the egg or sperm contained aberrant chromosome numbers.

47. How do natural killer (NK) cells contribute to the body's defense?

    (A) NK cells attack infected cell membranes.

    (B) NK cells ingest invading organisms.

    (C) NK cells release histamine.

    (D) NK cells produce T cells.

48. For *E. coli* to utilize lactose as a carbon and energy source, the protein β-galactosidase must be translated. In the presence of both lactose and glucose, *E. coli* will preferentially utilize glucose, conserving the resources necessary to produce β-galactosidase. However, when glucose is absent, lactose will functionally induce the expression of β-galactosidase. This most strongly suggests

    (A) lactose represses expression of the *lac* operon.

    (B) lactose is sufficient to cause an increase in β-galactosidase.

    (C) glucose decreases expression of the *lac* operon in the presence of lactose.

    (D) glucose is necessary for β-galactosidase production.

49. A certain species of bird has beaks of variable length, with an average length of 10 centimeters and a standard deviation of 2 centimeters. It feeds primarily on a species of insect that burrows underground, with most of the insects located 8 to 10 centimeters beneath the surface. If the insect starts burrowing to a greater depth, which of the following population distributions for beak length is most likely to result?

    (A) A single mean at 10 centimeters, with a standard deviation of 4 centimeters

    (B) Two means, one at 10 centimeters and one at 15 centimeters, each with a standard deviation of 2 centimeters

    (C) A single mean at 12 centimeters, with a standard deviation of 2 centimeters

    (D) A single mean at 10 centimeters, with a standard deviation of 1 centimeter

50. An object known as the *Murchison meteorite* struck the Earth in Australia in 1969. Analysis of the rock shows it contains at least 50 amino acids, 19 of which are found in living organisms on Earth, as well as several nucleotides. Assume such amino acid-containing meteorites have existed throughout the history of the Solar System. Of the following statements, it is most reasonable to conclude that one or more meteorites

    (A) are the source of all life on Earth.

    (B) cannot be the origin of the amino acids on Earth.

    (C) are the likely source of the precursors of life.

    (D) contributed some of the precursors of life, but never contained any living organisms.

GO ON TO THE NEXT PAGE

51. The phylogenetic tree above depicts five newly-discovered bacterial species. The branch point separating species 1 and 2 from 3, 4, and 5 is the presence of a gene for enzyme X, required for the metabolism of glucose. The branch point separating species 1 from species 2, and species 3 and 4 from species 5 is the presence of protein Y, a pump that removes a certain antibiotic from cells.

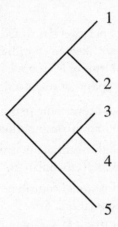

Which of the following findings, if true, would require the most extensive re-drawing of the tree?

(A) Molecular studies indicate that protein Y evolved 250,000 years before enzyme X.

(B) Molecular studies indicate that enzyme X evolved 1,000,000 years before enzyme Y.

(C) Species 3 expresses protein Z, a pump that regulates entry of sodium ions into the cell; none of the other species does.

(D) A newly discovered species 6 expresses protein Y but not enzyme X.

52. The *cecum* is a portion of the large intestine located near the junction of the small intestine and the large intestine. The following table lists the diets of several vertebrates, as well as the average length of the cecum as measured in 20 individuals of that species.

Based on the following information, which of the following conclusions is most plausible?

Species	Average Cecum Length	Diet
A	40.1 cm	herbivore
B	5.7 cm	omnivore
C	6.8 cm	carnivore
D	30.2	ruminant

(A) The cecum can become a vestigial structure in carnivores, since it is shorter in species B and C than in species A and D.

(B) The cecum evolved to have an important role in the digestion of protein, since it is shorter in species B and C than in species A and D.

(C) Species A is more closely related to species D than to species B, because A and D are both herbivores.

(D) Species B and C have a common ancestor, since the cecum in species B and C is approximately the same length.

53. Which of the following can regulate transcription?

(A) Ligase

(B) Histone acetylation

(C) Codons

(D) Okazaki fragments

GO ON TO THE NEXT PAGE

54. The hydrolysis of table sugar (sucrose) to glucose and fructose is exergonic, occurring spontaneously with a release of free energy. Which of the following best explains why a solution of sucrose dissolved in water will sit for years at room temperature with no appreciable hydrolysis?

   (A) The reaction is so rapid that it is undetectable.

   (B) The free energy of the products is greater than the free energy of the reactants.

   (C) The reaction requires a catalyst such as sucrase to proceed.

   (D) The energy required to reach the transition state is prohibitively high.

Questions 55–57 refer to the following image.

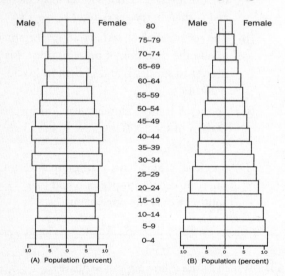

55. Which most accurately describes age-structures A and B?

   (A) A is more likely to have a higher fertility rate

   (B) B is more likely in a period of population growth

   (C) There is a relatively greater proportion of postreproductive individuals in B

   (D) B is more likely to represent a developed country in North America or Western Europe

56. Which social issue would you most expect to see in a country with an age-structure like A?

   (A) High unemployment for working-age people

   (B) Delayed retirement

   (C) Increased child mortality rate

   (D) Decreased work-related injury rate

57. Would a population structured like A or B be more likely to contribute to the global problems of overpopulation and overconsumption of resources?

   (A) A would contribute to overpopulation and B to overconsumption.

   (B) A would contribute to overconsumption and B to overpopulation.

   (C) A and B would contribute equally to overpopulation.

   (D) A and B would contribute equally to overconsumption.

GO ON TO THE NEXT PAGE

Questions 58–60 refer to the following image.

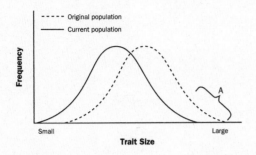

58. The region indicated by the A in this figure depicts

(A) the variability of sizes in the population.

(B) the mean size of the population.

(C) the phenotypes that were successful in the given environmental conditions.

(D) the phenotypes that did poorly in recent environmental conditions.

59. What type of selection does this figure depict?

(A) Disruptive

(B) Directional

(C) Stabilizing

(D) Forceful

60. Starting with the current population distribution, what would you expect to happen if, over the next several generations, the smallest and largest individuals had lower reproductive rates than moderately sized individuals? What would the new population curve look like?

(A) The curve would resemble the original population curve in shape and location.

(B) The curve would be shifted to the left of the current one.

(C) The curve would have two peaks, one at smaller sizes and another at larger sizes.

(D) The curve would be more narrow and higher than the current one; the mean size would remain the same.

Questions 61–63 to the following information.

Sickle cell anemia is caused by mutant hemoglobin DNA, which is more common in humans with African ancestry than in those with European ancestry. The sickle cell allele creates an altered mRNA codon that produces hemoglobin containing valine rather than glutamic acid. If a person inherits both alleles for the sickle cell trait, his hemoglobin will polymerize under low oxygen conditions (i.e., elevated physical activity). This can result in brain damage, paralysis, kidney failure, and other very serious physiological problems.

61. The described mutation is an example of

(A) a base-pair substitution.

(B) a frame-shift mutation.

(C) a silent mutation.

(D) a mutagen.

62. Heterozygotes for the sickle cell trait have increased resistance to malaria. If malaria were eradicated and effective treatment for sickle cell anemia made universally available, what would be the expected effect on the sickle cell trait?

(A) The frequency of the trait would remain constant.

(B) The frequency of the trait would decrease.

(C) The frequency of the trait would increase.

(D) The frequency of homozygous individuals would decrease, and the frequency of heterozygous individuals would remain constant.

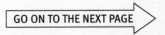

GO ON TO THE NEXT PAGE

63. Genetic mutations have a greater impact on the genetic diversity of bacteria populations than on that of human populations because

(A) human sexual reproduction recombines existing alleles.

(B) bacteria reproduce more rapidly than humans.

(C) new bacteria are generated through sexual reproduction.

(D) genetic mutation is rarer in humans than in bacteria.

GO ON TO THE NEXT PAGE

GRID-IN ITEMS

Directions: The following questions require numeric answers. Calculate the correct answer for the question and then enter the answers on the provided grids. You may use a simple four-function calculator and the included formula sheet.

64. In a particular species of guppy, tails can either be long or short and either feathered or straight. A mating between a short feather-tailed female and a short straight-tailed male produces 30 short straight-tailed guppies, 42 short feather-tailed guppies, 10 long straight-tailed guppies, and 14 long feather-tailed guppies. Calculate the chi-squared value for the null hypothesis that the short feather-tailed guppy was heterozygous for the feather-tailed gene. Give your answer to the nearest hundredth.

65. A pea plant that is heterozygous for alleles that control pea color will express the dominant phenotype and produce yellow peas. If the heterozygous plant is cross-pollinated with one that produces the recessive phenotype (green peas) and 800 offspring are produced, how many would you expect to have green peas?

66. When glucose is broken down into carbon dioxide, water releases about 360 kilocalories of energy, most of which is lost as heat. If 5 kilocalories are conserved per ATP molecule, what is the efficiency of cellular respiration? Give your answer as a percentage.

67. The data table shown displays the population growth for a population of bacteria that reproduces every 12 hours. Calculate the mean rate of population growth (individuals per hour) between 36 and 48 hours. Give your answer to the nearest whole number.

Hours	Number of Individuals
0	20
12	60
24	180
36	540
48	1,620
60	4,860
72	14,580
84	43,740
96	44,299
108	44,800

68. A paleontologist discovers a fossil and wants to determine its age. Using carbon dating, she determines that the fossil has $\frac{1}{8}$ of its original C-14 remaining. If the half-life of C-14 is 5,730 years, how old is the fossil?

69. On your summer break, you conducted a mark and recapture investigation to determine the size of the population of crickets in your yard. On your first attempt, you capture, mark, and release 107 crickets. On your second attempt, you capture 81 crickets, 17 of which are marked. What is your estimate of the total population size?

IF YOU FINISH BEFORE TIME IS CALLED, YOU MAY CHECK YOUR WORK ON THIS SECTION ONLY. DO NOT TURN TO ANY OTHER SECTION IN THE TEST.

10-Minute Reading Period

Take the next 10 minutes to glance over the four questions that comprise Section II of this test. You can take notes in the margins, and begin writing your responses, but it is advisable that you use this time to develop your responses.

When the 10 minute period is over, you will have 80 minutes to complete the eight free-response questions.

Section II: Free-Response Questions

Time: 90 Minutes
2 Long Questions
6 Short Questions

Part A Directions: Answer each part of the following free-response questions with complete sentences. Answers should be in essay form. Make sure to provide a detailed response to each part of each question. Diagrams and other figures may be included to demonstrate knowledge of a topic, but they should not be the only answer you provide.

1. Mammals have a unique ability to heal when they are invaded by a variety of foreign organisms. Humans have an entire immune system dedicated to protecting the body from foreign invaders such as viruses and bacteria. People have also developed artificial means of protecting the body through the use of vaccines and medicine.

   (A) **Describe** how the mammalian immune system responds to (i) a laceration, such as an injury from falling on a barbed-wire fence, and (ii) a viral infection.

   (B) **Explain** how the flu vaccination works and why it has a fairly low success rate.

2. Cellular respiration is a process that occurs in many organisms. The rate of cellular respiration can be measured using several different methods.

   (A) **Design** a specific experiment to measure the rate of cellular respiration in an organism of your choice. Identify, isolate, and test one experimental variable in your experiment. Identify any variables that you must control, and explain the apparatus that you will use to measure the cellular respiration.

   (B) **Construct** a graph to report your results, and graph the results that you would expect in your experiment.

   (C) **Explain** the significance of your results.

GO ON TO THE NEXT PAGE ⟩

Part B Directions: Answer each of the following short free-response questions with complete sentences. Answers should be in paragraph form. Diagrams and other figures may be included to demonstrate knowledge of a topic, but they should not be the only answer you provide.

3. Define trophic levels. Provide a specific example of four species and describe their trophic-level relationships.

4. **Compare** and **contrast** allopatric and sympatric speciation.

5. Sickle cell anemia is caused by mutant hemoglobin DNA, which is more common in humans with African ancestry than in those with European ancestry. The sickle cell allele creates an altered mRNA codon that produces hemoglobin containing valine rather than glutamic acid. If a person inherits both alleles for the sickle cell trait, his hemoglobin will polymerize under low oxygen conditions (i.e., elevated physical activity). This can result in brain damage, paralysis, kidney failure, and other very serious physiological problems.

   Describe the type of mutation that causes sickle cell anemia.

6. Why do larger organisms tend to have MORE cells (and not LARGER) cells than smaller organisms?

7. Define and describe an example of competitive inhibition of enzymatic activity.

8. Describe the endosymbiotic theory. Give evidence for this theory.

PRACTICE TEST 2 ANSWER KEY

1.	C	25.	D	49.	C
2.	C	26.	D	50.	D
3.	C	27.	B	51.	A
4.	C	28.	D	52.	A
5.	B	29.	B	53.	B
6.	D	30.	B	54.	D
7.	A	31.	A	55.	B
8.	D	32.	D	56.	B
9.	C	33.	B	57.	B
10.	D	34.	B	58.	D
11.	D	35.	A	59.	B
12.	A	36.	D	60.	D
13.	A	37.	C	61.	A
14.	B	38.	B	62.	A
15.	D	39.	A	63.	B
16.	D	40.	C	64.	2.66
17.	C	41.	D	65.	400
18.	B	42.	D	66.	50% or 0.50
19.	C	43.	A	67.	90
20.	D	44.	A	68.	17,190
21.	A	45.	C	69.	510
22.	D	46.	A		
23.	A	47.	A		
24.	D	48.	C		

ANSWERS AND EXPLANATIONS

SECTION I: MULTIPLE-CHOICE QUESTIONS AND GRID-IN ITEMS

1. C
Although (A) states something true, this fact tells us little if anything about how an animal lives or what it eats. Choice (B) is incorrect because broad, hard surfaces used for grinding are characteristic of herbivores' teeth; similarly, (D) describes carnivores' teeth, not herbivores' teeth.

2. C
The first two processes described in the question occur during meiosis I, while fertilization is the joining of two haploid gametes. None of these processes create gametes, which are haploid, not diploid. Asexually reproducing organisms, (D), do not go through meiosis.

3. C
A protein that changes the rate of a reaction is an enzyme or catalyst, (B). This is done by lowering activation energy, (A), not by changing the free-energy change ΔG between reactants and products, (C). Enzymes are affected by pH, (D).

4. C
One concern about genetically engineered crop plants is the possibility that the manipulated genes in crops could be transferred to other plants through hybridization. This could give the other plants benefits (such as pest resistance or drought tolerance) that could make them pestilential or invasive.

5. B
This question requires knowledge of community ecology terminology and concepts. Competitive exclusion, (A), refers to the extinction of one species because another has won the competition for an area's limited resources. Symbiosis, (C), is a close mutualistic relationship between two species. In interference competition, (D), a species prevents another from gaining access to a portion of habitat.

6. D
Subunits of both free and bound ribosomes are made in the nucleolus. Free ribosomes are found in the cytosol and tend to produce proteins that will be used in the cytosol, (B) and (C). Bound ribosomes are bound to the endoplasmic reticulum, or nuclear envelope, and tend to produce proteins for membranes, used in other organelles or for secretion. There is no difference in the structure of the two types, (A); a ribosome may serve as either free or bound, depending on the cell's metabolic needs.

7. A
Linkage maps are based on recombination frequencies, which increase the farther apart genes are on a chromosome.

8. D
A honeycombed internal structure makes the bones light but strong. Much energy is needed to fly—birds have air sacs that work with the lungs for efficient gas exchange, (C), and increased pectoral muscles, (B), to power the wings. Being toothless, which decreases a bird's weight, is compensated for by having a gizzard that grinds the food before it enters the stomach, (A).

9. C
Aposematic coloration is used to warn predators that an organism is poisonous or distasteful. In Müllerian mimicry, (A), multiple species with noxious predator-defense mechanisms have evolved to resemble each other. In Batesian mimicry, (B), a species without noxious predator-defense mechanisms resembles a noxious species. Cryptic coloration, (D), means that an organism blends into its environment, making it difficult to detect.

10. D
Plasmodesmata are not found in animal cells. They do all of the things listed in (A), (B), and (C), and are an important structural feature that unifies plant cells into an entire plant organism.

11. D

Although other pigment molecules are excited by photons, only the chlorophyll molecule at the reaction center donates an electron.

12. A

Synapsis occurs during prophase I in meiosis I.

13. A

The membrane between the vacuole and the cytosol, the tonoplast, serves to keep cytosol outside of the vacuole. The vacuole is filled with cell sap.

14. B

Increasing the concentration of substrate will increase the likelihood that substrate molecules, rather than inhibitor molecules, will bind to the enzyme's active site. All of the other choices would reduce the effectiveness of an enzyme, which is what a competitive inhibitor does.

15. D

Exergonic reactions are spontaneous and release free energy. At equilibrium, $\Delta G = 0$. ΔG does not reveal whether the system is open or closed.

16. D

Proteins can be organized from smaller (amino acids) to larger structural components (grouping of subunits). Proteins consisting of only one polypeptide chain will not have a quaternary structural level.

17. C

Cells normally show density-dependence inhibition, meaning that when cells become crowded (e.g., more than a single layer in a culture dish), they stop dividing. Normal cells will also stop dividing if they are not attached to substrate, either the culture dish or extracellular tissue matrix (anchorage dependence). The other answers are true statements but do not address the reason why clumping or overlapping cells are abnormal.

18. B

Enzymes move a phosphate group from ATP to another molecule. This phosphorylated molecule has the capacity to perform work, during which it releases inorganic phosphate. When ATP loses its phosphate group, it is converted to ADP. ADP can be phosphorylated during cellular respiration, which regenerates the cell's supply of ATP.

19. C

Prokaryotic cells and plant cells (eukaryotic) contain cell walls. Only animal cells contain centrioles.

20. D

Deep-diving animals are least likely to be buoyant, (D), or tending to float, since they depend on being able to stay underwater. Myoglobin, (B), stores oxygen in the muscle. Deep divers tend to have high concentrations of this protein, which allows much greater storage of O_2 in their muscles than commonly seen in non-deep-diving animals. Choices (A) and (C) are also modifications that are found in deep divers.

21. A

The electron transport chain releases energy in steps, which allows the cell to take advantage of the energy more efficiently.

22. D

Zooplankton (consumers) rapidly deplete the phytoplankton (producers) biomass, which allows the zooplankton to have a larger standing crop. This is sustainable because phytoplankton reproduce very quickly and can replenish consumed biomass. However, because the zooplankton are continually grazing down the phytoplankton, the phytoplankton standing crop population size remains small. Choice (B) is unlikely because it makes a two-part blanket statement (about water and about caloric needs), both parts of which would have to be true for this example. Although there may be a measurement error, (C), involved, it would have to be huge to create a pattern as extreme as an inverted biomass pyramid.

23. A

Liquids need to absorb heat to escape the liquid phase and be converted into a gas (heat of vaporization). Water's high heat of vaporization results in evaporative cooling because the liquid that is converted into gas when sweat evaporates into the environment absorbs and removes heat from the remaining molecules.

24. D

As water cools to $0°C$, water molecules bond to a maximum of four neighboring water molecules and become trapped in a lattice as their motion slows to form solid ice. The hydrogen bonds prevent the molecules from getting as close to each other in solid form as they do in liquid form, which makes ice less dense than water and, thus, able to float.

25. D

G_1, S, and G_2 are subphases of interphase. Mitosis (duplication and separation of chromosomes) and cytokinesis (division of the cell into two cells) make up the mitotic phase (M phase). The phases proceed in the order listed in (D).

26. D

Viruses that infect bacteria (bacteriophages) can insert their DNA into the bacterial chromosome (lysogenic: doesn't kill the host cell; phage DNA is passed along when the cell replicates) or can immediately start producing phage progeny, which causes the bacterial cell to lyse and release the new phages (lytic).

27. B

Light reactions, which convert solar energy into chemical energy (ATP, NADPH) in the thylakoid membranes, fuel the Calvin cycle's production of sugars in the stroma.

28. D

Primary succession occurs on newly created substrate, like new volcanic islands or land scoured by a retreating glacier, which lack soil (and living organisms). Secondary succession, (B), occurs when an existing community is exterminated in a disturbance, but soil remains. The climax community, (A), is the final, stable community in the succession process. Choice (C) is incorrect because *r*-selected species tend to outcompete *K*-selected species when habitats are new and resources are abundant.

29. B

Choices (A), (C), and (D) are all terms for types of change in chromosome structure.

30. B

Macroevolution refers to evolution on a large scale, such as the emergence of new species. Darwin's view is based on the idea that natural selection modifies populations through differential survival and reproduction of individuals with traits better suited to the environment, and this mechanism of natural selection caused common ancestral species to be modified over time into the diversity of species on Earth today. Prior to Darwin, Lamarck propounded an incorrect view of evolution based on inheritance of acquired characteristics, (D).

31. A

When an egg is fertilized, an intracellular release of calcium is triggered. This "calcium signal" causes the fertilization envelope to form. If the calcium signal is inhibited or stopped, the fertilization envelope will not form.

32. D

Because sexual reproduction joins two haploid gametes, it produces unique combinations of parental genetic material in the offspring. Offspring resulting from asexual reproduction will be genetically identical (except for mutations) to the parent that produced them. Choices (A) and (B) are not true. Choice (C) is an advantage of asexual reproduction.

33. B

Retroviruses reverse the order of information flow from RNA to DNA, using reverse transcriptase to synthesize DNA from viral RNA. This DNA is incorporated in the host cell's DNA as a provirus, which is then transcribed into viral proteins.

34. B

Only prokaryotes have operons, which consist of clusters of genes with related functions. In eukaryotes, genes with related functions are often scattered throughout the genome, and each gene is individually transcribed.

35. A

Survivorship curves, which plot number of survivors on the *y*-axis against age of individuals, have been generalized to take on three main shapes (Type I, II, or III). The Type I curve depicts high survivorship until late in life (e.g., humans, elephants, and other long-lived organisms usually having parental care). The Type II curve depicts a constant decrease in survivorship throughout the life span (some birds, small mammals, and reptiles). The Type III curve represents species where there is high mortality (low survivorship) early in life until some critical age is reached (typically organisms that produce millions of offspring, of which few survive to adulthood; e.g., many fish, annual plants, and insects).

36. D

Density-dependent factors are those that have a greater effect on population size, as well as on an individual, as population density increases (e.g., limited food resources or increased pathogen spread). Density-independent factors occur at intensities and frequencies that are unrelated to population density. Climatic events do not become more or less severe or change in frequency based on population density.

37. C

Oxidative phosphorylation of ADP in ATP synthase is powered by the exergonic flow of hydrogen ions as they move down the H^+ gradient across the inner mitochondrial membrane. This gradient is also known as proton-motive force. Fermentation, (B), refers to another type of cellular respiration.

38. B

Oxidation of pyruvate, which produces a lot of ATP, occurs in respiration but not fermentation. The statements in (A), (C), and (D) are about glycolysis, which occurs in both respiration and fermentation.

39. A

The Golgi apparatus stores, modifies, and transports products made by the ER in eukaryotic cells. Lysosomes, (B), digest macromolecules. Mitochondria, (C), are centers of energy production for the cell. Ribosomes, (D), are the sites of protein synthesis.

40. C

NAD^+ serves as an electron shuttle when it is reduced to NADH. The reaction consists of the removal of two hydrogen ions (H^- and H^+) from sugar or some other fuel by the enzyme dehydrogenase. H^- is accepted by NAD^+ and the proton is released into the environment. This redox reaction (NADH can be oxidized to NAD^+) is critical for cellular respiration.

41. D

Osmosis is the diffusion of a liquid through a permeable membrane from an area of low solvent concentration to an area of high solvent concentration. It is a vital process for both saltwater and freshwater fish, as their skin is a permeable membrane. Freshwater fish have blood that is saltier than the water they live in, so osmosis forces water into their bodies continuously. The opposite is true for saltwater fish.

42. D

Sodium and potassium movement across the axon membrane polarizes and depolarizes the membrane, which initiates and moves action potentials along the axon.

43. A

Alternation of generations refers specifically to the situation where generations alternate between multicellular diploid individuals (sporophytes) and multicellular haploid individuals (gametophytes).

44. A

Plants close their stomata in response to hot, dry conditions, which conserves water. Closed stomata, however, prevent CO_2 from entering the leaf, which is necessary for photosynthesis. Desert and other plants that encounter hot, dry conditions regularly have evolved photosynthetic adaptations that fix carbon in organic acids for later use in the Calvin cycle. This allows photosynthesis to occur with the stomata closed or partially closed during hot, dry conditions.

45. C

Norms of reaction are the phenotypic effects of different environmental conditions on the same genotype. For example, an environment with predators will cause some *Daphnia* species to grow a protective spike on their heads. Individuals with the same genetic makeup do not develop this predator protection (phenotypic trait) if predators are absent during development (which presents a different environmental condition).

46. A

The acrosomal reaction allows a sperm to penetrate the egg's jelly coat and fuse to the egg's plasma membrane. This depolarizes the membrane and prevents other sperm from fusing to the membrane. By the time the membrane voltage has returned to normal, the cortical reaction functions to block additional sperm (an increase in Ca^{2+} ions triggers the development of the fertilization envelope, which resists sperm).

47. A

NK cells lyse infected cells.

48. C

Process of elimination is probably best here. (A) can be ruled out because when lactose is present (without glucose) then the operon is expressed. (B) can be ruled out, because lactose (in the presence of glucose) does not cause an increase in β-galactosidase. (C) seems to match the scenario. (D) is most nearly an opposite.

49. C

The question stem gives us the original distribution of bird beak length, and asks us what would happen if the insect on which the bird feeds begins burrowing deeper into the soil. This suggests that there would be selection *for* birds with longer beaks and *against* birds with shorter beaks. Over time, we would expect the mean beak length to increase. Only (C) does so. (A) would require an increase in the proportion of short-beak birds, which would be unlikely. (B) is an example of *disruptive* selection which selects against individuals near the median. This would be unlikely, since it would require selection against some birds with longer beaks. (D) would result from selection against *both* birds with short beaks *and* birds with long beaks.

50. D

Current hypotheses state that life on Earth was preceded by the generation of the precursors of life, such as amino acids and nucleotides. That generation thus took place under *abiotic* conditions. Of the stated choices, the most reasonable conclusion is (D): some of the amino acids from earlier meteorites might have served as precursors of life. (A) implies that life on Earth originated on meteorites. This is a much stronger conclusion than we can infer from the limited data in the question stem. (B) might seem tempting—after all, there are *20* amino acids in proteins, so the Murchison meteorite must be missing at least one of the amino acids found in proteins. That said, it is not reasonable to conclude that if meteorites didn't provide *all* the necessary amino acids, then they didn't provide *any* of the amino acids. The statements here are insufficient to conclude that meteorites are the *likely* source of precursors of life; they simply suggest that they are *plausible* candidates, so (C) can be eliminated.

51. A

A phylogenetic tree shows the evolution of organisms. Here, the *x*-axis of such a diagram represents time. Thus, to argue that the diagram shown here is correct, we must assume that enzyme X evolved *before* protein Y. If the reverse were true, as (A)suggests, we would need to redraw the entire tree, because the branch points would be in the wrong order. (B) would support the tree as drawn. (C) is a plausible result, since Species 3 is shown as having separated from Species 4 *after* the evolution of protein Y. (D) can be ruled out, because information about a new species would not necessarily require redrawing the whole tree (especially since there is likely to be a place for it on the existing tree).

52.   A
The data here suggest that the cecum is four to five times longer in species that consume plant matter for their diets (A and D) compared to those that consume at least some meat (B and C). Of the statements listed, the only one that logically follows is (A): in carnivores, the cecum becomes unnecessary, and starts to shorten. (B) is the opposite of what we expect; it would be logical if it played a prominent role in digesting *carbohydrates*. (C) The data on cecum length and diet are insufficient to draw conclusions about the relatedness of species, so rule out (C) and (D).

53.   B
Of the answer choices, only histone acetylation has been shown to regulate transcription.

54.   D
The reaction is spontaneous, meaning the reaction will proceed without the input of additional energy. However, this thermodynamic information does not provide any details about the kinetics of the reaction. The best explanation is that the activation energy is prohibitively high, which matches (D). (A), (B) and (C) contradict information in the question stem.

55.   B
The answer is (B) because of the higher proportion of young people. Fertility rate refers to the number of live births per female. A smaller percentage of the population is made up of children (bars below age 15) in structure A compared with B. Furthermore, a structure like A, where the population is fairly evenly distributed among all the age classes and where there is a large proportion of postreproductive

individuals, is common for North America and Western Europe. In other regions of the world, there is often a much smaller proportion of the population that survives to older age classes, and the majority of the population consists of children and adults of reproductive age.

56.   B
This answer is the most logical given the age distribution of structure A, with a larger proportion of older individuals who tend to live longer. Choices (A) and (C) would be most consistent with age-structure B. Choice (D) does not follow logically from the question (why is this a social issue?) or from information available from the diagrams (how do you relate injury rate to age distribution without making a lot of assumptions?).

57.   B
Typically, industrialized nations, which usually have age-structures similar to A, have fewer children but spend much more per person on lifestyle expenses, which contributes greatly to the global problem of resource depletion. In contrast, many countries with age-structures similar to B have very rapidly growing populations, which increases the global population. Although individuals in these countries typically spend far less than the average person in an A-type population, the sheer number of individuals requiring even a small share of resources is a global issue.

58.   D
Individuals with the larger sizes of this hypothetical trait (right side of graph) have been selected against, which caused the average size of the trait to decrease (the hump of the new population is at a lower size than the hump of the original population).

59.   B
This graph depicts directional selection, which moves the mean trait value of a population toward more extreme values (smaller or larger, in this case). Disruptive (or diversifying) selection, (A), favors extreme phenotypes so that intermediate phenotypes become less common and extreme ones are more common. Stabilizing selection, (C), favors

intermediate phenotypes, which become more common (larger and smaller phenotypes become less common). Forceful selection, (D), is a made-up term.

60. D

This situation describes stabilizing selection, which favors intermediate phenotypes over extremes. Choices (A) and (B) depict directional selection, while (C) describes disruptive (diversifying) selection.

61. A

The sickle cell anemia allele has adenine substituted for thymine, which codes for valine rather than the normal glutamic acid in the hemoglobin protein. Clues that this scenario describes a base-pair substitution include: it involves a single codon and the protein is still produced, but with a different amino acid. Usually frame-shift mutations will have a much greater effect on the resulting protein. Silent mutations have no effect on the protein because of redundancy in the genetic code. Mutagens are agents that cause mutations.

62. A

If malaria were eliminated, selection for the sickle cell allele would stop. Individuals with the allele would no longer have a greater chance of survival than those without it. At the same time in this scenario, homozygous individuals would receive treatment that would ensure their survival through reproductive age and beyond. As it would no longer be selected for or against, the sickle cell allele's frequency would remain constant.

63. B

Bacteria, which have short generation times, reproduce asexually and very rapidly. This allows rare individual genetic mutations that create new alleles to have a large effect on the overall genetic diversity observed in a population. Much of the genetic variation among humans is the result of sexual recombination of existing alleles.

64. 2.66

The expected progeny would have phenotypes according to the following frequencies: three short straight-tailed, three short feather-tailed, one long straight-tailed, and one long feather-tailed. There are a total of 96 progeny as a result of this cross, so the null hypothesis would be that we would expect to see 36 short straight-tailed, 36 short feather-tailed, 12 long straight-tailed, and 12 long feather-tailed. We construct the chi-squared analysis as follows:

$$\chi^2 = \frac{(30-36)^2}{36} + \frac{(42-36)^2}{36} + \frac{(10-12)^2}{12} + \frac{(14-12)^2}{12} = 2.66$$

65. 400

The Punnett square for the described cross would show Yy crossed with yy. The offspring produced would be 50 percent Yy and 50 percent yy, so half of 800, or 400, would show the recessive phenotype.

66. 50% or .50

The missing number in this calculation is 36, the number of ATP molecules produced by cellular respiration. This means 36 × 5 kilocalories are conserved. Divide 180 by 360 to get 50 percent.

67. 90

From 36 to 48 hours, the population of bacteria grew from 540 to 1,620 individuals. The mean growth rate is the change in population over the change in time.

$$\frac{1,620 - 540 \text{ bacteria}}{48 - 36 \text{ hours}} = \frac{1,080 \text{ bacteria}}{12 \text{ hours}}$$
$$= 90 \text{ bacteria / hour}$$

68. 17,190

The fossil has undergone three half-lives. Because each half-life is 5,730 years long, the fossil is 5,730 × 3 = 17,190 years old.

69. 510

You can use the Lincoln-Peterson method to estimate the population size of crickets.

$$N = \frac{MC}{R}$$

where N is the total population size, M is the number of crickets captured and marked on your first attempt, C is the number of crickets captured on your second attempt, and R is the number of crickets that were recaptured on your second attempt.

$$N = \frac{(107)(81)}{17}$$
$$N = 510 \text{ crickets}$$

SECTION II: FREE-RESPONSE QUESTIONS

PART A

1. Questions about the immune system frequently appear on the exam, so it's a good idea to become familiar with its components and the ways they interact. The methodology for attacking each part of the question is the same. Read and reread each part so you know what the question is asking. Be aware that the two sections of Part (A) will draw on similar information about immune responses, so it shouldn't be too difficult to score points here. Part (B) asks for a biological explanation of a familiar phenomenon: the ability of pathogens to evolve into new forms and the ways the immune system adapts itself to counter them.

   (A) (i) Key points to include: roles of phagocytic blood cells, antimicrobial proteins, the inflammatory response, chemokines, clotting, natural killer cells, histamine

Here is a possible response:

The first level of the human immune system is the skin (the innate immune system), which provides a protective barrier around the body. When this protective barrier is broken, the body responds with an inflammatory response, releasing the chemical histamine. In an inflammatory response, the body sends specialized phagocytic blood cells and antimicrobial proteins to protect against foreign invading organisms. Damage to skin cells results in the release of chemicals from these cells, which signals blood vessels to dilate. This dilation causes a rush of blood to the area, which causes swelling and recognition of foreign invading bodies through the chemical complement, antibodies, and cytokines.

The body's immune response helps to protect the rest of the body from infection. When a laceration is made in the skin, white blood cells called neutrophils surround and engulf any foreign organisms in a process called phagocytosis. The neutrophils then attack the engulfed organisms with a variety of chemicals and enzymes, breaking them down into pieces to be excreted and processed by cells called lymphocytes. B lymphocytes produce antibodies to invading organisms, which stick to the invading organisms and prevent them from sticking to the host cell or releasing chemicals that can damage the host cell. The B lymphocytes mark the invading microbes with antimicrobial proteins, so that if they continue to attack or attack again, the body will know how to fight them off. B lymphocytes pass the information about the invading organisms on to T cells. The T cells activate macrophages, which digest pieces of the antigen. B lymphocytes and T cells are natural killer cells, which attack any foreign body that enters a person. That is why it is important for an organ donor to have the same blood type as the organ recipient—the recipient's body will attack the new organ if the blood cells are not the same type. The entire immune response causes clotting at the site of the laceration, as white blood cells and lymphocytes die and blood coagulates on the surface of the skin, preventing organisms from further entering the bloodstream.

(ii) Key points to include: lymphocytes, specific immunity, antigens, B cells, T cells, antibodies, memory cells, primary and secondary immune response

Here is a possible response:

When a human body gets a virus, it is being invaded by foreign organisms, as it is in a laceration. The difference is that the cells of the skin layer have not been damaged and do not release histamine in one particular area of the body. A virus is not specific—it will attack all host cells within the body. Each virus is recognized by a certain protein. If an organism has been invaded by the virus before, it will "remember" how to fight off the virus because the specific protein has been encoded in the body's memory cells. The body then has specific immunity toward the virus that has already invaded.

When a virus first invades a body, lymphocytes recognize the virus as a foreign invader and engulf the virus so that it cannot invade any other cells. The B cells, which are a type of lymphocyte, produce antibodies to the virus, which is tagged with antimicrobial proteins. These tagged viruses are then passed on to the T cells, which hold information on how to fight off invading viruses. The primary immune response is when lymphocytes and white blood cells rush to engulf the invading virus so that it is cut off from harming other cells in the rest of the body. The secondary immune response is the production of antibodies, which allow the B cells and T cells to continue fighting off the same invading organism.

(B) Key points to include: active immunity, secondary immune response, the problem of predicting evolution of a quickly reproducing virus, the "arms race" between host and pathogen, different strains

Here is a possible response:

The flu vaccine works by stimulating the secondary immune response of cells in the body: production of antibodies against invading organisms. This vaccine contains parts of commonly known strains of inactivated flu virus. The parts of these strains float around the body until they come in contact with proteins in the body that "match" the chemical makeup of the inactive viruses. When the viruses and proteins "match," the proteins activate the production of cells and antibodies that remove the foreign organism (the inactivated virus strains) from the body. When the body has the antibodies and antimicrobial proteins necessary to fight off a given virus, the body has active immunity against that virus. The body already knows how to fight off that particular invading organism.

Flu vaccines can protect against certain strains of the flu virus that are known. The problem with the vaccine is that the flu virus can rapidly multiply into different strains, which are resistant to the vaccine. The vaccine is only effective two or more weeks after a person receives the vaccine. That amount of time gives a person's body the chance to develop immunity to the strains of flu virus found in the vaccine. There is no way for the manufacturer of a vaccine to predict how the flu virus will mutate, so by the time the vaccine is made, it may be ineffective against a variety of new strains of flu or may only partially work against them.

An "arms race" is thereby constantly occurring between host and pathogen. The host is constantly trying to fight off pathogens through the development of active immunity, and the pathogen (virus) is constantly trying to evolve so that it is effectively able to invade and attack the host organism. In people with compromised immune systems, such as people with cancer and AIDS, there are not as many B lymphocytes and T cells to fight off invading organisms. That is why people with those diseases often develop an overgrowth of thrush or the *Candida* organism, which naturally occurs at low levels in the human body (kept at bay by the body's symbiotic defense system), or pass away from a secondary infection such as pneumonia.

2. You should immediately realize that this is a question about cellular respiration. You should draw on your knowledge of this topic to answer the question. Use the organism that you used when completing this lab (commonly germinating peas).

(A) Key points to include: the organism you chose, clear identification of the experimental variable, clear explanation for the variables that you must control, an explanation of the apparatus that you will use to measure the cellular respiration

Here is a possible response:

Cellular respiration can most easily be measured by consumption of O_2 or by the production of CO_2. In this experiment, the volume of O_2 consumed by germinating peas will be measured. The experiment will test peas that have been germinating one day vs. peas that have been germinating for three days. Therefore, the number of days the peas have been germinating will be my experimental variable. I will use an apparatus called a respirometer to measure the amount of O_2 consumed. This device will be submerged underwater with a pipette attached to the end. I will be able to measure the amount of water drawn into the pipette by comparing where the water mark begins and where it ends. I will eliminate the production of CO_2 as a variable by using potassium hydroxide (KOH) to fix CO_2 into a solid form: potassium carbonate (K_2CO_3). KOH will be added to an absorbent cotton ball and placed on the bottom of the respirometer with a non-absorbent cotton ball in between, so the KOH will not interfere with the experiment. Because volume must be controlled, I will use glass beads to control the volume differences between the two germinating pea samples.

I will place each respirometer in the same tub of water to control temperature between the two germinating pea samples.

Hypothesis: Measuring cellular respiration for 30 minutes at intervals of 10 minutes at a time will demonstrate that peas that have been germinating for three days will consume more oxygen through cellular respiration than peas that have been germinating one day.

Procedure: I will place 20 peas that have been germinating for one day in one respirometer and 20 peas that have been germinating for three days in another respirometer. After a 10-minute equilibration period, I will begin to measure the amount of oxygen consumed at 10-minute intervals for 30 minutes. I will record results measured by the graduated intervals on the pipette attached to the respirometer.

(B) Key points to include: labels on each axis, regular intervals on the graph, a specific title on the graph, points plotted on the graph, and a line connecting the appropriate points. If there is more than one plot on the graph, you should use a dotted line for one line and a solid line for the other. Alternatively, you can simply write a short phrase above each line for identification.

Here is a sample graph:

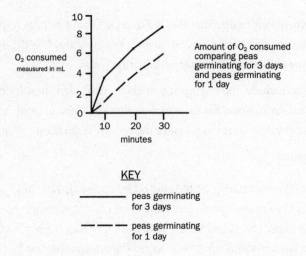

(C) Key points to include: clear explanation of the graph and clear explanation of the significance of the results

Here is a possible response:

The graph shows that peas that have been germinating for three days consume more oxygen during a 30-minute period than peas that have been germinating for one day. The peas that have only been germinating for one day are not as well developed. Therefore, these peas are not undergoing as much cellular respiration as the more developed peas that have been germinating for three days.

PART B

3. Key points to include: define trophic levels; sample community should contain a primary
   producer, primary consumer, secondary consumer, tertiary consumer, and/or decomposer

Here is a possible response:

Trophic levels are the levels of a food chain or hierarchy in which each organism plays a
particular role based on the foods it eats. Each organism fits into a specific level of the
food chain. The first level of the food chain includes primary producers. Primary producers
are autotrophic organisms (plants) that form organic compounds from inorganic
compounds. Tiny blue-green algae, called cyanobacteria, are photosynthetic producers.

Organisms at the second level of the food chain, primary consumers, feed on the plants
in the first level. They are herbivores. Certain fish, for example, feed on algae and lakeweed.
These fish are eaten by a secondary consumer, such as a bear. A secondary consumer
feeds on a primary consumer, but they are not always solely carnivorous. Some secondary
consumers feed on primary producers (such as bears eating berries), but their main source
of food is primary consumers.

These secondary consumers are then fed upon by another predator, the tertiary
consumer. The tertiary consumer occupies the end of the food chain. Humans are tertiary
consumers. When a person hunts and eats a bear, he or she is the tertiary consumer of
the bear. When a human being dies, the body is decomposed in the soil by microbes, and the
nutrients left from the decomposed body are absorbed by organisms such as fungi and
bacteria.

4. Key points to include: allopatric speciation—role of geographic barriers, more common
   than sympatric; sympatric speciation—reproductive isolation without geographic barriers

Here is a possible response:

Allopatric speciation occurs due to geographic isolation. It is more common than
sympatric speciation, because species often develop when a few individuals move to
a different area, whether by choice or force (for example, a storm or food shortage).
Sympatric speciation does not occur due to geographic isolation, since groups of similar
organisms can coexist in the same geographic or overlapping geographic area without
interbreeding. Species diverge in sympatric speciation due to behavioral, temporal, and
resource-based mechanisms of reproductive isolation.

5. Key points to include: substitution, amino acids

Here is a possible response:

Sickle cell anemia has adenine substituted for thymine, which codes for valine rather than
the normal glutamic acid in the hemoglobin protein. Clues that this scenario describes
a base-pair substitution include that it involves a single codon and the protein is still
produced, but with a different amino acid. Frameshift mutation would have a much greater

effect on the resulting protein because it would change all of the amino acids that follow the mutation. Silent mutations have no effect on the protein.

6. Key points to include: surface-to-volume ratio, plasma membrane

Here is a possible response:

Chemical exchange (oxygen, nutrients, and waste) across the cell's plasma membrane can occur only so quickly. Larger things have smaller surface-to-volume ratios (in this case, surface = plasma membrane and volume = cell interior)—i.e., there is less plasma membrane available for each unit of cell interior the larger a cell gets. Because rates of chemical exchange are limited across the membrane surface, there is a limit to cell size. Cells that are too large will not be able to get enough oxygen, and so on, to survive.

7. Key points to include: substrate, active site

Here is a possible response:

In competitive inhibition, a molecule that resembles the normal substrate of an enzyme competes for the active site. Both the substrate and the inhibitor can bind the enzyme at the active site, resulting in fewer actual reactions proceeding. Many prescription drugs use competitive inhibition to prevent unhealthy behavior. For example, lithium is used to treat bipolar disorder by competing with the substrate that binds to the enzyme that causes manic behavior.

8. Key points to include: engulf, mitochondria, chloroplasts

Here is a possible response:

The endosymbiotic theory suggests that mitochondria and chloroplasts evolved when prokaryotes were engulfed by eukaryotic cells. Three primary pieces of evidence have led scientists to this conclusion. First, mitochondria and chloroplasts have unique DNA that is not found in the cell's nucleus. Second, these two organelles are produced via division and not via synthesis by the cell. Third, the inner mitochondrial and chloroplast membranes are similar to those in prokaryotes.

APPENDIX A

AP BIOLOGY EQUATIONS AND FORMULAS

STATISTICAL ANALYSIS AND PROBABILITY	
Standard Error	Mean
$SE_{\bar{x}} = \dfrac{s}{\sqrt{n}}$	$\bar{x} = \dfrac{1}{n}\sum\limits_{i=1}^{n} x_i$
Standard Deviation	Chi-Square
$S = \sqrt{\dfrac{\sum(x_i - \bar{x})^2}{n-1}}$	$\chi^2 = \sum \dfrac{(o-e)^2}{e}$

s = sample standard deviation (i.e., the sample-based estimate of the standard deviation of the population)

$\bar{x}$ = mean

n = size of the sample

o = observed individuals with observed genotype

e = expected individuals with observed genotype

Degrees of freedom equal the number of distinct possible outcomes minus one.

CHI-SQUARE TABLE

Degrees of Freedom

p	1	2	3	4	5	6	7	8
0.05	3.84	5.99	7.82	9.49	11.07	12.59	14.07	15.51
0.01	6.64	9.32	11.34	13.28	15.09	16.81	18.48	20.09

LAWS OF PROBABILITY

If A and B are mutually exclusive, then P(A or B) = P(A) + P(B)

If A and B are independent, then P(A and B) = P(A) x P(B)

HARDY-WEINBERG EQUATIONS

$p^2 + 2pq + q^2 = 1$

$p + q = 1$

p = frequency of the dominant allele in a population

q = frequency of the recessive allele in a population

METRIC PREFIXES		
Factor	Prefix	Symbol
10^9	giga	G
10^6	mega	M
10^3	kilo	k
10^{-2}	centi	c
10^{-3}	milli	m
10^{-6}	micro	μ
10^{-9}	nano	n
10^{-12}	pico	p

Mode = value that occurs most frequently in a data set

Median = middle value that separates the greater and lesser halves of a data set

Mean = sum of all data points divided by number of data points

Range = value obtained by subtracting the smallest observation (sample minimum) from the greatest (sample maximum)

RATE AND GROWTH		
Rate dY/dt **Population Growth** $dN/dt = B-D$ **Exponential Growth** $$\frac{dN}{dt} = r_{max}N$$ **Logistic Growth** $$\frac{dN}{dt} = r_{max}N\left(\frac{K-N}{K}\right)$$	dY = amount of change t = time B = birth rate D = death rate N = population size K = carrying capacity r_{max} = maximum per capita growth rate of population	**Water Potential (Ψ)** $\Psi = \Psi p + \Psi s$ Ψp = pressure potential Ψs = solute potential The water potential will be equal to the solute potential of a solution in an open container, since the pressure potential of the solution in an open container is zero. **The Solute Potential of the Solution**
Temperature Coefficient Q$_{10}$ $$Q_{10} = \left(\frac{k_2}{k_2}\right)^{\frac{10}{t_2-t_1}}$$ **Primary Productivity Calculation** mg O_2/L x 0.698 = mL O_2/L mL O_2/L x 0.536 = mg carbon fixed/L	t_2 = higher temperature t_1 = lower temperature k_2 = metabolic rate at t_2 k_1 = metabolic rate at t_1 Q_{10} = the factor by which the reaction rate increases when the temperature is raised by ten degrees	$\Psi s = -iCRT$ i = ionization constant (For sucrose, this is 1 because sucrose does not ionize in water.) C = molar concentration R = pressure constant (R = 0.0831 liter bars/mole K) T = temperature in Kelvin (273 + °C)

SURFACE AREA AND VOLUME		
Volume of a Sphere $V = 4/3\pi r^3$ **Volume of a Cube (or Square Column)** $V = lwh$ **Volume of a Column** $V = \pi r2h$ **Surface Area of a Sphere** $A = 4\pi r^2$ **Surface Area of a Cube** $A = 6a$ **Surface Area of a Rectangular Solid** $A = \Sigma$ (surface area of each side)	r = radius l = length h = height w = width A = surface area V = volume Σ = sum of all a = surface area of one side of the cube	**Dilution—used to create a dilute solution from a concentrated stock solution** $C_iV_i = C_fV_f$ i = initial (starting) C = concentration of solute f = final (desired) V = volume of solution **Gibbs Free Energy** $\Delta G = \Delta H - T\Delta S$ ΔG = change in Gibbs free energy ΔS = change in entropy ΔH = change in enthalpy T = absolute temperature (in Kelvin) pH = $-$ log [H+]

GLOSSARY

abiotic
Nonliving, as in the physical environment

absorption
The process by which water and dissolved substances pass through a membrane

acoelomate
An animal that lacks a coelom, exhibits bilateral symmetry, and has one internal space, the digestive cavity

action potential
The change in electrical potential across a nerve or muscle cell when stimulated, as in a nerve impulse

active immunity
Protective immunity to a disease in which the individual produces antibodies as a result of previous exposure to the antigen

adaptation
A behavioral or biological change that enables an organism to adjust to its environment

adaptive radiation
The production of a number of different species from a single ancestral species

adenine
A purine base that pairs with thymine in DNA and uracil in RNA

adenosine phosphate
Adenosine diphosphate (ADP) and adenosine triphosphate (ATP), which are energy-storage molecules

adipose
Fatty tissue, fat-storing tissue, or fat within cells

adrenaline (epinephrine)
An "emergency" hormone stimulated by anger or fear; increases blood pressure and heart rate to supply the emergency needs of the muscles

adventitious roots
Plant roots that develop in an unusual place

aerobe
An organism that requires oxygen for respiration and can live only in the presence of oxygen

aerobic
Requiring free oxygen from the atmosphere for normal activity and respiration

aerobic catabolism
Metabolic breakdown of complex molecules into simple ones through the use of oxygen; results in the release of energy

agonistic response
Response of aggression or submission between two organisms

alimentary canal
The digestive tract

allele
One of two or more types of genes, each representing a particular trait; many alleles exist for a specific gene locus

allopatric speciation
Evolution of species that occurs in separate geographic areas

alternation of generations
The description of a plant life cycle that consists of a diploid, asexual, sporophyte generation and a haploid, sexual, gametophyte generation

anaerobe
An organism that does not require free oxygen to respire

anaerobic
Living or active in the absence of free oxygen; pertaining to respiration that is independent of oxygen

anaerobic catabolism
Metabolic breakdown of complex molecules into simple ones without the use of oxygen; results in the release of energy

analogous
Describes structures that have similar function but different evolutionary origins (e.g., a bird's wing and a moth's wing)

anaphase
The stage in mitosis that is characterized by the migration of chromatids to opposite ends of the cell; the stage in meiosis during which homologous pairs migrate (anaphase I); and the stage in meiosis during which chromatids migrate to different ends of the cell (anaphase II)

androgen
A male sex hormone (e.g., testosterone)

angiosperm
A flowering plant; a plant of the class Angiospermae that produces seeds enclosed in an ovary and is characterized by the possession of fruit and flowers

Animalia
Kingdom that includes all extinct and living animals

antibiotic
An antipathogenic substance (e.g., penicillin)

antibody
Globular proteins produced by tissues that destroy or inactivate antigens

antigen
A foreign protein that stimulates the production of antibodies when introduced into the body of an organism

appendage
A structure that extends from the trunk of an organism and is capable of active movements

Archaea
Domain comprised of an ancient group of microorganisms (prokaryotes) that are metabolically and genetically different from bacteria; they came before the eukaryotes

artery
A blood vessel that carries blood away from the heart

asexual reproduction
The production of daughter cells by means other than the sexual union of gametes (as in budding and binary fission)

ATPase
Adenosine triphosphatase; enzyme that catalyzes the hydrolysis of ATP to ADP, thereby releasing energy

autonomic nervous system
The part of the nervous system that regulates the involuntary muscles, such as the walls of the alimentary canal; includes the parasympathetic and sympathetic nervous systems

autosomal genes
Non-sex-linked genes

autosome
Any chromosome that is not a sex chromosome

autotroph
An organism that utilizes the energy of inorganic materials such as water and carbon dioxide or the sun to manufacture organic materials; plants are examples of autotrophs

axon
A nerve fiber

bacteria
Kingdom of single-celled organisms that reproduce by fission and can be spiral, rod, or spherical shaped; often pathogenic organisms that rapidly reproduce

base-pair substitution
When one base pair is incorrectly reproduced and substituted for another base pair

bilateral symmetry
The equal division of an organism into a left and right half

bile
An emulsifying agent secreted by the liver

bile salts
Compounds in bile that aid in emulsification

binary fission
Asexual reproduction; in this process, the parent organism splits into two equal daughter cells

binomial nomenclature
The system of naming an organism by its genus and species names

biological species concept (BSC)
Definition of a species as a naturally interbreeding population of organisms that produces viable, fertile offspring

biome
A habitat zone, such as desert, grassland, or tundra

biotic
Living, as in living organisms in the environment

Calvin cycle
Cycle in photosynthesis that reduces fixed carbon to carbohydrates through the addition of electrons (also known as the "dark cycle")

CAM (crassulacean acid metabolism)
Storage of carbon dioxide at night in the form of organic acids

carbohydrate
An organic compound to which hydrogen and oxygen are attached; the hydrogen and oxygen are in a 2:1 ratio; examples include sugars, starches, and cellulose

carbon cycle
The recycling of carbon from decaying organisms for use in future generations

carbon fixation
Conversion of carbon dioxide into organic compounds during the Calvin cycle, the second stage of photosynthesis; known as a "dark reaction"

carnivore
A flesh-eating animal; an animal that subsists on other animals or parts of animals

carrying capacity
The number of organisms an environment can support

catabolism
Metabolic breakdown of complex molecules into simple ones, releasing energy

cell

Smallest structural unit of an organism

cell wall

A wall composed of cellulose that is external to the cell membrane in plants; it is primarily involved in support and in the maintenance of proper internal pressure; fungi have cell walls made of chitin, and some protists also have cell walls

central nervous system (CNS)

Encompasses the brain and the spinal cord

chemiosmosis

The coupling of enzyme-catalyzed reactions

chi-squared analysis

Test to see if a theory is backed up by experimental results

chlorophyll

A green pigment that performs essential functions as an electron donor and light entrapper in photosynthesis

chloroplast

A plastid containing chlorophyll

chromatid

One of the two strands that constitute a chromosome; chromatids are held together by the centromere

chromatin

A nuclear protein of chromosomes that stains readily

chromosome

A short, stubby rod consisting of chromatin that is found in the nucleus of cells; contains the genetic or hereditary component of cells (in the form of genes)

chromosome map

The distribution of genes on a chromosome, derived from crossover frequency experiments

circadian rhythms

Daily cycles of behavior

circulatory system

System that circulates blood throughout the body; includes the heart, blood, and blood vessels

cleavage

The division in animal cell cytoplasm caused by the pinching in of the cell membrane

clotting

The coagulation of blood caused by the rupture of platelets and the interaction of fibrin, fibrinogen, thrombin, prothrombin, and calcium ions

codominant

The state in which two genetic traits are fully expressed and neither dominates

codon

Three adjacent nucleotides in the genetic code that signal ribosomes to insert an amino acid or end protein synthesis

coelom

The space between the mesodermal layers that form the body cavity of some animal phyla

coelomates

Organisms that contain a coelom

coenzyme

An organic cofactor required for enzyme activity

commensal

Describes an organism that lives symbiotically with a host; this host neither benefits nor suffers from the association

communities

Groups of interacting organisms that live in the same geographic area under similar environmental conditions

complementary base pairs

Pairing of purines and pyrimidines in DNA and RNA

concentration gradient
Difference in concentration of a solute between two areas of a solution

conditioning
The association of a physical, visceral response with an environmental stimulus with which it is not naturally associated; a learned response

cone
A cell in the retina that is sensitive to colors and is responsible for color vision

conifers
Phylum of cone-bearing gymnosperm trees and shrubs that are primarily needle- and scale-leaved

connective tissue
Highly vascular matrix that forms the supporting and connecting structures of the body

consumer
Organism that consumes food from outside itself instead of producing it; consumers are classified as primary, secondary, and tertiary

convergence
Adaptive evolution of similar structures, such as wings

coupled reaction
Chemical reaction in which energy is transferred from one side of the reaction to the other through a common intermediate

cristae
Inward folds of the mitochondrial membrane

crossing over
The exchange of parts of homologous chromosomes during meiosis

cytokinesis
A process by which the cytoplasm and the organelles of the cell divide; the final stage of mitosis

cytoplasm
The living matter of a cell, located between the cell membrane and the nucleus

cytosine
A nitrogen base that is present in nucleotides and nucleic acids; it is paired with guanine

cytoskeleton
The organelle that provides mechanical support and carries out motility functions for the cell

dark (Calvin) reactions
Processes that occur after the light reactions of photosynthesis (during carbon fixation), without the presence of light

Darwin, Charles Robert (1809–1882)
Naturalist who came up with the theory of evolution based on natural selection

decomposers
Organisms that feed on and break down dead plant or animal matter

degree of freedom (d.f.)
Independent statistical category; the number of categories of observation minus one

deletion
The loss of all or part of a chromosome

dendrite
The part of the neuron that transmits impulses to the cell body

density-dependent factors
Effects that increase with population density and smaller population size

density-independent factors
Effects that are independent of population size

deoxyribose
A five-carbon sugar that has one oxygen atom less than ribose; a component of DNA (deoxyribonucleic acid)

determinate cleavage
Irreversible division of an egg into specific areas for further development

deuterostomes
Means "second mouth"; mouth forms from the second opening of the digestive tract in embryos; these organisms have a mouth, radial cleavage, anus, coelom, and indeterminate cleavage in common

differentiation
A progressive change from which a permanently more mature or advanced state results; for example, a relatively unspecialized cell's development into a more specialized one

diffusion
The movement of particles from one place to another as a result of their random motion

digestion
The process of breaking down large organic molecules into smaller ones

digestive system
The alimentary canal and glands, which ingest, digest, and absorb food

dihybrid
An organism that is heterozygous for two different traits

dihybrid cross
A hybridization between two traits, each with two alleles

diploid
Describes cells that have a double set of chromosomes in homologous pairs (2N)

directional selection
Favors organisms that have extreme variation of traits within a population

disruptive (diversifying) selection
Sudden changes in the environment cause organisms with extreme variation of traits in a population to be favored

DNA
Deoxyribonucleic acid; found in the cell nucleus, its basic unit is the nucleotide; contains coded genetic information; can replicate on the basis of heredity

domains
Biological classification of prokaryotes and eukaryotes into Bacteria, Archaea, and Eukarya

dominance
A dominant allele suppresses the expression of the other member of an allele pair when both members are present; a dominant gene exerts its full effect regardless of the effect of its allelic partner

ecological succession
The orderly process by which one biotic community replaces another until a climax community is established

ecology
The study of organisms in relation to their environment

ecosystem
Ecological community and its environment

ectoderm
The outermost embryonic germ layer that gives rise to the epidermis and the nervous system

egg (ovum)
The female gamete; it is nonmotile, large in comparison with male gametes, and stores nutrients

electrochemical gradient
Diffusion gradient of an ion, including potential and kinetic energy of the ion

electron transport chain
A complex carrier mechanism located on the inside of the inner mitochondrial membrane of the cell; releases energy and is used to form ATP

endemic
Pertaining to a restricted locality; ecologically, occurring only in one particular region

endocrine gland
A ductless gland that secretes hormones directly into the bloodstream

endocrine (hormone) system
Collection of ductless glands that secrete hormones into the bloodstream with various effects on the body (includes thyroid gland, pituitary gland, etc.)

endocytosis
A process by which the cell membrane is invaginated to form a vesicle that contains extracellular medium

endoderm
The innermost embryonic germ layer that gives rise to the lining of the alimentary canal and to the digestive and respiratory organs

endoplasm
The inner portion of the cytoplasm of a cell or the portion that surrounds the nucleus

endoplasmic reticulum
A network of membrane-enclosed spaces connected with the nuclear membrane; transports materials through the cell; can be smooth or rough

energy flow
The movement of energy throughout the trophic levels of an ecosystem

enzyme
An organic catalyst and protein

epidermal tissue
The outer or integumentary layer of the body, including sebum, adipose, and skin cells

epidermis
The outermost surface of an organism

epithelium
The cellular layer that covers external and internal surfaces of animal bodies

Eukarya
Domain containing all eukaryotic organisms

eukaryote
Organism consisting of one or more cells with genetic material in membrane-bound nuclei

evolution, theory of
Theory that organisms have developed over time to produce current biota

excretion
The elimination of metabolic waste matter

exocrine
Pertaining to a type of gland that releases its secretion through a duct; e.g., the salivary gland or the liver

exocytosis
A process by which a vesicle in the cell fuses with the cell membrane and releases its contents to the outside

exons
DNA that is transcribed to RNA and codes for protein synthesis

exoskeleton
Describes the skeletal or supporting structures that are outside the skin of arthropods and other animals

extracellular matrix
Material occurring outside of the cell

F1
The first filial generation (first offspring)

F2

The second filial generation; offspring resulting from the crossing of individuals of the F1 generation

fats

Solid, semi-solid, or liquid organic compounds composed of glycerol, fatty acids, and organic groups

feedback mechanism

The process by which a certain function is regulated by the amount of the substance it produces

ferns

Seedless, flowerless vascular plants that reproduce by spores

fertilization

The fusion of the sperm and egg to produce a zygote

fitness

The ability of an organism to contribute its alleles and, therefore, its phenotypic traits to future generations

food web

The interaction of feeding levels in a community, including energy flow throughout the community

functional groups

Chemical groups attached to carbon skeletons that give compounds their functionality

Fungi

Kingdom of eukaryotic organisms that lack vascular tissues and chlorophyll, possessing chitinous cell walls; reproduction occurs through spores

gamete

A sex or reproductive cell that must fuse with another of the opposite type to form a zygote, which subsequently develops into a new organism

gametogenesis

The formation of gametes

gametophyte

The haploid sexual stage in the life cycle of plants (alternation of generations)

gas exchange

The exchange of gases such as oxygen and carbon dioxide through respiratory surfaces, gills, lungs, or tracheae

gastrula

A stage of embryonic development characterized by the differentiation of the cells into the ectoderm and endoderm germ layers and by the formation of the archenteron; can be two-layer or three-layer

gastrulation

Formation of a gastrula

gene

The portion of a DNA molecule that serves as a unit of heredity; found on the chromosome

gene expression

Conversion of information from a gene to mRNA to a protein

gene frequency

A decimal fraction that represents the presence of an allele for all members of a population that have a particular gene locus

genetic code

A four-letter code made up of the DNA nitrogen bases A, T, G, and C; each chromosome is made up of thousands of these bases

genetic drift

Random evolutionary changes in the genetic makeup of a (usually small) population

genotype

The genetic makeup of an organism without regard to its physical appearance; a homozygous dominant and a heterozygous organism may have the same appearance but different genotypes

genus

In taxonomy, a classification level between species and family; a group of very closely related species (e.g., *Homo, Felis*)

geographic barrier

Any physical feature that prevents the ecological niches of different organisms (not necessarily different species) from overlapping

geographic isolation

Isolation due to geographic factors; islands are geographically isolated

geotropism

Any movement or growth of a living organism in response to the force of gravity

gills

Respiratory organ of aquatic animals

glycolysis

The anaerobic respiration of carbohydrates

Golgi apparatus

Membranous organelles involved in the storage and modification of secretory products

gravitropism

Directional growth according to the gravitational field; roots grow downward with gravity, while the shoots of plants grow up toward the sunlight

growth curve

Growth of an organism or population plotted over time

guanine

A purine (nitrogenous base) component of nucleotides and nucleic acids; it links up with cytosine in DNA

gymnosperm

A plant that belongs to the class of seed plants in which the seeds are not enclosed in an ovary; includes the conifers

habitat

The environment a community or organism lives in

haploid

Describes cells (gametes) that have half the chromosome number typical of the species (*n* chromosome number)

haploid sporophytes

Spore-producing phase of a plant that contains a single set of chromosomes, allows the plant to reproduce asexually

Hardy-Weinberg equilibrium

In a randomly breeding population, gene frequency and genotype ratios remain constant over generations of organisms

herbivore

A plant-eating animal

hermaphrodite

An organism that possesses both male and female reproductive organs

heterotroph

An organism that must get its inorganic and organic raw materials from the environment; a consumer

heterozygous

Describes an individual that possesses two contrasting alleles for a given trait (e.g, *Tt*)

homologous

Describes two or more structures that have similar forms, positions, and origins despite the differences between their current functions; examples are the arm of a human, the flipper of a dolphin, and the foreleg of a horse

homozygous

Describes an individual that has the same gene for the same trait on each homologous chromosome (e.g., *TT* or *tt*)

homozygous dominant

Having two dominant alleles of the same gene; dominant alleles are expressed in a heterozygous as well as a homozygous genotype

homozygous recessive

Having two recessive alleles of the same gene; recessive alleles are only expressed when a gene is homozygous recessive

hormone

A chemical messenger that is secreted by one part of the body and carried by the blood to affect another part of the body, usually a muscle or gland

hormone-receptor system

Chemical messengers (hormones) travel throughout the body and are read by receptor proteins, which respond to the message each hormone codes for

hybrid

An offspring that is heterozygous for one or more gene pairs

hybridization

Cross-breeding organisms to form base pairs between two strands of DNA that weren't originally paired

hydrophilic

Having an affinity for water

hydrophobic

Repelling water

hypertonic

Describes a fluid that has a higher osmotic pressure than another fluid it is compared with; it exerts greater osmotic pull than the fluid on the other side of a semipermeable membrane; hence, it possesses a greater concentration of particles and acquires water during osmosis

hypotonic

Describes a fluid that has a lower osmotic pressure than a fluid it is compared with; it exerts lesser osmotic pull than the fluid on the other side of a semipermeable membrane; hence, it possesses a lesser concentration of particles and loses water during osmosis

immunity

A resistance to disease developed through the immune system

incomplete dominance

Genetic blending; each allele exerts some influence on the phenotype (for example, red and white parents may yield pink offspring)

Independent Assortment, Law of

Genes independently sort and do not affect the sorting of other genes in the formation of gametes in diploid organisms

indeterminate cleavage

Early divisions of a cell that produce blastomeres

indeterminate growth

Growth without a termination point

induction

Initiating enzyme production or genetic transcription

ingestion

The intake of food from the environment into the alimentary canal

insertion

Addition of one or more nucleotides to a chromosome, usually by mutation

integument

Refers to protective covering, such as the covering of an ovule, that develops into the seed coat, or an animal's skin

integumentary system
The skin, hair, nails, sebaceous glands, and sweat glands

intermembrane space
Space between the outer and inner membranes of a mitochondrion

interphase
The cellular phase between meiotic or mitotic divisions

intestines
Part of the alimentary canal that extends from the stomach to the anus

introns
Part of a gene that is located between exons and is removed before the translation of mRNA; does not code for protein synthesis

isolation
The separation of some members of a population from the rest of their species; prevents interbreeding and may lead to the development of a new species

isomer
One of a group of compounds that is identical in atomic composition, but different in structure or arrangement

isotonic
Describes a fluid that has the same osmotic pressure as a fluid it is compared with; it exerts the same osmotic pull as the fluid on the other side of a semipermeable membrane; hence, it neither gains nor loses net water during osmosis and possesses the same concentration of particles before and after osmosis occurs

kinesis
Movement of an organism in response to light

kingdom
Second-highest taxonomic level of classification of organisms, after domain

Krebs cycle
Process of aerobic respiration that fully harvests the energy of glucose; also known as the citric acid cycle

K-selected
Organisms in more stable environments that tend to produce fewer offspring and invest more energy in rearing offspring

lactic acid fermentation
A type of anaerobic respiration found in fungi, bacteria, and human muscle cells

Lamarck, Jean Baptiste de (1744–1829)
Naturalist who studied evolution and classified invertebrate organisms; Lamarck produced several theories of evolution including *Philosophie zoologique*; best known for his incorrect theory of inheritance of acquired characteristics

light reactions
Photosynthetic reactions that occur in the presence of light

linkage
Occurs when different traits are inherited together more often than they would have been by chance alone; it is assumed that these traits are linked on the same chromosome

lipase
A fat-digesting hormone

lipid
A fat or oil

lumen
The inner cavity of a tubular organ, such as an intestine

lungs
Saclike respiratory organs of most vertebrates

lymph
A body fluid that flows in its own circulatory system in lymphatic vessels separate from blood circulation

lymphocyte
A kind of white blood cell in vertebrates that is characterized by a rounded nucleus; involved in the immune response

lysosome
An organelle that contains enzymes that aid in intracellular digestion

meiosis
A process of cell division whereby each daughter cell receives only one set of chromosomes; the formation of gametes

membrane
Thin structure connecting or separating structures or regions of an organism

Mendelian laws
Laws of classical genetics established through Mendel's experiments with peas; include segregation and independent assortment

mesoderm
Embryonic germ layer from which body systems and tissues develop; located between the ectoderm and endoderm

metabolism
A group of life-maintaining processes that includes nutrition, respiration (the production of usable energy), and the synthesis and degradation of biochemical substances

metaphase
A stage of mitosis; chromosomes line up at the equator of the cell

micronutrients
Vitamins or minerals essential for growth and metabolism in an organism

minerals
Naturally occurring inorganic elements essential in the nutrition of organisms

mitochondria
Cytoplasmic organelles that serve as sites of respiration; rod-shaped bodies in the cytoplasm known to be the center of cellular respiration

mitosis
A type of nuclear division that is characterized by complex chromosomal movement and the exact duplication of chromosomes; occurs in somatic cells

mitotic divisions
Cell division during mitosis

molecular clock hypothesis
Hypothesis that genetic mutations occur in a genome at a linear rate

Monera
The kingdom of bacteria

monohybrid
An individual that is heterozygous for only one trait

monohybrid cross
Cross involving a single trait and two alleles

monosaccharide
A simple sugar; a five- or six-carbon sugar (e.g., ribose or glucose)

mosses
Green nonvascular plants of the division Bryophyta

mRNA (messenger RNA)
RNA that transfers genetic information from the cell nucleus to ribosomes, serving as a template for protein synthesis

muscle tissue
Bundles of contractile fibers that allow movement; can be cardiac, skeletal, or smooth

mutagenic agent
Agent that induces mutations; typically carcinogenic

mutation
Changes in genes that are inherited

mutualism
A symbiotic relationship from which both organisms involved derive some benefit

myelin
Fatty lipid material that forms an insulating sheath around nerve fibers

NAD⁺
An abbreviation of nicotinamide-adeninedinucleotide, also called DPN; a respiratory oxidation-reduction molecule

NADP
An abbreviation of nicotinamide-adeninedinucleotide-phosphate, also called TPN; an organic compound that serves as an oxidation-reduction molecule

natural selection
Process by which organisms best adapted to their environment survive to pass their genes on through offspring; idea pioneered by Charles Darwin

negative feedback loop
Serves to stabilize a system, e.g., maintain temperature, concentration, direction

nerve
A bundle of nerve axons

nerve tissue
Tissue composed of nerve cells, dendrites, neuroglia, and nerve fibers

neuron
A nerve cell

neurotransmitters
Messenger molecules that affect the behavior of neurons

niche
The functional role and position of an organism in an ecosystem; embodies every aspect of the organism's existence

nitrogenous bases
The five purine and pyrimidine bases found in nucleic acid—adenine, thymine (in DNA only), cytosine, guanine, and uracil (in RNA only)

novel structures
Cellular structures used primarily for reproduction

nuclear membrane
A membrane that envelops the nucleus and separates it from the cytoplasm; present in eukaryotes

nucleolus
A dark-staining small body within the nucleus; composed of RNA

nucleotide
An organic molecule consisting of joined phosphate, five-carbon sugar (deoxyribose or ribose), and a purine or a pyrimidine (adenine, guanine, uracil, thymine, or cytosine)

nucleus
An organelle that regulates cell functions and contains the genetic material of the cell

nutrient
A substance that can be metabolized by an organism to provide energy and build tissue

ontogeny
The origin and subsequent growth of an organism, from embryo to adult

operon
A unit of genetic material that controls production of mRNA through the use of an operator gene, promoter, and two or more structural genes

organ
A group of tissues that perform a specific function in the body

organ system
A group of organs that work together to perform a specific task in the body, such as the circulatory system distributing blood throughout the body

organelle
A specialized structure that carries out particular functions for eukaryotic cells; examples include the plasma membrane, the nucleus, and ribosomes

organic molecules
Molecules that contain carbon (C)

organogenesis
Formation and development of organs

The Origin of Species
Charles Darwin's book in which he expressed his theory that species evolve through natural selection (survival of the fittest)

osmosis
The diffusion of water through a semipermeable membrane, from an area of greater concentration to an area of lesser concentration

osmotic pressure
Pressure exerted by the flow of water through a semipermeable membrane that separates two solutions of differing concentrations

oxidation
The removal of hydrogen or electrons from a compound or addition of oxygen; half of a redox (oxidation or reduction) process

oxidative phosphorylation
Formation of ATP from energy released during the oxidation of various substances and substrates, as in aerobic respiration and the Krebs cycle

oxygen
A nonmetallic element essential in animal and plant respiration

pairing (synapsis)
An association of homologous chromosomes during the first meiotic division

parapatric speciation
Occurs when limited interbreeding and negligible genetic exchange takes place between two populations

pedigree
A family tree depicting the inheritance of a particular genetic trait over several generations

pH
A symbol that denotes the relative concentration of hydrogen ions in a solution: the lower the pH, the more acidic a solution; the higher the pH, the more basic a solution; pH is equal to $-\log [H^+]$

phagocyte
Any cell capable of ingesting another cell

phenotype
The physical appearance of an individual, as opposed to its genetic makeup

phloem
The vascular tissue of a plant that transports organic materials (photosynthetic products) from the leaves to other parts of the plant

phosphoenol pyruvic acid (PEP)
An intermediate enzyme that fixes carbon dioxide into a four-carbon molecule in a mesophyll cell of C4 plants

phospholipids
Phosphorus-containing lipids composed of a phosphate group, organic molecule, and fatty acids

photophosphorylation
The synthesis of ATP using radiant energy absorbed during photosynthesis

photosynthesis
The process by which light energy and chlorophyll are used to manufacture carbohydrates out of carbon dioxide and water; an autotrophic process using light energy

phototropism
Plant growth stimulated by light (stem: +, toward light; root: −, away from light)

phylogeny
The study of the evolutionary descent and interrelations of groups of organisms

phylum
A category of taxonomic classification that is ranked above class; kingdoms are divided into phyla

physiology
The study of all living processes, activities, and functions

Plantae
Kingdom containing all extinct and living plants

plasma
The liquid part of blood

plasma membrane
The cell membrane

pollen
The microspore of a seed plant

pollination
The transfer of pollen to the micropyle or to a receptive surface that is associated with an ovule (such as a stigma)

polymer
A large molecule that is composed of many similar molecular units (e.g., starch)

polymorphism
The individual differences of form among the members of a species

polyploidy
A condition in which an organism may have a multiple of the normal number of chromosomes (4N, 6N, etc.)

polysaccharide
A carbohydrate that is composed of many monosaccharide units joined together, such as glycogen, starch, and cellulose

population
All the members of a given species inhabiting a certain locale

postzygotic barriers
Mechanisms that prevent the development of a zygote into a fertile adult offspring

prezygotic barriers
Mechanisms that prevent the formation of a zygote, leading to reproductive isolation

primary consumers
Eat primary producers; herbivores

primary producers
First level of the food chain, use light to produce energy through photosynthesis

progeny
Offspring

prokaryote
Unicellular organism lacking organelles, specifically a nucleus

promoter
Initial binding site on an operon for RNA polymerase

prophase
A mitotic or meiotic stage in which the chromosomes become visible and during which the spindle fibers form; synapsis takes place during the first meiotic prophase

protein
One of a class of organic compounds that is composed of many amino acids; contains C, H, O, and N

protein synthesis
The creation of proteins, coded for by nucleic acids

protobionts
Metabolically active protein clusters that inaccurately reproduce; possible evolutionary precursors to prokaryotic cells

Protoctista
Kingdom composed of eukaryotic microorganisms and their immediate descendants, such as slime molds and protozoa

protostomes
Organisms that have a "mouth first" during gastrulation

pseudocoelomates
Organisms with fluid-filled body cavity surrounding the gut

punctuated equilibrium
Evolution characterized by long periods of virtual standstill

Punnett square
Diagram used to show a cross between genes of breeding organisms

radial symmetry
Symmetrical arrangement of radiating parts about a central point

receptor cells
Cells throughout the body that contain receptor proteins and are activated by chemical signals, producing a systemic response

receptor proteins
Proteins that bind to specific molecules such as hormones and cytokines

recessive
Pertains to a gene or characteristic that is masked when a dominant allele is present

reduction
A change from a diploid nucleus to a haploid nucleus, as in meiosis

regeneration
The ability of certain animals to regrow missing body parts

repressor proteins
Proteins that bind to genes called operators to slow down transcription in cells

reproduce vegetatively/vegetative reproduction
Asexual reproduction

reproductive system
System of organs associated with reproduction; includes the gonads, ducts, and external genitalia

respiration
A chemical action that releases energy from glucose to form ATP

respiratory system
System of organs involved in the intake and exchange of oxygen and carbon dioxide between an organism and its environment

resting membrane potential
Electrical state in an excitable cell where the membrane potential is more negative inside the cell than outside

ribose
A pentose sugar that occurs in nucleotides, nucleic acids, and riboflavin

ribosome
An organelle in the cytoplasm that contains RNA; serves as the site of protein synthesis

RNA
An abbreviation of ribonucleic acid, a nucleic acid in which the sugar is ribose; a product of DNA transcription that serves to control certain cell activities; acts as a template for protein translation; types include mRNA (messenger), tRNA (transfer), and rRNA (ribosomal)

r-selected
Species that monopolize rapidly changing environments and produce many offspring in a short amount of time

rubisco
Plant protein that accepts oxygen in place of carbon dioxide and fixes carbon in photosynthetic organisms

secondary consumers
Organisms that eat primary consumers

Segregation, Law of
Genes come in pairs in diploid organisms and each gamete gets one gene at random from each gene pair

selective advantages
Characteristics that are good for courtship and mating

selective disadvantages
Secondary representations of fitness that are good for sexual selection, but make an organism easily seen by predators, such as the tail feathers of peacocks

selectively permeable
Membranes that allow some substances and particles to pass through, but not others

self-pollination
The transfer of pollen from the stamen to the pistil of the same flower

senescence
The process of growing old

sex chromosome
There are two kinds of sex chromosomes, X and Y; XX signifies a female and XY signifies a male; there are fewer genes on the Y chromosome than on the X chromosome

sex linkage
Occurs when certain traits are determined by genes on the sex chromosomes

sex-linked cross
A cross between sex-linked genes

sexual reproduction
Reproduction by the fusion of a male and female gamete to form a zygote

sexual selection
Selection driven by the competition for mates, in relation to natural selection

skeletal system
Body system consisting of the bones, cartilage, and joints

soma
The whole body of an organism or the cell body, exclusive of the germ cells

somatic cell
Any cell that is not a reproductive cell

species
A group of populations that can interbreed to produce fertile, viable offspring

sperm
A male gamete

spermatogenesis
The process of forming the sperm cells from primary spermatocytes

spicules
Small needlelike structures that support the soft tissue of some invertebrates, such as sponges

spindle
A structure that arises during mitosis and helps separate the chromosomes; composed of tubulin

spore
A reproductive cell that is capable of developing directly into an adult

sporophyte
An organism that produces spores; a phase in the diploid-haploid life cycle that alternates with a gametophyte phase

stabilizing selection
Selection that maintains the same distribution mean in a phenotypic distribution by removing individuals from both phenotypic ends

sterols
Polycyclic compounds (lipids) such as cholesterol that play an important role in lipid metabolism

stomach
The portion of the alimentary canal in which some protein digestion occurs; its muscular walls churn food so that it is more easily digested; its low-pH environment activates certain protein-digesting enzymes

stomata
Pores in a leaf through which gas and water vapor pass; a small opening on the surface of a membrane; a mouthlike opening in an organism

substrate
A substance that is acted upon by an enzyme

symbiosis
The living together of two organisms in an intimate relationship; includes commensalism, mutualism, and parasitism

sympathetic
Pertaining to a subdivision of the autonomic nervous system

sympatric speciation
Speciation due to behavioral, temporal, or ecological factors; species live in the same geographic area but do not interbreed

synapse
The junction or gap between the axon terminal of one neuron and the dendrites of another neuron

synapsis
The pairing of homologous chromosomes during meiosis

synergistic
Describes organisms that are cooperative in action, such as hormones or other growth factors that reinforce each other's activity

system
A group of complementary organs that together perform a designated bodily function

taxis
The responsive movement of an organism toward or away from a stimulus such as light

taxonomy
The science of classification of living things

telophase
A mitotic stage in which nuclei reform and nuclear membrane reappears

terminator
Sequence of nucleotides that signals the end of synthesis of a protein or nucleic acid, as well as the end of translation or transcription

terrestrial plants
Plants that live and grow on land

test cross
The breeding of an organism with a homozygous recessive individual to determine whether an organism is homozygous dominant or heterozygous dominant for a given trait

tetrad
A pair of chromosome pairs present during the first metaphase of meiosis

thermoregulation
The ways in which organisms regulate their internal heat

thylakoid space
An inner compartment formed from the connected spaces between flattened sacs called thylakoids in chloroplasts

thymine
A pyrimidine component of nucleic acids and nucleotides; pairs with adenine in DNA

tissue
A mass of cells that have similar structures and perform similar functions

tracheal openings
Opening to the windpipe or trachea, which is a cartilaginous tube that brings oxygen to the lungs

transcription
The first stage of protein synthesis, in which the information coded in the DNA base is transcribed onto a strand of mRNA

translation
The final stages of protein synthesis in which the genetic code of nucleotide sequences is translated into a sequence of amino acids

translocation
The transfer of a piece of chromosome to another chromosome

transpiration
The evaporation of water from leaves or other exposed surfaces of plants

tRNA (transfer RNA)
RNA molecules that transport amino acids to ribosomes

uracil
A pyrimidine found in RNA (but not in DNA); pairs with adenine

urea
An excretory product of protein metabolism

vacuole
A space in the cytoplasm of a cell that contains fluid

vitamin
An organic nutrient required by organisms in small amounts to aid in proper metabolic processes; may be used as an enzymatic cofactor; because it is not synthesized, it must be obtained prefabricated in the diet

wood
Xylem that is no longer being used; gives structural support to the plant

xylem
Vascular tissue of the plant that aids in support and carries water

zygote
A cell resulting from the fusion of gametes